VOLUME FIVE HUNDRED AND THIRTY FOUR

METHODS IN ENZYMOLOGY

Endosome Signaling Part A

METHODS IN ENZYMOLOGY

Editors-in-Chief

JOHN N. ABELSON and MELVIN I. SIMON
Division of Biology
California Institute of Technology
Pasadena, California

ANNA MARIE PYLE
Departments of Molecular, Cellular and Developmental
Biology and Department of Chemistry
Investigator, Howard Hughes Medical Institute
Yale University

GREGORY L. VERDINE
Department of Chemistry and Chemical Biology
Harvard University

Founding Editors

SIDNEY P. COLOWICK and NATHAN O. KAPLAN

VOLUME FIVE HUNDRED AND THIRTY FOUR

METHODS IN ENZYMOLOGY

Endosome Signaling Part A

Edited by

P. MICHAEL CONN

Senior Vice President for Research
Associate Provost
Professor of Internal Medicine and Cell Biology
Texas Tech University Health Sciences Center
Lubbock, TX 79430, USA

AMSTERDAM • BOSTON • HEIDELBERG • LONDON
NEW YORK • OXFORD • PARIS • SAN DIEGO
SAN FRANCISCO • SINGAPORE • SYDNEY • TOKYO

Academic Press is an imprint of Elsevier

Academic Press is an imprint of Elsevier
525 B Street, Suite 1800, San Diego, CA 92101-4495, USA
225 Wyman Street, Waltham, MA 02451, USA
Radarweg 29, PO Box 211, 1000 AE Amsterdam, The Netherlands
The Boulevard, Langford Lane, Kidlington, Oxford, OX5 1GB, UK
32 Jamestown Road, London NW1 7BY, UK

First edition 2014

Copyright © 2014 Elsevier Inc. All Rights Reserved.

No part of this publication may be reproduced, stored in a retrieval system or transmitted in any form or by any means electronic, mechanical, photocopying, recording or otherwise without the prior written permission of the publisher

Permissions may be sought directly from Elsevier's Science & Technology Rights Department in Oxford, UK: phone (+44) (0) 1865 843830; fax (+44) (0) 1865 853333; email: permissions@elsevier.com. Alternatively you can submit your request online by visiting the Elsevier web site at http://elsevier.com/locate/permissions, and selecting *Obtaining permission to use Elsevier material*

Notice
No responsibility is assumed by the publisher for any injury and/or damage to persons or property as a matter of products liability, negligence or otherwise, or from any use or operation of any methods, products, instructions or ideas contained in the material herein. Because of rapid advances in the medical sciences, in particular, independent verification of diagnoses and drug dosages should be made

For information on all Academic Press publications
visit our website at store.elsevier.com

ISBN: 978-0-12-397926-1
ISSN: 0076-6879

Printed and bound in United States of America
14 15 16 17   11 10 9 8 7 6 5 4 3 2 1

CONTENTS

Contributors   xiii
Preface   xix

Section I
Compartments

1. Monitoring Phosphatidylinositol 3-Phosphate in Multivesicular Endosome Biogenesis   3

Catherine Sem Wegner, Kay Oliver Schink, Harald Stenmark, and Andreas Brech

1. Introduction	4
2. Localization of PtdIns3P and EGFR	5
3. Correlative Light and Electron Microscopy of MVEs	16
4. Concluding Remarks	21
Acknowledgments	22
References	22

2. Methods to Discriminate the Distribution of Acidic Glycohydrolases Between the Endosomal–Lysosomal Systems and the Plasma Membrane   25

Alessandro Magini, Alice Polchi, Brunella Tancini, Lorena Urbanelli, Manlio Di Cristina, Roberta Mannucci, Ildo Nicoletti, and Carla Emiliani

1. Introduction	27
2. Purification of Lipid Microdomains from Cell Membranes and Glycohydrolases Activity Determination	28
3. Discrimination of Cell Surface Lipid Microdomain-Associated Glycohydrolases from the Intracellular Counterparts	31
4. Immunology Capture of Lipid Microdomains Containing Glycohydrolases	35
5. *In Vivo* Assay of Cell Surface Glycohydrolases	35
6. Fluorescence Microscopy Analysis of Hex Intracellular Trafficking	38
7. Summary	43
Acknowledgment	44
References	44

3. Visualizing of Signaling Proteins on Endosomes Utilizing Knockdown and Reconstitution Approach   47

Myoungkun Jeoung and Emilia Galperin

1. Introduction   48
2. Description of Methods   49
3. Conclusions   61
Acknowledgments   62
References   62

4. Virus-Induced Signaling Influences Endosome Trafficking, Virus Entry, and Replication   65

David J. Marchant, Leanne Bilawchuk, Gordon Nish, and Richard G. Hegele

1. Introduction   66
2. Isolation of Fibroblasts from Wild-Type or MyD88$^{-/-}$ Mice and Purification of Virus   66
3. Screening Assays for the Determination of Principal Host Kinases Involved During Virus Replication   68
4. Virus–Host Cell Receptor Interactions and Kinase Activation   70
5. Detection of Virus-Induced Cell Signaling and Virus Entry into Endosomes Using Immunofluorescence Confocal Microscopy   73
6. Summary   75
References   75

5. Methods to Evaluate Zinc Transport into and out of the Secretory and Endosomal–Lysosomal Compartments in DT40 Cells   77

Taiho Kambe

1. Introduction   78
2. Intracellular Zinc Transporters Localized to the Secretory and Endosomal–Lysosomal Compartments Play Crucial Roles   79
3. Establishment of DT40 Cells Deficient in Zinc Transporters Genes   82
4. Experimental Procedures Used in Studies of DT40 Cells Deficient in Zinc Transport   84
5. Functional Analysis of Zinc Mobilization into or out of the Secretory and Endosomal–Lysosomal Compartments   86
6. Concluding Remarks   89
Acknowledgment   89
References   89

6. Interactions Between Endosomal Maturation and Autophagy: Analysis of ESCRT Machinery During *Caenorhabditis elegans* Development — 93

Marion Manil-Ségalen, Emmanuel Culetto, Renaud Legouis, and Christophe Lefebvre

1. Introduction	94
2. Strains and Reagents	95
3. Fluorescent-Tagged Protein Construction and Transgenesis	98
4. Analysis of Developmental Phenotypes in ESCRT Mutants	100
5. Analysis of Vesicular Compartments	104
6. Analysis of Autophagy	110
7. Methods to Visualize Amphisome, the Fusion Organelle Between Endosomes and Autophagosomes	114
8. Conclusions	116
Acknowledgments	116
References	116

7. Assessment of Cation Trapping by Cellular Acidic Compartments — 119

François Marceau, Caroline Roy, and Johanne Bouthillier

1. Introduction	120
2. Quinacrine Uptake by Cells	123
3. Macroautophagic Accumulation in Cells That Have Accumulated Cations	128
4. Summary	129
Acknowledgments	130
References	130

8. Signaling Initiated by the Secretory Compartment — 133

Carmen Ruggiero, Jorge Cancino, Monica Giannotta, and Michele Sallese

1. Introduction	134
2. ER-to-Golgi Traffic-Synchronization Protocols	136
3. Read-Outs for Traffic-Generated Signaling	141
4. KDELR Signaling	147
5. The KDELR Transduction Machinery	149
6. Conclusions	151
Acknowledgments	151
References	152

9. Image-Based and Biochemical Assays to Investigate Endosomal Protein Sorting   155
Sophia Y. Breusegem and Matthew N.J. Seaman

1. Introduction   156
2. Antibody-Uptake Assays   157
3. Detailed Characterization of Endosomes   169
4. Endosome Recruitment and/or Association   171
5. Summary   177
Acknowledgments   177
References   177

Section II
Transport and Transfer

10. Cytokines, Polarity Proteins, and Endosomal Protein Trafficking and Signaling—The Sertoli Cell Blood–Testis Barrier System *In Vitro* as a Study Model   181
Xiang Xiao, Elissa W.P. Wong, Pearl P.Y. Lie, Dolores D. Mruk, Chris K.C. Wong, and C. Yan Cheng

1. Introduction   182
2. Endocytosis Assay   184
3. Materials   187
4. Buffers   187
5. Methods   188
6. Cell Staining to Assess Endocytosis   190
7. Results   190
8. Summary   191
Acknowledgments   192
References   192

11. Methods of Analysis of the Membrane Trafficking Pathway from Recycling Endosomes to Lysosomes   195
Takahide Matsui and Mitsunori Fukuda

1. Introduction   196
2. Degradation of TfR in Lysosomes   197
3. Screening Methods for Rab Proteins Involved in Lysosomal Degradation of TfR   200
4. Effect of Rab12 Knockdown on an EGFR Endocytic Pathway and a Tf Recycling Pathway   203

5. Concluding Remarks	205
Acknowledgments	205
References	205

12. Measurement of Intercellular Transfer to Signaling Endosomes   207
Kristopher D. Marjon and Jennifer M. Gillette

1. Introduction	208
2. Labeling of Transferred Signals	209
3. Measuring ICT Using Fluorescence Techniques	214
4. Signaling Endosomes	216
5. Experimental Example: ICT to SARA-Positive Signaling Endosomes	217
6. Summary	218
References	219

13. Liposome-Based Assays to Study Membrane-Associated Protein Networks   223
Christian Niehage, Christoph Stange, Mihaela Anitei, and Bernard Hoflack

1. Introduction	224
2. Isolation of Core Machineries Required for Carrier Biogenesis on Synthetic Membranes	226
3. Identification of Core Machineries by Mass Spectrometry-Based, Label-Free Quantitative Proteomics	234
4. Visualization of Protein Dynamics on Giant Unilamellar Vesicles by Fluorescence Microscopy	238
5. Summary	241
Acknowledgments	242
References	242

14. Mouse Models of PI(3,5)P_2 Deficiency with Impaired Lysosome Function   245
Guy M. Lenk and Miriam H. Meisler

1. Introduction	246
2. Design of Mouse Models	247
3. A Spontaneous Null Mutation of *Fig4*: The *Pale Tremor* Mouse	249
4. Tissue-Specific *Fig4* Transgenes: Neurons Versus Astrocytes	251
5. Conditional Knockout of *Fig4* in Neurons	251
6. The Human Disease Mutation FIG4-I41T in Transgenic Mice	252
7. A Spontaneous Missense Mutation of *Vac14* in the *ingls* Mouse	252
8. A Null Gene-Trap Allele of *Vac14*	253

9.	A Hypomorphic Gene-Trap Allele of *Pikfyve* (*Fab1*)	253
10.	A Conditional Knockout of Pikfyve	254
11.	Genetic Interactions: *Fig4*, *Vac14*, and *Mtmr2*	254
12.	Genetic Effects of Strain Background	255
13.	Future Applications of Mouse Models of PI(3,5)P$_2$ Deficiency	255
	Acknowledgments	257
	References	257

15. Monitoring Endosomal Trafficking of the G Protein-Coupled Receptor Somatostatin Receptor 3 — 261

Cristy Tower-Gilchrist, Melanie L. Styers, Bradley K. Yoder, Nicolas F. Berbari, and Elizabeth Sztul

1.	Introduction	262
2.	Development of Cell Lines Stably Expressing SSTR3	264
3.	Live Imaging of SSTR3 and RABS in Mammalian Kidney Cells	269
4.	Dynamics of SSTR3 Transit Relative to RABS	272
5.	Effects of Dominant Negative Rabs on SSTR3 Trafficking	276
6.	Summary	278
	Acknowledgments	278
	References	278

Section III
Proteins

16. Genetic Circuitry Modulating Notch Signals Through Endosomal Trafficking — 283

Kazuya Hori, Anindya Sen, and Spyros Artavanis-Tsakonas

1.	Introduction	284
2.	Genetic Screen Using the Exelixis Collection	285
3.	Notch Localization in Endosomes	289
4.	Optical Approaches	292
5.	Ubiquitination Status of Notch	295
6.	Conclusion	297
	Acknowledgments	297
	References	297

17. Monitoring Notch/Delta Endosomal Trafficking and Signaling in Drosophila — 301

Sylvain Loubéry and Marcos González-Gaitán

1.	Introduction	302
2.	Antibody Uptake Assays to Monitor Notch and Delta Trafficking	305

3.	Correlative Imaging of Dividing SOP Cells and of Their Progeny	313
4.	Concluding Remarks	318
	Acknowledgements	319
	References	320

18. Toll-Interacting Protein Pathway: Degradation of an Ubiquitin-Binding Protein   323

Miho Shimizu, Asami Oguro-Ando, Eri Ohoto-Fujita, and Yoriko Atomi

1.	Introduction	324
2.	Expression of GFP-htt Protein and RFP-Tollip and How to Count Aggregation	325
3.	Cytoskeleton-Dependent Transport of Tollip to htt^{pQ}-Aggresome	327
4.	Determinants of Early-/Late Endosomal Localization in the Cells	328
5.	Summary	328
	Acknowledgment	329
	References	329

19. Measuring Interactions of FERM Domain-Containing Sorting Nexin Proteins with Endosomal Lipids and Cargo Molecules   331

Rajesh Ghai, Mehdi Mobli, and Brett M. Collins

1.	Introduction	332
2.	Secondary Structure-Based Domain Classification	333
3.	Production of Recombinant PX-FERM Proteins	335
4.	Measuring Interactions of PX-FERM Proteins with Phosphoinositide Lipids and Peptide Cargo Motifs by Isothermal Titration Calorimetry	338
5.	Analysis of SNX17 PX Domain-PI3P Interactions by Nuclear Magnetic Resonance Spectroscopy	341
6.	Summary	347
	Acknowledgments	347
	References	347

Author Index	*351*
Subject Index	*369*

CONTRIBUTORS

Mihaela Anitei
Biotechnology Center, Dresden University of Technology, Dresden, Germany

Spyros Artavanis-Tsakonas
Department of Cell Biology, Harvard Medical School, Boston, Massachusetts, USA

Yoriko Atomi
Graduate School of Information Science and Technology, and Radioisotope Center, Cell to Body Dynamics Laboratory 1, The University of Tokyo, Tokyo, Japan

Nicolas F. Berbari
Department of Cell, Developmental and Integrative Biology, University of Alabama at Birmingham, Birmingham, Alabama, USA

Leanne Bilawchuk
Li Ka Shing Institute of Virology, Department of Medical Microbiology and Immunology, University of Alberta, Edmonton, Alberta, Canada

Johanne Bouthillier
Centre de recherche en rhumatologie et immunologie, Centre Hospitalier Universitaire de Québec, Québec, Canada

Andreas Brech
Centre for Cancer Biomedicine, Faculty of Medicine, University of Oslo, and Institute for Cancer Research, The Norwegian Radium Hospital, Oslo University Hospital, Oslo, Norway

Sophia Y. Breusegem
Department of Clinical Biochemistry, Cambridge Institute for Medical Research, University of Cambridge, Addenbrooke's Hospital, Cambridge, United Kingdom

Jorge Cancino
Department of Life Sciences, Institute of Protein Biochemistry, National Research Council and Telethon Institute of Genetics and Medicine, Naples, Italy

C. Yan Cheng
The Mary M. Wohlford Laboratory for Male Contraceptive Research, Center for Biomedical Research, Population Council, New York, USA

Brett M. Collins
Institute for Molecular Bioscience, University of Queensland, St. Lucia, Queensland, Australia

Emmanuel Culetto
Centre de Génétique Moléculaire, CNRS UPR3404, Université Paris-Sud, 91198 Gif-sur-Yvette Cedex, France

Manlio Di Cristina
Department of Experimental Medicine and Biochemical Sciences, Section of Biochemistry and Molecular Biology, University of Perugia, Perugia, Italy

Carla Emiliani
Department of Experimental Medicine and Biochemical Sciences, Section of Biochemistry and Molecular Biology, University of Perugia, Perugia, Italy

Mitsunori Fukuda
Laboratory of Membrane Trafficking Mechanisms, Department of Developmental Biology and Neurosciences, Graduate School of Life Sciences, Tohoku University, Sendai, Miyagi, Japan

Emilia Galperin
Department of Molecular and Cellular Biochemistry, University of Kentucky, Lexington, Kentucky, USA

Rajesh Ghai
Institute for Molecular Bioscience, University of Queensland, St. Lucia, Queensland, Australia

Monica Giannotta
FIRC Institute of Molecular Oncology Foundation, Milan, Italy

Jennifer M. Gillette
Department of Pathology, University of New Mexico Health Sciences Center, Albuquerque, New Mexico, USA

Marcos González-Gaitán
Department of Biochemistry, Faculty of Sciences, and Department of Molecular Biology, Faculty of Sciences, Geneva University, Geneva, Switzerland

Richard G. Hegele
Department of Laboratory Medicine and Pathobiology, University of Toronto, Toronto, Ontario, Canada

Bernard Hoflack
Biotechnology Center, Dresden University of Technology, Dresden, Germany

Kazuya Hori
Department of Cell Biology, Harvard Medical School, Boston, Massachusetts, USA

Myoungkun Jeoung
Department of Molecular and Cellular Biochemistry, University of Kentucky, Lexington, Kentucky, USA

Taiho Kambe
Graduate School of Biostudies, Kyoto University, Kyoto, Japan

Christophe Lefebvre
Centre de Génétique Moléculaire, CNRS UPR3404, Université Paris-Sud, 91198 Gif-sur-Yvette Cedex, France

Renaud Legouis
Centre de Génétique Moléculaire, CNRS UPR3404, Université Paris-Sud, 91198 Gif-sur-Yvette Cedex, France

Guy M. Lenk
Department of Human Genetics, University of Michigan, Ann Arbor, Michigan, USA

Pearl P.Y. Lie
The Mary M. Wohlford Laboratory for Male Contraceptive Research, Center for Biomedical Research, Population Council, New York, USA

Sylvain Loubéry
Department of Biochemistry, Faculty of Sciences, and Department of Molecular Biology, Faculty of Sciences, Geneva University, Geneva, Switzerland

Alessandro Magini
Department of Experimental Medicine and Biochemical Sciences, Section of Biochemistry and Molecular Biology, University of Perugia, Perugia, Italy

Marion Manil-Ségalen
Centre de Génétique Moléculaire, CNRS UPR3404, Université Paris-Sud, 91198 Gif-sur-Yvette Cedex, France

Roberta Mannucci
IMAGE Analysis Laboratory, University of Perugia, Perugia, Italy

François Marceau
Centre de recherche en rhumatologie et immunologie, Centre Hospitalier Universitaire de Québec, Québec, Canada

David J. Marchant
Li Ka Shing Institute of Virology, Department of Medical Microbiology and Immunology, University of Alberta, Edmonton, Alberta, Canada

Kristopher D. Marjon
Department of Pathology, University of New Mexico Health Sciences Center, Albuquerque, New Mexico, USA

Takahide Matsui
Laboratory of Membrane Trafficking Mechanisms, Department of Developmental Biology and Neurosciences, Graduate School of Life Sciences, Tohoku University, Sendai, Miyagi, Japan

Miriam H. Meisler
Department of Human Genetics, University of Michigan, Ann Arbor, Michigan, USA

Mehdi Mobli
Centre for Advanced Imaging & School of Chemistry and Molecular Biosciences, The University of Queensland, St. Lucia, Queensland, Australia

Dolores D. Mruk
The Mary M. Wohlford Laboratory for Male Contraceptive Research, Center for Biomedical Research, Population Council, New York, USA

Ildo Nicoletti
IMAGE Analysis Laboratory, University of Perugia, Perugia, Italy

Christian Niehage
Biotechnology Center, Dresden University of Technology, Dresden, Germany

Gordon Nish
Li Ka Shing Institute of Virology, Department of Medical Microbiology and Immunology, University of Alberta, Edmonton, Alberta, Canada

Asami Oguro-Ando
Department of Neuroscience and Pharmacology, Rudolf Magnus Institute of Neuroscience, University Medical Centre Utrecht, Utrecht, The Netherlands

Eri Ohoto-Fujita
Graduate School of Information Science and Technology, and Radioisotope Center, Cell to Body Dynamics Laboratory 1, The University of Tokyo, Tokyo, Japan

Alice Polchi
Department of Experimental Medicine and Biochemical Sciences, Section of Biochemistry and Molecular Biology, University of Perugia, Perugia, Italy

Caroline Roy
Centre de recherche en rhumatologie et immunologie, Centre Hospitalier Universitaire de Québec, Québec, Canada

Carmen Ruggiero
Department of Cellular and Translational Pharmacology, Fondazione Mario Negri Sud, Unit of Genomic Approaches to Membrane Traffic, Santa Maria Imbaro (CH), Italy

Michele Sallese
Department of Cellular and Translational Pharmacology, Fondazione Mario Negri Sud, Unit of Genomic Approaches to Membrane Traffic, Santa Maria Imbaro (CH), Italy

Kay Oliver Schink
Centre for Cancer Biomedicine, Faculty of Medicine, University of Oslo, and Institute for Cancer Research, The Norwegian Radium Hospital, Oslo University Hospital, Oslo, Norway

Matthew N.J. Seaman
Department of Clinical Biochemistry, Cambridge Institute for Medical Research, University of Cambridge, Addenbrooke's Hospital, Cambridge, United Kingdom

Anindya Sen
Department of Cell Biology, Harvard Medical School, Boston, Massachusetts, USA

Miho Shimizu
Graduate School of Information Science and Technology, and Radioisotope Center, Cell to Body Dynamics Laboratory 1, The University of Tokyo, Tokyo, Japan

Christoph Stange
Biotechnology Center, Dresden University of Technology, Dresden, Germany

Harald Stenmark
Centre for Cancer Biomedicine, Faculty of Medicine, University of Oslo, and Institute for Cancer Research, The Norwegian Radium Hospital, Oslo University Hospital, Oslo, Norway

Melanie L. Styers
Department of Biology, Birmingham-Southern College, Birmingham, Alabama, USA

Elizabeth Sztul
Department of Cell, Developmental and Integrative Biology, University of Alabama at Birmingham, Birmingham, Alabama, USA

Brunella Tancini
Department of Experimental Medicine and Biochemical Sciences, Section of Biochemistry and Molecular Biology, University of Perugia, Perugia, Italy

Cristy Tower-Gilchrist
Department of Cell Biology, Emory University, Atlanta, Georgia, USA

Lorena Urbanelli
Department of Experimental Medicine and Biochemical Sciences, Section of Biochemistry and Molecular Biology, University of Perugia, Perugia, Italy

Catherine Sem Wegner
Centre for Cancer Biomedicine, Faculty of Medicine, University of Oslo, and Institute for Cancer Research, The Norwegian Radium Hospital, Oslo University Hospital, Oslo, Norway

Chris K.C. Wong
Department of Biology, Hong Kong Baptist University, Hong Kong, China

Elissa W.P. Wong
The Mary M. Wohlford Laboratory for Male Contraceptive Research, Center for Biomedical Research, Population Council, New York, USA

Xiang Xiao
The Mary M. Wohlford Laboratory for Male Contraceptive Research, Center for Biomedical Research, Population Council, New York, USA, and Department of Reproductive Physiology, Zhejiang Academy of Medical Sciences, Hangzhou, Zhejiang, China

Bradley K. Yoder
Department of Cell, Developmental and Integrative Biology, University of Alabama at Birmingham, Birmingham, Alabama, USA

PREFACE

Endosomes are membrane-bound compartments that transport internalized material from the plasma membrane to the lysosome and elsewhere. These compartments, often about 500 nm, but ranging in size, have the capability to sort molecules, routing some contents to the lysosomes for degradation, and recycling other materials back to the plasma membrane. The Golgi apparatus also provides molecules to the endosome, some of which are delivered to lysosomes and others are recycled back to the Golgi. Because of this ability to differentially deliver molecules, the endosome is viewed as a presorting structure.

Endosomes are categorized by size, enzymatic content, morphology, and by other criteria such as the length of time it takes internalized material to reach them. Endosomes may provide platforms for cross talk between signaling systems, and this consideration has provided them elite status among cellular components that contribute to signaling.

This volume provides descriptions of the range of methods used to analyze and evaluate these important compartments. The authors explain how these methods are able to provide important biological insights in the context of particular models.

Authors were selected based on both their research contributions and on their ability to describe their methodological contributions in a clear and reproducible way. They have been encouraged to make use of graphics, comparisons to other methods, and to provide tricks and approaches not revealed in prior publications that make it possible to adapt their methods to other systems.

The editor wants to express appreciation to the contributors for providing their contributions in a timely fashion, to the senior editors for guidance, and to the staff at Academic Press for helpful input.

P. MICHAEL CONN
Lubbock, TX, USA

SECTION I

Compartments

CHAPTER ONE

Monitoring Phosphatidylinositol 3-Phosphate in Multivesicular Endosome Biogenesis

Catherine Sem Wegner[*,†], **Kay Oliver Schink**[*,†], **Harald Stenmark**[*,†], **Andreas Brech**[*,†,1]

[*]Centre for Cancer Biomedicine, Faculty of Medicine, University of Oslo, Oslo, Norway
[†]Institute for Cancer Research, The Norwegian Radium Hospital, Oslo University Hospital, Oslo, Norway
[1]Corresponding author: e-mail address: Andreas.Brech@rr-research.no

Contents

1. Introduction 4
2. Localization of PtdIns3P and EGFR 5
   2.1 PtdIns3P and EGFR localization by confocal microscopy 7
   2.2 PtdIns3P and EGFR localization by super-resolution light microscopy 9
   2.3 PtdIns3P and EGFR localization by immuno-EM 10
   2.4 PtdIns3P localization by pre-embedding immuno-EM 13
3. Correlative Light and Electron Microscopy of MVEs 16
   3.1 Seeding of cells on gridded coverslips 17
   3.2 Fixation, staining, and mounting 18
   3.3 Observation of cells with light microscopy and registration of cell position 18
   3.4 Epon embedding, locating registered cells, and serial sectioning 19
   3.5 Electron microscopy 21
   3.6 Correlation and image processing 21
4. Concluding Remarks 21
Acknowledgments 22
References 22

Abstract

The endocytic pathway comprises a variety of intracellular compartments that regulate sorting of internalized plasma membrane constituents as well as extracellular material. A major sorting station on this route is the early endosome, where internalized receptors destined for degradation are trafficked from the limiting membrane into the interior of the endosome by formation of intraluminal vesicles (ILVs). This invagination and budding process leads to the biogenesis of multivesicular endosomes (MVEs). The formation of ILVs depends on the sequential action of protein complexes that are partly recruited in a phosphatidylinositol 3-phosphate (PtdIns3P)-dependent manner. The underlying mechanisms of the biogenesis of MVEs are still not completely understood and it is therefore of great interest to study the sorting of PtdIns3P in this process. We are describing several methods to track these sorting events by both light and electron microscopy and combination of both methods.

1. INTRODUCTION

Multivesicular endosomes (MVEs) are important organelles in the endocytic pathway where they serve to regulate signaling through sorting and subsequent lysosomal degradation of receptors and their ligand (Gruenberg & Stenmark, 2004). Receptors such as the epidermal growth factor receptor (EGFR) are sorted into intraluminal vesicles (ILVs) of the MVE and thereafter degraded upon fusion with the lysosome. The charged phospholipid phosphatidylinositol 3-phosphate (PtdIns3P) has been shown to be necessary for the formation of ILVs (Futter, Collinson, Backer, & Hopkins, 2001). PtdIns3P is found in the endocytic pathway and most abundantly on early endosomes (EEs) and internal vesicles of MVEs (Gillooly et al., 2000). PtdIns3P generated on EEs can be further phosphorylated to produce phosphatidylinositol 3,5-bisphosphate [PtdIns(3,5)P_2], which is thought to reside predominantly on late endosomes (De Matteis & Godi, 2004).

Effectors of PtdIns3P directly bind the phospholipid, mostly via one of two PtdIns3P-binding domains, the FYVE domain (Burd & Emr, 1998; Gaullier et al., 1998; Patki, Lawe, Corvera, Virbasius, & Chawla, 1998; Raiborg, Schink, & Stenmark, 2013) or the PX (Phox homology) domain (Simonsen & Stenmark, 2001). The FYVE domain is the only phosphoinositide-binding domain found so far to only bind a single phosphoinositide (namely, PtdIns3P) (De Matteis & Godi, 2004). The FYVE domain is also the smallest of the phosphoinositide-binding domains and the one with the highest sequence conservation (Simonsen & Stenmark, 2001). Several of the FYVE domain-containing proteins are involved in vesicle trafficking, signal transduction, and cytoskeletal functions.

A function of the FYVE domain was first suggested when it was found to be responsible for targeting early endosomal antigen1 (EEA1) to the EE (Stenmark, Aasland, Toh, & D'Arrigo, 1996). EEA1 has later been shown to be recruited to the EE by PtdIns3P and the small GTPase Rab5. Here EEA1 acts on homo- and heterotypic fusion of endosomes. Other proteins localizing to the endosomal pathway have also been shown to contain FYVE domains, such as Rabenosyn5 and hepatocyte growth factor-regulated tyrosine kinase substrate (Hrs) (De Matteis & Godi, 2004).

Hrs is the main component of ESCRT-0 (endosomal sorting complex required for transport). There are four ESCRT complexes, which are involved in sorting of receptors in the endosomal pathway as well as the formation of ILVs leading to the biogenesis of MVEs (Gruenberg & Stenmark, 2004). However, an ESCRT-independent mechanism has also been proposed (Stuffers, Sem Wegner, Stenmark, & Brech, 2009; Theos et al., 2006; Trajkovic et al., 2008). The ESCRT complexes have been suggested to be recruited and function sequentially. The ESCRT-0 component Hrs contains in addition to its FYVE domain, an ubiquitin-interacting motif (UIM), which can bind to ubiquitinated cargo, such as the EGFR. Both of these domains are necessary for the function of Hrs in endosomal sorting (Gruenberg & Stenmark, 2004). Components of both ESCRT-I and ESCRT-II also contain UIMs, which serve to promote inward vesiculation of the endosome membrane and help sort the ubiquitinated membrane proteins into ILVs. The ESCRT-III complex is thought to be responsible for the final abscission of the neck of the forming ILV, while Vps 4 (sometimes referred to as the fifth ESCRT complex) is required for the disassembly of the ESCRT subunits (Gruenberg & Stenmark, 2004; Henne, Buchkovich, Zhao, & Emr, 2012).

Even though a quite established model for MVE biogenesis exists, there are still several unanswered questions regarding the detailed mechanisms of this process. In the following procedures, we describe a diverse set of microscopy methods to study MVEs and their biogenesis with special reference to PtdIns3P.

2. LOCALIZATION OF PTDINS3P AND EGFR

The EGFR is in unstimulated cells located to the plasma membrane and also to some extent in the limiting membrane of endosomes. Upon stimulation with EGF, the receptor is internalized via endocytosis and will be found on the limiting membrane of EE. Following ubiquitination, the EGFR is sorted into the lumen of the endosomes through the action of PtdIns3P and the four ESCRT complexes. The receptor will then be found on the membrane of ILVs and signaling is no longer possible. Studies into this distribution have mainly been executed, employing tagged EGF and antibodies against the receptor.

Determination of PtdIns3P localization has been performed using FYVE or PX domains as probes. This can be done either by treating the cells with the probe in much the same way as when labeling with antibodies or by transfecting cells with a tagged version of the probe. Isolated FYVE domains have a rather weak affinity for PtdIns3P (Gillooly et al., 2000; Sankaran, Klein, Sachdeva, & Lemmon, 2001). *In vivo*, homodimerization of the proteins containing a FYVE domain together with additional structural features such as the UIM in Hrs seems to be necessary for their efficient membrane recruitment (Gruenberg & Stenmark, 2004). To increase the binding affinity of genetically engineered probes, it has therefore been a common approach to employ two FYVE domains together when studying the localization of PtdIns3P (Gillooly et al., 2000; Hayakawa et al., 2004; Stuffers et al., 2010). The FYVE domain can either be expressed as monomers, which can be homodimerized by adding a cell permeable dimerizer (Hayakawa et al., 2004; Stuffers et al., 2010), or by expressing a recombinant 2xFYVE probe (Gillooly et al., 2000), in both cases tagged with green fluorescent protein (GFP). These dimerized domains have a very high avidity for PtdIns3P. Transient or stably transfected cells expressing these probes can then be studied directly by confocal microscopy or by electron microscopy (EM) following labeling against the GFP tag of the FYVE probe. When using protein–lipid interaction domains to study the localization of phosphoinositides, one has to be aware of the fact that overexpression of these probes might displace endogenous PtdIns3P-binding proteins and thereby induce alteration in endosomal morphology and biology. High-level overexpression of fluorescently tagged 2xFYVE and PX domains probes leads to enlargement and clustering of endosomes, indicating defects in the endosomal sorting pathway.

Fluorescently labeled PtdIns3P and other phosphoinositides have been described and are commercially available; however, their intracellular distribution is heavily influenced by the dye conjugated to the lipid, rendering them unsuitable for monitoring phosphoinositides in living cell. Therefore, the most suitable approach is still the use of probes based on protein–lipid interaction domains like the FYVE or PX domain (Schultz, 2010). We have found that it is crucial to have low expression levels of the probe in order to prevent dominant-negative effects. The most reasonable approach to achieve this is the generation of stable cell lines, either by selecting a weakly expressing clone or by sorting a weakly expressing population of cells using flow cytometry. Furthermore, we found that weak promoters, for example, the PGK promoter, results in low expression levels with minimal perturbance of endosomal morphology (unpublished data).

2.1. PtdIns3P and EGFR localization by confocal microscopy

For studying the localization of PtdIns3P, we have generated stably transfected U2OS cells with a GFP-tagged 2xFYVE. U2OS cells were transfected with pEGFP-2xFYVE using FuGENE6 (Roche) and seeded out at low density. Stable cell clones were selected by 800 μg/ml G418 selection and characterized by fluorescent microscopy. Care was taken to select clones with a moderate expression level.

1. U2OS-GFP2xFYVE cells are seeded out on coverslips (4×10^4 cells/well) and grown in DMEM containing 100 U/ml Penicillin, 100 μg/ml Streptomycin, 10% FCS, and 2 mM stable glutamine (this medium is used throughout this chapter unless specified otherwise).
2. After 16–24 h cells are either incubated with 50 ng/ml Alexa Fluor 647 EGF (Life Technologies, UK) for 15 min or fixed directly.
3. For fixation cells are placed on ice, washed once in cold PBS, and fixed in 3% Formaldehyde (FA)/1× PBS (from 10% Formalin solution, Sigma-Aldrich, St. Louis, MO) for 15 min.
4. Wash three times in PBS.
5. Prepare a solution of PBS containing 0.05% saponin. This will be used for diluting antibodies and washing.
6. Incubate cells in PBS/saponin for 5 min to permeabilize the cells. For some antibodies, conducting this step before fixation and thereby washing out some of the cytosol can improve the visualization of membrane-associated proteins.
7. Using some drops of water, stick parafilm on the bench and apply small drops (30–50 μl) of the primary antibody, Sheep anti-EGFR (1/1000 in PBS/saponin) (Fitzgerald Industries International, Acton, MA). Coverslips are placed (cells facing downward) on each drop. Cover with a lid and incubate for 20 min.
8. Replace in wells and wash three times in PBS/saponin.
9. Incubate with secondary antibody, Alexa Fluor 555 anti-sheep (1/200) as in point 7, for 20 min in the dark.
10. Wash once in PBS/saponin, then twice in PBS.
11. It is possible at this step to also stain the nuclei with a brief incubation in Hoechst as described in Section 3.2.

12. Dip each coverslip in dH$_2$O before mounting on a microscope slide using 10 μl of Mowiol. Let solidify for 30 min at 37 °C.
13. Cells are imaged using a Confocal scanning microscope (LSM 780, Zeiss, Oberkochen, GER or similar) and images are deconvolved with appropriate software (in our case Huygens Essential Software). Results are shown in Fig. 1.1C.

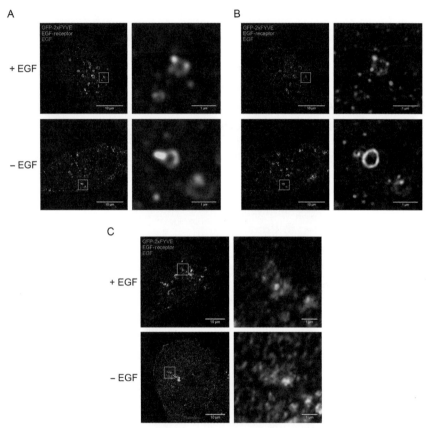

Figure 1.1 Localization of GFP2xFYVE, EGF, and its receptor studied using super-resolution and confocal microscopy. U2OS-GFP2xFYVE cells seeded on coverslips. After about 20 h the cells were either fixed directly (A, B, C, lower) or following a 15 min pulse with 50 ng EGF-Alexa647 per ml medium (A, B, C, upper). Cells were fixed in 4% FA/0.1% GA (A and B) or 3% FA (C) and labeled with Sheep anti-EGFR (1/1000) followed by a secondary Alexa Fluor 555 antibody. (A) Deconvolved widefield SIM images, (B) reconstructed SIM images, and (C) deconvolved confocal images. Bars: 10 μm (overviews, A, B, C, left), 1 μm (enlarged areas, A, B, C, right). (See color plate.)

2.2. PtdIns3P and EGFR localization by super-resolution light microscopy

For studying more closely the relationship between PtdIns3P and the EGFR, we have employed a 3D-SIM microscope (DeltaVision OMX Microscope Applied Precision, GE Healthcare, Issaquah, WA). The 3D structured illumination technique doubles the optical resolution of the light microscope and greatly facilitates imaging of subdomains of endosomes, which are close to the resolution limit of a confocal microscope. This technology can be used to study live cells or fixed cells on coverslips prepared in much the same way as for conventional confocal microscopy, using the same fluorescent antibodies and dyes. However, it is important that the fluorophores are bleach resistant especially if the signal is weak, to allow proper reconstruction of the images. Flurophores such as Alexa dyes and DyLight work well. However, even if direct imaging of GFP is often possible, week signals may need enhancement by antibody staining.

Samples prepared for confocal microscopy can also be studied using SIM microscopy, which is the advantage of this technique as compared to other super-resolution techniques such as STED or STORM/PALM. However, we have noticed that some modifications of both sample preparation and microscopy procedures can significantly enhance the result. First, when studying endosomes using the SIM, a higher degree of fixation will give a better sample preservation, resulting in rounder vesicles. 4% FA/0.1% glutaraldehyde (GA) (Fluka, Sigma-Aldrich) in PBS for 45 min can be used without giving off high autofluorescence or inhibiting antibody binding, which will result from using higher concentrations of GA.

Labeling with antibodies is then performed as described in the previous section. Mounting of the coverslip on microscope slides can be done using regular Mowiol (as above) or by mounting with ProLong® Gold Antifade Reagent (Life Technologies) and sealing around the edges of the coverslip using nail polish. It is important that the mounting media has properly solidified before microscopy.

We performed 3D SIM imaging on an OMX V4 system equipped with two cooled sCMOS cameras and 405, 488, 568, and 642 nm laser lines. Using the appropriate excitation wavelengths of the used fluorophores, cells were illuminated with a grid pattern and for each image plane, 15 images were collected (five phases for three rotations of the illumination pattern). To ensure proper reconstruction, at least seven image planes with z-spacing of 125 nm were collected from each cell. Super-resolution images were reconstructed from the raw image files using Softworx software (Applied Precision, GE Healthcare) and processed in ImageJ for presentation. For comparison, wide field images were generated from the raw SIM files by averaging the five phases of one image rotation. The resulting wide field image was then deconvolved using Softworx software and processed in ImageJ. Results are shown in Fig. 1.1A and B.

2.3. PtdIns3P and EGFR localization by immuno-EM

For preparation of cells for immuno-EM, we grow 4×10^5 U2OS-GFP2xFYVE or HEp2 cells in 5 cm dishes for 16–24 h. Cells are either stimulated with 50 ng/ml EGF for 60 min or fixed directly in 4% FA/0.1% GA (from 16% EM grade FA, Polysciences, Warington, PA) in phosphate buffer (pH 7.4) for 45 min at room temperature. This will allow detection of the FYVE probe using a Rabbit anti-GFP, kindly provided by Terje Johansen (Tromsø University, Tromsø, Norway) and detection of the EGFR using Sheep anti-EGFR (Fitzgerald Industries International). When antibodies to be used bind with low affinity, the GA in the fixation step can be left out, but this will result in poorer structural preservation. To avoid any stress on the cells before fixation, the fixation mixture is made at double concentration and then added directly to the cells in the same amount of medium. Cells are further processed according to the following protocol:

1. After fixation, the cells are washed thrice in buffer and scraped in approximately 1 ml of 1% gelatin in PBS from the dish. They are then centrifuged for 2 min at 2000 rpm in 1.5 ml microcentrifuge tubes. The greater part of the gelatin is removed and replaced with 12% gelatin/PBS, the cells are resuspended and left for infiltration for 10 min at 37 °C, and pelleted at 10,000 rpm for 10 min.
2. Excess gelatin is removed and the samples are transferred onto ice for at least 20 min to solidify before blocks of approximately 0.2 mm^3 are prepared from the pellet (see Fig. 1.2 for details).
3. The blocks are infused with 2.3 M sucrose for at least 1 h at 4 °C on a rotator. Allowing infusion up to 16 h will facilitate sectioning. Blocks are then mounted on sample pins and frozen in liquid nitrogen.
4. Ultrathin cryosections are cut at -110 °C on a cryo-microtome (Leica Ultracut UCT, Leica Microsystems or similar) and collected with a 1:1 mixture of 2.2% methyl cellulose and 2.3 M sucrose.
5. Sections are transferred to formvar/carbon-coated mesh-grids (GilderGrids, Lincolnshire, UK).

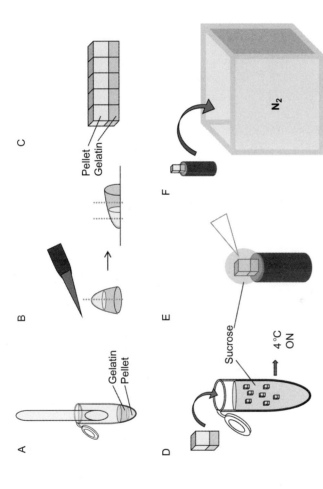

Figure 1.2 (A) After the pellet infused in 12% gelatin has hardened on ice for at least 20 min, it is removed from the tube by employing a straw, cut at an angle and dipped in 2.3 M sucrose. This will facilitate loosening the pellet from the sides of the tube. (B) The pellet, with a small amount of sucrose, is then placed on a cutting board (Petri dish), bottom facing upward. With a sharp, pointed knife, cut the pellet in half. For each half lay it down and cut off excess gelatin (right side) and if a large pellet also the tip of this (left). (C) The resultant half disk is cut into two or three strips depending on pellet size and further sectioned into blocks, approximately $0.2^2 \times 0.3$ mm. (D) The blocks are immersed in 2.3 M sucrose and incubated overnight (ON) at 4 °C on a rotator. (E) The blocks are mounted on sample pins and excess sucrose removed using filter paper. (F) Samples are immersed in liquid nitrogen. Proceed with trimming and sectioning of the frozen cells on the cryo-ultramicrotome. (For color version of this figure, the reader is referred to the online version of this chapter.)

6. The gelatin in the sections is removed by floating the grids on water (e.g., in a 4-well plate) at 37 °C and labeled with primary antibodies for 20 min at room temperature followed by a bridging secondary antibody if necessary, and protein A–gold conjugates (Peters, Neefjes, Oorschot, Ploegh, & Geuze, 1991; Slot, Geuze, Gigengack, Lienhard, & James, 1991). The Rabbit anti-GFP antibody is diluted 1/150 in PBS/0.5% BSA, while the Sheep anti-EGFR is diluted 1/500 in PBS only. The use of BSA as blocking agent in combination with sheep antibodies can cause background on the sections. All antibody dilutions are centrifuged at 10,000 rpm for 2 min before labeling. Protein A–gold conjugates (ECM, Utrecht) are otherwise used according to the manufacturer's recommendations and incubation should not exceed 20 min.
7. After embedding in 2.2% methyl cellulose/0.4% uranyl acetate, sections are observed at 60–80 kV in a TEM.

Results shown in Fig. 1.3 confirm previous findings on localization of PtdIns3P and EGFR. Labeling of anti-GFP is seen on internal vesicles of MVEs and on small vesicles surrounding the MVEs. However, we also find slight labeling of GFP (2xFYVE) on the limiting membrane of the MVE. This can be due to high level of expression of the GFP-2xFYVE probe which might retain PtdIns3P in the limiting membranes of the endosomes for a longer time or suggest the presence of PtdIns3P on the limiting membrane of MVEs. The EGFR is upon stimulation with EGF also found on ILVs.

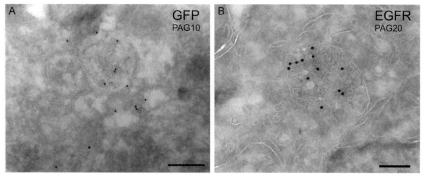

Figure 1.3 Localization of GFP2xFYVE and the EGFR studied by cryo-immuno-EM. (A) U2OS-GFP2xFYVE cells were grown in 5 cm Petri dishes for about 20 h before fixation. Sections were labeled with Rabbit anti-GFP (1/150) and 10 nm protein A–gold (PAG). (B) Hep2 cells were grown for about 20 h before being stimulated with EGF (50 ng/ml) for 60 min. Sections were labeled with Sheep anti-EGFR (1/500), Rabbit anti-Goat (1/100) bridging antibody, and 20 nm PAG. Bars: 200 nm.

2.4. PtdIns3P localization by pre-embedding immuno-EM

When conducting experiments using EM as with any other method, it is important to bear in mind that the experimental method and preparation of the specimen have great influence on the observed distribution of PtdIns3P. This is very well demonstrated by repeating the above experiment using the alternative method of pre-embedding labeling. This method can be more sensitive as antibodies are able to recognize antigens throughout the whole cell volume, although permeabilization may lead to loss of cell constituents. Another influencing factor is that penetration of antibodies into cell organelles might be hampered due to varying penetration through these membranes. When using the Tokuyasu method as described above, one should bear in mind that only antigens on the section surface are recognized.

2.4.1 Pre-embedding labeling

We seed out U2OS-GFP2xFYVE cells on coverslips in 24-well plates at a density of 4×10^4 cells/well. We routinely prepare four to eight coverslips per experimental setting when doing pre-embedding immuno-labeling and subsequent monolayer embedding, as data acquisition is possible only from a limited number of cells per coverslip. This is in contrast to confocal microscopy where one can study the whole coverslip, or Epon-embedded cell pellets, where numbers of cells are usually not a limitation. In addition, coverslips can easily be damaged during the preparation and subsequent sectioning of the monolayer is more prone to errors as compared to sectioning of pelleted cells.

Subsequent labeling was carried out with the same antibody used in the previous section (Rabbit anti-GFP); however, the concentration could be reduced to 1:2000 when following the protocol described below:

1. Cells are fixed in 4% FA/0.1% GA in 0.1 M PHEM for 45 min at room temperature and washed three times in buffer.
2. The cells are permeabilized for 10 min in 0.01% Saponin/0.5% BSA/ 0.1 M PHEM buffer (60 mM PIPES/25 mM HEPES/10 mM EGTA/ 2 mM MgCl$_2$ with pH of 6.9) (from here referred to as buffer A).
3. Incubate with primary antibody (Rabbit anti-GFP) diluted in buffer A for 30 min and subsequent washing in buffer A three times for 5 min.
4. The secondary antibody which is conjugated directly to nanoGold (Alexa Fluor 594 FluoroNanogold Fab'-rabbit) is used at

manufacturer's recommendations (Life Technologies) and the cells are incubated for 30 min. Further three times for 5-min washing in buffer A follows, with subsequent steps of washing in PHEM buffer.
5. Cells are then postfixed in 1% GA in PHEM buffer for 10 min and washed twice in PHEM.
6. Quench in 50 nM glycine for 3 min, further three washing steps in PHEM followed by 3×5 min in dH_2O.
7. To increase the size and thereby ensure visibility of the nanoGold, we perform silver enhancement and subsequent gold toning as follows: Two premade solutions are kept at $-20\,°C$ and in the dark.
    - SE mix: 23% gum arabic, 1% hydroquinone in 23.5 mM HEPES
    - 0.7% silver lactate
8. Let the mixes thaw and reach room temperature (a couple of hours) before continuing. Immediately before incubation, mix 170 µl SE mix + 30 µl silver lactate for each set of four coverslips. In a dark chamber, apply drops of approximately 40 µl on parafilm and quickly place one coverslip (cells facing down) on each drop. Place the lid on top and incubate for 5–15 min. The duration of this incubation step will determine the size of the particles. Several time points should therefore be tested upon labeling a new antigen.
9. After washing twice in dH_2O for 5 min, gold toning is performed by the following incubations steps:
    - 2% Na-acetate three times for 5 min at room temperature
    - 0.05% AuCl for 10 min on ice
    - 0.3% $Na_2S_2O_3$ twice for 10 min on ice
10. Wash thoroughly three times with dH_2O before continuing with Epon embedding.

2.4.2 Epon embedding

The following procedure can be employed for either monolayers of cells on coverslips (as above) or for pelleted cells. The main difference is the time of incubation in different solutions due to the longer penetration time for pelleted cells compared to a monolayer. The whole procedure is carried out at room temperature and under a vented hood.
1. After initial fixation, followed or not by labeling, cells are postfixed with 1% OsO_4 and 0.15% KFeCN in dH_2O for 1 h in the dark.

2. Thorough washing is carried out by three sequential incubations with dH$_2$O lasting 15 min. The cells are stained with 4% uranyl acetate for 30 min in the dark.
3. Cells are then dehydrated by incubations lasting 2 min (10 min for pelleted cells) in sequentially increasing concentration of ethanol as follows, 70%, 80%, 85%, 90%, 96%, and 100%, ending with two further incubations in 100% ethanol. For pelleted, cells continue to point 9.
4. After dehydration, place the coverslips (cells facing upward) on, for instance, Petri dish lids or plastic weighing dishes and quickly add a small drop of Epon solution to each (Sigma-Aldrich). Leave for 1 h at room temperature.
5. Gelatin capsules (0.5 ml, Davcaps Ltd., Cheltenham, UK) labeled with a specimen name and filled with Epon are placed, opening downward on top of each coverslip, after removal of excess Epon from the coverslip.
6. The Epon is polymerized at 60 °C for 48–60 h.
7. To remove the coverslip from the hardened Epon, the samples are placed on a hot plate at 100 °C for 2 min and subsequently immersed in liquid nitrogen; if necessary repeat until the coverslip is loose and can be removed from the specimen. Continue at point 13.
8. The pellet immersed in 100% ethanol is cut into smaller pieces. The pieces are incubated in Epon on a rotator for 16–20 h at room temperature (still inside a fume hood).
9. The Epon is exchanged for fresh and incubated further on a rotator at room temperature for at least 1 h.
10. Each piece of the pellet is placed in a gelatine capsule, which is then filled with embedding medium. Label the specimen by inserting a small numbered paper slip in the gelatine capsule before filling with Epon.
11. The Epon is polymerized at 60 °C for 48–60 h.
12. Remove the gelatine capsule by incubating in water at 60 °C for 30 min while stirring.
13. Ultrathin sections are cut on a microtome (Leica Ultracut UCT, Leica Microsystems or similar) and collected on formvar/carbon-coated mesh-grids (GilderGrids) by immersing the grid in the water bath on the knife and picking up sections from underneath. They are then contrasted using 1% Pb-citrate for 2 min.
14. Sections are observed at 60–80 kV in a TEM. Images are processed using Photoshop. Results are shown in Fig. 1.4.

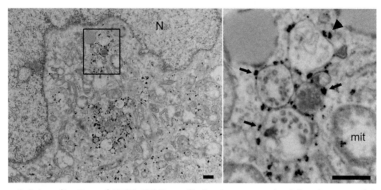

Figure 1.4 Localization of GFP2xFYVE studied using pre-embedding immuno-labeling and Epon embedding. U2OS-GFP2xFYVE cells were grown on coverslips for about 20 h before being fixed. Then labeled with Rabbit anti-GFP (1/2000) followed by Alexa Fluor 594 FluoroNanogold Fab anti-rabbit, silver enhanced and embedded in Epon. When highly expressed, 2xFYVE results in clustering and a slight enlargement of endosomes. Rabbit anti-GFP is seen on the limiting membranes of both EE (arrowhead) and MVEs (arrow). Nucleus (N) and mitochondria (mit) are indicated. Bars: 2 μm.

It is important to conduct the whole procedure up to the point of polymerization, under a vented hood and avoid skin contact with solutions used. Take special care when using fixation solutions as they are toxic and strong allergens. Different solutions must be collected in separate containers and disposed of according to local guidelines. OsO_4 can be neutralized by adding surplus amount of olive oil or vitamin C.

3. CORRELATIVE LIGHT AND ELECTRON MICROSCOPY OF MVES

An increasingly more widespread method to obtain additional information about a cell and its components is to correlate directly findings from the light microscope with those from the EM. Even with the increasing resolution of the light microscope, the resolution obtained with EM is still 100-fold higher. Moreover, fluorescence microscopy lacks any structural information and organelles or structures that have not been labeled will not be visible. Correlative light and electron microscopy (CLEM) is normally conducted by first imaging the cells at the confocal level, live or fixed. They are then fixed and further processed for EM, by conventional chemical fixation or using high-pressure freezing, whereupon the same cell studied by light microscopy can then be studied at the ultrastructural level. This method can be very useful when studying short-lived events, such as specific stages of the cell cycle, or when having low transfection efficiencies in your cell cultures. If characterization of cellular structures by EM is hampered, due to low antibody sensitivity the CLEM method is particularly useful in identifying structures on the light microscopy level followed by EM. We show how this method can be employed to closely study the structure of specific endosomes.

Cell lines stably expressing fluorescently tagged marker proteins greatly facilitate the process of correlative microscopy. The fluorophore will facilitate the imaging at the confocal level, enable live cell microscopy if necessary, and the tags can often more easily be labeled against the EM level. However, as we have done here, it is also fully possible to fixate the cells, label with an antibody which can be imaged with a light microscope, and thereafter process the sample for EM, following the procedure outlined as follows:

1. Seeding of cells on gridded coverslips
2. Fixation, staining, and mounting
3. Observation of cells with light microscopy and registration of cell position
4. Epon embedding, locating registered cells and serial sectioning
5. Electron microscopy
6. Correlation and image processing

3.1. Seeding of cells on gridded coverslips

We have employed an inducible HeLa cell line, which after induction with $CdCl_2$ expresses constitutively active Rab5Q79L, resulting in enlarged endosomes (Stenmark et al., 1994). 2×10^4 HeLa-Rab5Q79L cells are seeded on sterilized, photoetched, gridded coverslips (#72265-12, Electron Microscopy Sciences, Hatfield, PA). It is important that the cells are subconfluent, as the pattern of cells will aid in the localization of the cell of interest and cells packed too densely will mask the grid pattern on the coverslip. Each square in the grid on the coverslip is 0.5 mm and contains a number from 01 to 100. Ensure that the coverslip has the etched side facing up. The numbers will be shown the right way and not inverted. This is necessary in order to transfer the grid and numbers onto the Epon sample made at Section 3.4.

3.2. Fixation, staining, and mounting

After about 20 h, cells are fixed in 4% FA/0.1% GA/0.1 M PHEM for 45 min at room temperature. Endosomes are labeled using a mouse anti-human CD63 (obtained from the Developmental Studies Hybridoma Bank of the University of Iowa (Iowa City, IA)) followed by the secondary antibody Alexa Fluor 488 anti-mouse (1/500) following the protocol detailed in Section 2.1. Before mounting, the cells are stained with Hoechst 2 µg/ml for 10 min to assist correlation and washed twice in buffer. Thereafter the coverslip is dipped in dH_2O, mounted on microscope slides using Mowiol and left to set for 30 min at 37 °C. To ensure coverslips are kept separate when detaching them from the microscope slides, coverslip should be mounted on individual slides.

3.3. Observation of cells with light microscopy and registration of cell position

There are two main reasons to employ CLEM, which will determine the choice of light microscopy most appropriate to the procedure. One either wants to study in more detail the exact same structures viewed with the light microscope or when studying a rare event the light microscope is an aiding tool to help find the correct cell or structure to be studied by EM. For the latter purpose, a confocal microscope is well suited whereas with the first motive using a super-resolution system to obtain that extra level of detail at this point of the procedure will aid the actual correlation of that seen at the light microscope with the detailed micrograph obtained with the EM. Below the two alternatives are detailed.

3.3.1 Confocal microscopy

Using a 20× objective, find a cell of interest (or if possible two to three cells in the vicinity of each other). Image, using all necessary fluorescent channels in addition to face contrast, an overview of the cells. Make sure the imaged area covers at least one grid square with number. If necessary take a 3 × 3 tile. The focus plane showing the grid pattern will also show the part closest to the substrate and largest outline of cells, which will facilitate finding back the correct cells with the EM. In the overview image, note a number on each cell of interest. Save the position of the cells in the microscope software, switch to a 63× objective and revisit the cell of interest. Take high-resolution images of the cell in all necessary z planes.

3.3.2 Super-resolution light microscopy

The light microscopy step employing the SIM is executed in much the same two-step fashion as with confocal microscopy. First, a low magnification overview of cell and grid is acquired followed by a high-magnification study of the cell of interest. To most easily obtain a large overview, we employ a $20\times$ objective on a Delta Vision Deconvolution microscope (Applied Precision, GE Healthcare). An appropriate area for imaging is located and an overview image is obtained by acquiring a tile series of an area covering at least one square in the grid, showing all flurophores in addition to phase contrast. This image is then transferred to the SIM system. It is also important to register the x and y coordinates of the center of the imaged area.

The microscope slide is placed in the SIM in the same orientation as on the Delta Vision and x, y coordinates entered into the software. The SIM coordinates are calculated and area of interest revisited. For this to be possible it is important that the stages of the two microscopes are synchronized. When the correct area is located, the cell of interest is imaged using the stationary $60\times$ objective on the SIM. If imaging more than one cell on a coverslip, the order of acquisition of the cells is noted on the overview obtained using the Delta Vision microscope.

When all necessary images have been obtained, the coverslip is removed from the microscope slide. The Mowiol is dissolved by immersing microscope slide with coverslip in buffer for an hour at room temperature or overnight at 4 °C, until the coverslip has loosened. It is important not to damage the specimen by dragging the coverslip with forceps over the glass slide.

3.4. Epon embedding, locating registered cells, and serial sectioning

The cells are postfixed in OsO_4 and further processed for Epon embedding as described for monolayers of cells, in Section 2.4.2. The inverted grid pattern will be visible in the solidified block. The region of interest is located and trimmed down using a trimming device (Leica trimmer) and one corner (we as a rule use the upper-right corner) is trimmed back in order to orient the sections when observing them by EM. Ultrathin serial sections are cut on a microtome. We employ a non-coated slot grid as a tool to pick up sections, which are then placed on formvar/carbon-coated slot grids (GilderGrids) and contrasted using Pb-citrate (1%) for 2 min. For detailed instructions into serial sectioning see, for instance, Mironov and Beznoussenko (2012).

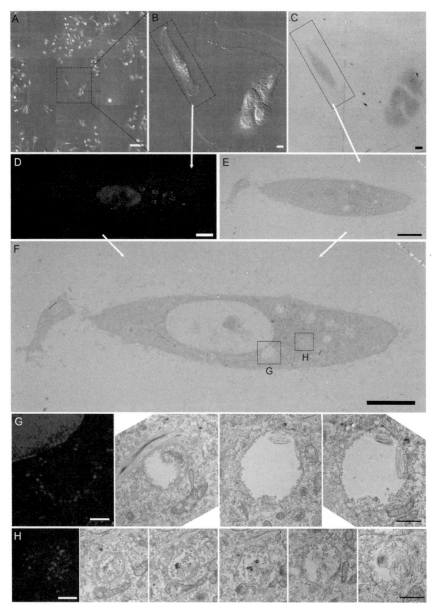

Figure 1.5 Study of ultrastructure of CD63 positive endosomes by CLEM. (A) An overview was obtained using the 20× objective on a Delta Vision Deconvolution microscope. The grid pattern in the coverslip is seen and the cell of interest is found on the left side of the number 55. An enlargement of the cell of interest with a neighboring cluster of four cells is shown in (B). The same area viewed at low magnification by EM is seen in (C). (D) Projection of the super-resolution image obtained with the SIM

3.5. Electron microscopy

Sections are observed at 60–80 kV in a TEM. It is important to acquire overview images at low magnification of both entire grid with sections as well as cell of interest with surrounding cells for orientation, in addition to the high-magnification images of structures of interest. The earliest sections are often best suited for relocating cell of interest, while other sections will contain the endosomes.

3.6. Correlation and image processing

Images from the SIM are processed by reconstruction, image alignment and alternatively quick projection. Using Photoshop images obtained with the light microscope are overlaid with those from the EM by copying the first into the second. The light microscopy layer is then made translucent by using "Blending options" and choosing "color" or "overlay". Results are shown in Fig. 1.5.

4. CONCLUDING REMARKS

We have described a variety of microscopical techniques useful for characterizing the process of MVE biogenesis using fluorescence and EM and the combination of both. When interpreting results using these techniques it should be taken into account that probes, such as 2xFYVE expressed in cells, might interfere and compete with endogenous effectors of PtdIns3P. We therefore encourage investigators employing these probes to apply several experimental approaches to verify their findings. We are, however, convinced that a combination of the above-described techniques are extremely useful in understanding the intracellular trafficking events governing the sorting of receptors in the endocytic pathway and in MVE biogenesis. Especially, the combination of light and electron microscopy in CLEM is a very powerful tool.

---

micosrocope, while (E) is a section from the EM of the same cell. This section is chosen because of being located centrally in the Z direction of the cell and all of the enlarged endosomes can be found in this section. An overlay of the SIM image (D) and the electron micrograph (E) are seen in (F). Two endosomes (G) and (H) are indicated and SIM projection and several sections of the same endosomes are shown. Bars: 100 μm (A), 10 μm (B–F), and 1 μm (G and H). (For color version of this figure, the reader is referred to the online version of this chapter.)

ACKNOWLEDGMENTS

This work was supported by grants from the Norwegian Research Council, the Norwegian Cancer Society, the European Science Foundation, the Bothner Foundation, and the European Research Council. Thanks to Marianne Smedstad and Lene Hermansen for technical support.

REFERENCES

Burd, C. G., & Emr, S. D. (1998). Phosphatidylinositol(3)-phosphate signaling mediated by specific binding to RING FYVE domains. *Molecular Cell, 2*(1), 157–162.
De Matteis, M. A., & Godi, A. (2004). PI-loting membrane traffic. *Nature Cell Biology, 6*(6), 487–492.
Futter, C. E., Collinson, L. M., Backer, J. M., & Hopkins, C. R. (2001). Human VPS34 is required for internal vesicle formation within multivesicular endosomes. *Journal of Cell Biology, 155*(7), 1251–1264.
Gaullier, J. M., Simonsen, A., D'Arrigo, A., Bremnes, B., Stenmark, H., & Aasland, R. (1998). FYVE fingers bind PtdIns(3)P. *Nature, 394*(6692), 432–433.
Gillooly, D. J., Morrow, I. C., Lindsay, M., Gould, R., Bryant, N. J., Gaullier, J. M., et al. (2000). Localization of phosphatidylinositol 3-phosphate in yeast and mammalian cells. *EMBO Journal, 19*(17), 4577–4588.
Gruenberg, J., & Stenmark, H. (2004). The biogenesis of multivesicular endosomes. *Nature Reviews Molecular Cell Biology, 5*(4), 317–323.
Hayakawa, A., Hayes, S. J., Lawe, D. C., Sudharshan, E., Tuft, R., Fogarty, K., et al. (2004). Structural basis for endosomal targeting by FYVE domains. *Journal of Biological Chemistry, 279*(7), 5958–5966.
Henne, W. M., Buchkovich, N. J., Zhao, Y., & Emr, S. D. (2012). The endosomal sorting complex ESCRT-II mediates the assembly and architecture of ESCRT-III helices. *Cell, 151*(2), 356–371.
Mironov, A. A., & Beznoussenko, G. V. (2012). Correlative light-electron microscopy a potent tool for the imaging of rare or unique cellular and tissue events and structures. *Methods in Enzymology, 504*, 201–219.
Patki, V., Lawe, D. C., Corvera, S., Virbasius, J. V., & Chawla, A. (1998). A functional PtdIns(3)P-binding motif. *Nature, 394*(6692), 433–434.
Peters, P. J., Neefjes, J. J., Oorschot, V., Ploegh, H. L., & Geuze, H. J. (1991). Segregation of MHC class II molecules from MHC class I molecules in the Golgi complex for transport to lysosomal compartments. *Nature, 349*(6311), 669–676.
Raiborg, C., Schink, K. O., & Stenmark, H. (2013). Class III phosphatidylinositol 3-kinase and its catalytic product PtdIns3P in regulation of endocytic membrane traffic. *FEBS Journal, 280*, 2730–2742.
Sankaran, V. G., Klein, D. E., Sachdeva, M. M., & Lemmon, M. A. (2001). High-affinity binding of a FYVE domain to phosphatidylinositol 3-phosphate requires intact phospholipid but not FYVE domain oligomerization. *Biochemistry, 40*(29), 8581–8587.
Schultz, C. (2010). Challenges in studying phospholipid signaling. *Nature Chemical Biology, 6*(7), 473–475.
Simonsen, A., & Stenmark, H. (2001). PX domains: Attracted by phosphoinositides. *Nature Cell Biology, 3*(8), E179–E182.
Slot, J. W., Geuze, H. J., Gigengack, S., Lienhard, G. E., & James, D. E. (1991). Immunolocalization of the insulin regulatable glucose transporter in brown adipose tissue of the rat. *Journal of Cell Biology, 113*(1), 123–135.

Stenmark, H., Aasland, R., Toh, B. H., & D'Arrigo, A. (1996). Endosomal localization of the autoantigen EEA1 is mediated by a zinc-binding FYVE finger. *Journal of Biological Chemistry, 271*(39), 24048–24054.

Stenmark, H., Parton, R. G., Steele-Mortimer, O., Lutcke, A., Gruenberg, J., & Zerial, M. (1994). Inhibition of rab5 GTPase activity stimulates membrane fusion in endocytosis. *EMBO Journal, 13*(6), 1287–1296.

Stuffers, S., Malerod, L., Schink, K. O., Corvera, S., Stenmark, H., & Brech, A. (2010). Time-resolved ultrastructural detection of phosphatidylinositol 3-phosphate. *Journal of Histochemistry and Cytochemistry, 58*, 1025–1032.

Stuffers, S., Sem Wegner, C., Stenmark, H., & Brech, A. (2009). Multivesicular endosome biogenesis in the absence of ESCRTs. *Traffic, 10*(7), 925–937.

Theos, A. C., Truschel, S. T., Tenza, D., Hurbain, I., Harper, D. C., Berson, J. F., et al. (2006). A lumenal domain-dependent pathway for sorting to intralumenal vesicles of multivesicular endosomes involved in organelle morphogenesis. *Developmental Cell, 10*(3), 343–354.

Trajkovic, K., Hsu, C., Chiantia, S., Rajendran, L., Wenzel, D., Wieland, F., et al. (2008). Ceramide triggers budding of exosome vesicles into multivesicular endosomes. *Science, 319*(5867), 1244–1247.

CHAPTER TWO

Methods to Discriminate the Distribution of Acidic Glycohydrolases Between the Endosomal–Lysosomal Systems and the Plasma Membrane

Alessandro Magini*,[1], Alice Polchi*,[1], Brunella Tancini*, Lorena Urbanelli*, Manlio Di Cristina*, Roberta Mannucci[†], Ildo Nicoletti[†], Carla Emiliani*,[2]

*Department of Experimental Medicine and Biochemical Sciences, Section of Biochemistry and Molecular Biology, University of Perugia, Perugia, Italy
[†]IMAGE Analysis Laboratory, University of Perugia, Perugia, Italy
[1]These authors contributed equally to this work
[2]Corresponding author: e-mail address: emiliani@unipg.it

Contents

1. Introduction — 27
2. Purification of Lipid Microdomains from Cell Membranes and Glycohydrolases Activity Determination — 28
   2.1 Purification of lipid microdomains by sucrose-density gradient centrifugation — 28
   2.2 Characterization of lipid microdomains and glycohydrolase Hex and Gal assay — 30
3. Discrimination of Cell Surface Lipid Microdomain-Associated Glycohydrolases from the Intracellular Counterparts — 31
   3.1 Biotinylation of plasma membrane proteins — 33
   3.2 Lipid microdomains concentration — 34
   3.3 Affinity chromatography — 34
4. Immunology Capture of Lipid Microdomains Containing Glycohydrolases — 35
5. *In Vivo* Assay of Cell Surface Glycohydrolases — 35
6. Fluorescence Microscopy Analysis of Hex Intracellular Trafficking — 38
   6.1 Preparation of HEXA cDNA — 38
   6.2 GFP construct, transfection, and cell selection — 38
   6.3 DEAE-chromatography — 38
   6.4 Cholesterol depletion and isolation of lipid microdomains — 40
   6.5 Fluorescence microscopy analysis — 40
   6.6 Colocalization studies by confocal microscopy analysis — 42

7. Summary	43
Acknowledgment	44
References	44

Abstract

The endosomal–lysosomal system plays important roles in cellular physiology. Beyond the well-known function as terminal degradative compartment, necessary to maintain the health of the cell, lysosomes are critical for many other cellular processes, such as termination of signaling mediated by cell surface receptors and processing of internalized peptides in antigen-presenting cells. Moreover, the intracellular membrane trafficking related to the endosomal–lysosomal system plays a pivotal role in diverse physiological and pathological processes, such as exocytosis, plasma membrane repair, and endocytosis. Increasing evidences suggest that several lysosomal glycohydrolases, together with nonlysosomal glycohydrolases, are associated with cell membranes in their active form, and they are localized into lipid microdomains. The role of these forms in physiological and pathological conditions, such as differentiation and aging, neurodegenerative diseases, and cancer spreading, is under investigation.

Here we provide general methods to purify lipid microdomain proteins and to discriminate cell surface lipid microdomains-associated glycohydrolases from those not exposed on cell surface. The methods reported here have been developed to characterize the membrane-associated forms of the acidic glycohydrolases β-hexosaminidase and β-galactosidase, but they may be applied to any other protein of interest.

ABBREVIATIONS

ASB-14 amidosulfobetaine-14
Brij 98 polyoxyethylene (20) oleyl ether
CT-B cholera toxin B subunit
EZ-link Sulfo-NHS-LC-Biotin Sulfosuccimidyl-6-(biotinamido)hexanoate
flot-2 flotillin-2
Gal β-galactosidase
GM1 Galβ1,3GalNAcβ1,4-(NeuAcα2,3)-Galβ1,4Glc-ceramide
GM2 3GalNAcβ1,4-(NeuAcα2,3)-Galβ1,4Glc-ceramide
GM3 NeuAcα2,3Galβ1,4Glc-ceramide
GSL glycosphingolipid
Hex β-hexosaminidase
LAMP-2 lysosomal-associated membrane protein 2
MbCD methyl-β-cyclodextrin
MUG 4-methylumbelliferyl-N-acetyl-β-D-glucosaminide
MUGal 4-methylumbelliferyl-β-D-galactopyranoside
MUGS 4-methylumbelliferyl-N-acetyl-β-D-glucosaminide-6-sulfate
Tween 20 Polyethylene glycol sorbitan monolaurate
TX-100 Triton X-100

1. INTRODUCTION

Lysosomes are not only terminal degradative organelles but are also involved in several cellular processes such as cell surface receptor regulation (Marchese, Paing, Temple, & Trejo, 2008) and autophagic pathway (Settembre & Ballabio, 2011). Besides, lysosomal exocytosis is critical for physiological events such as plasma membrane repair (Reddy, Caler, & Andrews, 2001) and cellular immune response (Bossi & Griffiths, 2005). Moreover, it has been reported that the induction of lysosomal exocytosis promotes cellular clearance in lysosomal storage diseases (Medina et al., 2011). There are several evidence on the presence of mature and active lysosomal glycohydrolases on the cell surface (Aureli et al., 2009; Magini et al., 2008; Mencarelli et al., 2005), thus indicating the existence of a not completely elucidated lysosome-to-plasma membrane delivery mechanism mediating the translocation of lysosomal enzymes to the cell surface. An example is represented by β-hexosaminidase (Hex, EC 3.2.1.52) and β-galactosidase (Gal, EC 3.2.1.23), as forms with the same properties of their lysosomal counterparts have been identified to be associated with the plasma membrane of human cells (Mencarelli et al., 2005), where they codistribute with the lipid microdomain marker flotillin-2 (flot-2) (Magini et al., 2012). Hex and Gal are both involved in the stepwise degradation of GM1–GM3 ganglioside. Hex is an acidic glycohydrolase that cleaves terminal β-linked N-acetylglucosamine or N-acetylgalactosamine residues from oligosaccharides, glycolipids, glycoproteins, and glycosaminoglycans (Mahuran, 1999), while Gal catalyzes the hydrolysis of terminal N-linked galactosyl moiety from oligosaccharides and glycosides (Okada & O'Brien, 1968). Genetic deficiency of both Hex and Gal leads to the occurrence of glycosphingolipid lysosomal storage diseases, a group of severe neurodegenerative disorders characterized by the intralysosomal accumulation of the deficient enzyme substrates, primarily in the CNS (Scriver, Beaudet, Sly, & Valle, 1995). Moreover, alteration of these two enzymes and their membrane-associated isoforms have been observed in pregnancy (Tancini et al., 2000), apoptosis (Rosati et al., 2007), neurodegenerative disorders (Emiliani et al., 2003; Orlacchio, Martino, Sarchielli, Gallai, & Emiliani, 1998), and cancer (Emiliani, Martino, Stirling, & Orlacchio, 1995).

Interestingly, Hex and Gal are not randomly distributed on the external leaflet of the plasma membrane, but they codistribute with the lipid microdomains (Magini et al., 2012). The presence of glycohydrolases in the mature form on specialized areas of the plasma membranes such as "lipid microdomains" supports their possible role in the modification of cell surface glycosphingolipids participating to the modulation of cell functions such as cell-to-cell interactions and signal transduction pathways (Magini et al., 2012; Sonnino, Aureli, Loberto, Chigorno, & Prinetti, 2010). In fact, lipid microdomains have been implicated in the regulation of numerous cellular events, including signal transduction, membrane traffic, and viral entry/infection, because they can change size and composition in response to intra- or extracellular stimuli and favor-specific protein–protein interactions, resulting in the activation of signaling cascades (Simons & Toomre, 2000).

Two major Hex isoenzymes exist in mammalian cells derived from the assembly of two subunits, α and β, encoded by closely related genes (Mahuran, 1999). Both isoenzymes Hex A (αβ-dimer) and Hex B (ββ-dimer) are able to hydrolyze neutral substrates, but only Hex A is capable to convert GM2 to GM3, in the presence of the GM2 activator protein (Mahuran, 1999). Lack of β-subunit activity, as is the case of Sandhoff disease, lead to an alteration in the terminal endocytic organelles of Sandhoff fibroblasts (Tancini et al., 2010), whereas the overexpression of Hex α-subunit induces the coordinate increased expression of other glycohydrolases participating in the glycosphingolipid catabolism, a regulatory effect that particularly involves the plasma membrane-associated isoforms (Tancini et al., 2012).

In this review we report protocols to purify total cell lipid microdomains and separate cell surface lipid microdomain-associated glycohydrolases. In addition, a further optimization of this protocol, based on the biotinylation of cell surface proteins, which allows to discriminate the component exposed on cell surface from that associated to internal cell membranes, is reported. A method based on immunoaffinity capture to isolate flot-2-decorated lipid microdomain vesicles containing glycohydrolase enzymes is also shown. The production of Hex A-GFP-overexpressing cell, which allows to analyze intracellular trafficking and membrane association of Hex by confocal microscope analysis, is also described.

2. PURIFICATION OF LIPID MICRODOMAINS FROM CELL MEMBRANES AND GLYCOHYDROLASES ACTIVITY DETERMINATION

2.1. Purification of lipid microdomains by sucrose-density gradient centrifugation

The biochemical method exploited to analyze the microdomain organization of membranes is the extraction using mild detergents. Even if detergent extraction disrupts most lipid–lipid interactions, a minor fraction of cell membranes is preserved and can be isolated by flotation exploiting their density. Detergent also disrupts lipid–protein interactions, and only few proteins retain their association with detergent-resistant membrane. Therefore, the association of a protein with detergent-resistant membrane is often indicative of a strong interaction with lipid microdomains. Here we report a lipid microdomains purification method based on that described by Rodgers and Rose (1996) with some modifications, with experimental procedures carried out with different types of detergents:

1. Resuspend 1×10^8 cells in 0.5 ml of 10 mM Tris, 150 mM NaCl, 5 mM EDTA, protease inhibitor cocktail (Sigma-Aldrich), pH 7.4 (TNE) containing either 1% (v/v) Triton X-100 (TX-100), 1% (v/v) Brij 98, 1% (v/v) Tween 20, or no detergent. We utilized Jurkat cells, a leukemic cell line growing in suspension, but the method can be successfully applied to adherent cells such as fibroblasts. In this case cell detachment from the flask by scraping should be preferred.
2. Carry out the protein extraction on ice for 30 min or incubate at 37 °C for 15 min the sample containing Brij 98 detergent.
3. Mechanically destroy cells using Dounce homogenization (10 strokes) and centrifuge the lysate at $1500 \times g$ for 5 min at 4 °C to remove nuclei and cellular debris.
4. Mix the supernatant with 0.5 ml of 85% (w/v) sucrose solution in TNE, transfer into a thin wall polyallomer centrifuge tube, then overlay with 2.75 ml of 35% (w/v) sucrose solution in TNE and with 5% (w/v) sucrose solution in TNE.
5. Ultracentrifuge at 4 °C for 18 h at 50,000 rpm using a MLS-50 rotor and an Optima Max ultracentrifuge.
6. Recover 11 fractions of equal volume (450 µl) from the top to the bottom of the discontinuous sucrose-density gradient.
7. Use 1/15 volume of each recovered fraction to carry out immunoblot analysis under reducing conditions to reveal lipid microdomain marker flot-2 using specific antibody.
8. Spot aliquots of each fraction (1.5 µl) in a nitrocellulose membrane to reveal the presence of GM1 by cholera toxin B.

As shown in Fig. 2.1, lipid microdomain extraction methods that use detergents give a lower diffusion of the microdomain markers GM1 and flot-2 compared to the detergent-free method. Among detergents, the TX-100 provides a more stringent method to exclude contamination by proteins not associated to lipid microdomains, due to its higher solubilizing

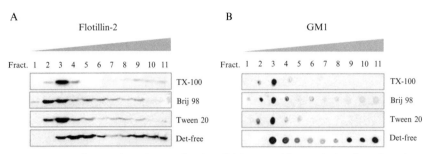

Figure 2.1 Lipid microdomains purification. (A) Immunoblot analysis of gradient fractions using anti-flotillin-2 antibody. (B) Dot blot analysis of gradient fractions to reveal the ganglioside GM1 by using cholera toxin B. Fractions are collected from the top to the bottom of the sucrose-density gradient. Fract, fraction number; Det-free, detergent free. Gray triangle indicates the sucrose concentration increase.

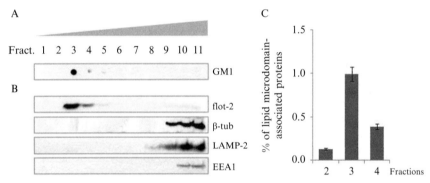

Figure 2.2 Characterization of TX-100 lipid microdomains. (A) Dot blot analysis of GM1. (B) Immunoblot analysis by using anti-flotillin-2 (flot-2), anti-β-tubulin (β-tub), anti-LAMP-2, and anti-Early endosome antigen 1 (EEA1) antibodies. (C) Protein percentage recovered in lipid microdomain fractions 2–4 with respect to the total proteins. Fractions are collected from the top to the bottom of the sucrose-density gradient. Gray triangle indicates the sucrose concentration increase.

capacity with respect to Brij 98 and Tween 20 (Schuck, Honsho, Ekroos, Shevchenko, & Simons, 2003).

2.2. Characterization of lipid microdomains and glycohydrolase Hex and Gal assay

Among the extraction methods reported above, the TX-100 is the preferred one to prevent nonlipid microdomain proteins contamination, due to its higher solubilizing capacity (Fig. 2.2).

1. Prepare TX-100 insoluble microdomains from Jurkat cells using discontinuous sucrose–density gradient centrifugation as described earlier.

2. Determine protein content of the recovered fractions using Bradford assay (Bradford, 1976).
3. Determine glycohydrolase enzyme activities (Fig. 2.3) using the final concentration of 2 mM of the fluorogenic substrate 4-methylumbelliferyl-N-acetyl-β-D-glucosaminide (MUG; Sigma-Aldrich) for Total Hex, 2 mM of 4-methylumbelliferyl-N-acetyl-β-D-glucosaminide-6-sulfate (MUGS; Toronto Research Chemicals Inc.) for Hex A and 1 mM 4-methylumbelliferyl-β-D-galactopyranoside (MUGal; Sigma-Aldrich) for Gal, in 0.1 M citric acid/0.2 M disodium phosphate buffer, pH 4.5. The enzymatic activity of the Hex system is usually determined by using the two substrates MUG, which is hydrolyzed by both α- and β-subunits forming Hex isoenzymes (Total Hex), and MUGS, which is hydrolyzed only by the α-subunit-containing isoenzyme (Hex A and Hex S). In this way with a simple activity assay we may also evaluate the amount of each isoenzyme (Magini et al., 2012). Gal activity is determined by using MUGal substrate.
4. Carry out the reaction assays in triplicate in 96-well black multiplates at 37 °C by mixing 20 μl of opportunely diluted sample and 40 μl of substrate. At the end of the reaction period add 0.290 ml of 0.4 M glycine–NaOH buffer, pH 10.4.
5. Measure the fluorescence of the liberated 4-methylumbelliferone at 360 nm excitation, 450 nm emission.

3. DISCRIMINATION OF CELL SURFACE LIPID MICRODOMAIN-ASSOCIATED GLYCOHYDROLASES FROM THE INTRACELLULAR COUNTERPARTS

Protein biotinylation is commonly used to recover proteins exposed on the external leaflet of cells, which has been standardized by our group for the purification of plasma membrane-associated glycohydrolases (Mencarelli et al., 2005). This method allows to recover native proteins which can be further characterized for their activity toward natural and synthetic substrates. We applied the *in vivo* cell surface biotinylation method. Subsequently, lipid microdomains are extracted to purify lipid microdomain proteins exposed on the external leaflet of the cell plasma membrane. Thereafter, lipid microdomain biotinylated proteins were recovered by affinity chromatography to obtain highly purified microdomain proteins exposed to the extracellular environment (Fig. 2.4).

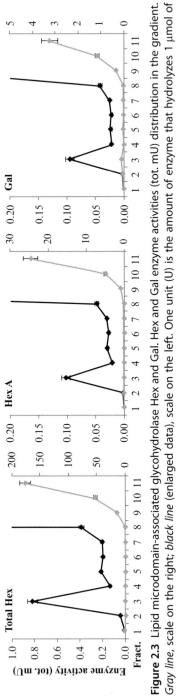

Figure 2.3 Lipid microdomain-associated glycohydrolase Hex and Gal. Hex and Gal enzyme activities (tot. mU) distribution in the gradient. *Gray line*, scale on the right; *black line* (enlarged data), scale on the left. One unit (U) is the amount of enzyme that hydrolyzes 1 µmol of substrate/min at 37 °C.

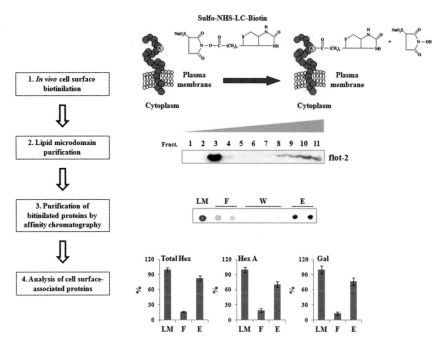

Figure 2.4 Purification of cell surface lipid microdomain proteins. (1) Schematic model of cell surface protein biotinylation. Cell surface proteins are labeled *in vivo* with EZ-Link Sulfo-NHS-LC-Biotin. (2) Lipid microdomains purification by sucrose-density gradient centrifugation. Fractions by sucrose-density gradient are analyzed by immunoblotting for flot-2. (3) Biotinylated protein purification by avidin affinity chromatography. Dot blot analysis by using HRP-conjugated streptavidin is performed to reveal biotinylated proteins purified by affinity chromatography. (4) Determination of Hex and Gal enzyme activities in the chromatographic fractions. Lipid microdomains in sucrose-density gradient fraction 3, LM. Flow through, F. Eluted biotinylated proteins, E. Gray triangle indicates the sucrose concentration increase. (For color version of this figure, the reader is referred to the online version of this chapter.)

3.1. Biotinylation of plasma membrane proteins

1. Wash cells three times with ice-cold 0.1 M sodium phosphate, 0.15 M NaCl, pH 7.4 (PBS) to eliminate culture medium. This method provides good results both with cells growing in suspension, such as Jurkat, and with adherent cells such as fibroblasts. In this last case, steps 2 and 3 are carried out directly in flask. After treatments, the cells are detached from the flask using a scraper (Mencarelli et al., 2005).

2. Incubate cells under gentle shaking for 30 min at room temperature (RT) in 20 ml of PBS containing EZ-Link Sulfo-NHS-LC-Biotin (Thermo Scientific) at the concentration of 1 mg/ml.
3. Wash cells three times with PBS containing 100 mM glycine to quench and remove excess biotin reagent and homogenize cells in 0.5 ml of TNE containing 1% (v/v) TX-100 and 20 µl of protease inhibitor cocktail (Sigma-Aldrich).
4. Isolate lipid microdomains (LM) as described above (Section 2.1).

3.2. Lipid microdomains concentration

1. Mix flot-2-positive fraction 3 with four volumes of TNE containing 0.5% (v/v) TX-100 to eliminate sucrose and concentrate lipid microdomain vesicles.
2. Ultracentrifuge the sample at 60,000 rpm at 4 °C for 2 h using a TLA-100.3 rotor and an Optima Max ultracentrifuge.
3. Resuspend the pellet in 100 µl of PBS containing 1% (v/v) TX-100 and disaggregate the lipid microdomain complexes by incubating for 10 min at 37 °C.

3.3. Affinity chromatography

1. Pack a column with 0.5 ml of Monomeric Avidin resin (Thermo Scientific).
2. Wash the column with 5 ml of PBS.
3. Wash the column with 2 ml of 2 mM biotin in PBS to block any non-reversible biotin binding site contained in avidin resin.
4. Remove the loosely bound biotin with 3 ml of 0.1 M glycine pH 2.8.
5. Equilibrate the column using 20 ml PBS.
6. Add four volumes of PBS to the sample and apply lipid microdomain preparation to the column.
7. Wash the column with five resin-bed volumes of PBS until all unbound proteins (W) are washed off and the A_{280} returned to baseline.
8. Elute the biotinylated proteins (E) with 5 mM D-biotin in PBS.
9. Spot aliquots (1.5 µl) of the recovered fractions on the nitrocellulose membrane.
10. Reveal the biotinylated proteins using 1:500 streptavidin HRP-conjugated (Thermo Scientific) for 1 h and develop by chemiluminescence.
11. Pool separately the flow through (F) and eluted (E) fractions and assay Hex and Gal activities.

4. IMMUNOLOGY CAPTURE OF LIPID MICRODOMAINS CONTAINING GLYCOHYDROLASES

The isolation of lipid microdomains containing glycohydrolases may be achieved by immunoprecipitation using specific antibodies against proteins associated to the microdomains. We adopted this method for Jurkat cells using lck antibodies (Magini et al., 2012). The tyrosine kinase lck is a Src-family kinase that connects the T-cell antigen receptor to Syk kinase ZAP-70. Upon stimulation, lck is massively translocated to the lipid microdomains. Lck-containing lipid microdomains are isolated by immunoprecipitation of lipid microdomain vesicles in flot-2-positive fraction 3 with anti-lck antibody.

1. Isolate lipid microdomains and concentrate as above.
2. Resuspend microdomain pellet from 5×10^7 cells in 200 µl TNE containing 1% (v/v) TX-100 or 200 µl TNE containing 1% (w/v) amidosulfobetaine-14 (ASB-14) and incubate overnight with 2 µg monoclonal anti-lck at 4 °C with rotation. ASB-14 is a zwitterionic ASB detergent with linear 14-C alkyl chains, which is able to solubilize the major part of lipid microdomain structures without affecting protein–protein interactions (Chevallet et al., 1998).
3. The following day, incubate the suspensions with 50 µl Dynabeads protein G (Invitrogen) at 4 °C for 3 h under constant agitation.
4. Perform controls using beads without antibody.
5. After magnetic application, recover the supernatants (IP super) and wash the beads five times with 500 µl of 200 mM NaCl.
6. Recover lipid microdomains (IP) by treating the bound complexes with 150 µl of TNE/1% (w/v) ASB-14 for 10 min at 37 °C.
7. Analyze the supernatant (IP super) and the eluted fractions (IP) by immunoblotting with anti-flot-2 antibody as control and measure of Hex and Gal activities (Fig. 2.5).

5. *IN VIVO* ASSAY OF CELL SURFACE GLYCOHYDROLASES

An alternative method to measure the Hex and Gal enzyme activities on the cell surface is the *in vivo* assay, based on the observation that the fluorogenic substrates commonly used for the *in vitro* assay of glycohydrolases activity are not taken up by living cells when cells are incubated in an

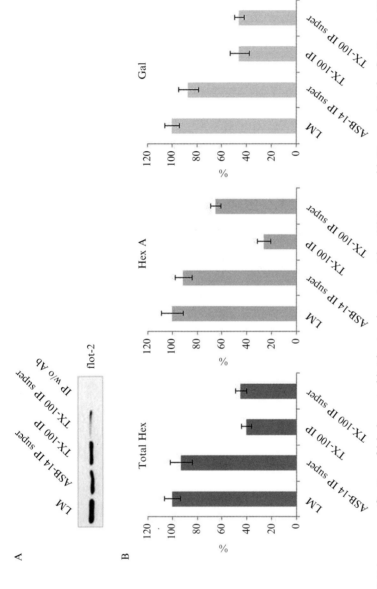

Figure 2.5 Immunopurification of lck-decorated-lipid microdomain vesicles. (A) Immunoblotting analysis for flot-2 and (B) Hex and Gal enzyme activities in immunoprecipitated lck-decorated-lipid microdomains. LM, lipid microdomains. Not immunoprecipitated vesicles resuspended with ASB-14 or TX-100 detergents, ASB-14 IP super or TX-100 IP super. Immunoprecipitated vesicles resuspended with TX-100, TX-100 IP.

isotonic medium containing the appropriate substrate (Aureli et al., 2009; Liang et al., 2006; Tancini et al., 2012). This is a useful method to assess quickly the amount of glycohydrolase exposed on the outer side of the cell, which does not require any further biochemical characterization. As control, to verify that the substrates are not hydrolyzed by enzymes eventually released into the medium during the incubation time, the cells are incubated with the reaction buffer without substrates. It is also important to determine the cell number before and after the incubation to evaluate the cell viability during the assay. In our experience, this assay does not alter cell viability.

1. Growth Jurkat cells up to the concentration of 1.5×10^6 cell/ml.
2. Recover the cells by centrifuge at $200 \times g$ for 5 min at RT.
3. Wash the cells three times by using PBS solution.
4. Centrifuge the cells at $200 \times g$ for 5 min at RT and carefully eliminate supernatant.
5. Resuspend the cell pellet (5×10^5 cell/ml) in DMEM-F12 medium without serum and phenol red pH 6.5, containing the appropriate substrate at the following concentrations: 1 mM MUG, 1 mM MUGS, and 0.5 mM MUGal and incubate at 37 °C.
6. At different times, recover the reaction mixture and stop by adding five volumes of 0.4 M Glycine–NaOH buffer (pH 10.4).
7. Measure the fluorescence of the liberated 4-methylumbelliferone at 360 nm excitation, 450 nm emission (Fig. 2.6).

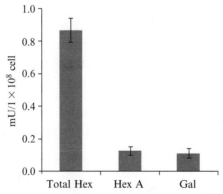

Figure 2.6 *In vivo* assay of cell surface-associated Hex and Gal. Plasma membrane-associated Hex and Gal activities are measured by incubating living cells with DMEM-F12 medium without serum and phenol red, pH 6.5, containing the appropriate substrate.

6. FLUORESCENCE MICROSCOPY ANALYSIS OF HEX INTRACELLULAR TRAFFICKING

The use of confocal microscopy and image analysis is the most effective method to evaluate the intracellular trafficking of glycohydrolases. The method reported here is based on the production of clones overexpressing the α-subunit of Hex fused with the green fluorescent protein (GFP). Hex α-subunit-GFP fusion protein is produced in HEK293 cells and leads to the formation of both Hex isoenzymes Hex A-GFP and Hex S-GFP.

6.1. Preparation of HEXA cDNA

1. Extract total RNA from 1×10^6 HEK293 cells using PureLink™ Total RNA Purification System (Invitrogen).
2. Synthesize cDNA by reverse transcription of total RNA with random hexamer primers and SuperScript™ II Reverse Transcriptase according to the manufacturer's procedure (Invitrogen).
3. Synthesize HEXA cDNA without stop codon by PCR using Phusion® Hot Start II (Finzymes) and the following primers: forward primer 5′-ATC TCG AGG CGG GCC ATG ACA AGC TC-3′ and reverse primer 5′-ATG AAT TCG GGT CTG TTC AAA CTC CTG CT-3′.

6.2. GFP construct, transfection, and cell selection

1. Clone the HEXA cDNA into the XhoI/EcoRI sites of pEGFP-N1 expression plasmid (BD Clontech) in the correct orientation and in frame with EGFP sequence.
2. One day before the transfection, seed 3×10^5 cells/well (6-well plate) in 2 ml of growth medium.
3. Transfect HEK293 cells with 1 μg/well of pEGFP-N1-HEXA or the empty pEGFP-N1 vector (Mock) using lipofectamine LTX reagent (Invitrogen) according to manufacturer's instructions.
4. Maintain the cells in standard medium and after 24 h add 1.2 mg/ml of G-418 to select stably transfected cell line (HEXA cells) (Fig. 2.7).

6.3. DEAE-chromatography

Separation and analysis of Hex isoenzymes and Gal from HEXA and mock cell extracts are performed by ion-exchange chromatography on DEAE-cellulose. The combination of chromatographic separation and enzymatic

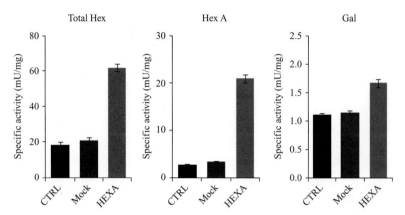

Figure 2.7 Glycohydrolases activity on control, mock, and HEXA cells. Specific activities of Hex and Gal are determined in the total cell lysates. The results are the means ± SD of three independent experiments. CTRL, control cells; Mock, pEGFP-N1 transfected cells; HEXA, pEGFP-N1-HEXA transfected cells.

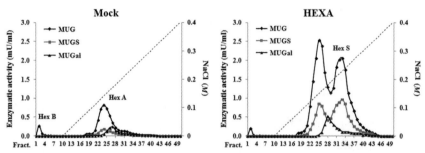

Figure 2.8 Chromatographic analysis of Hex and Gal on mock and HEXA cells. The separation of Hex isoenzymes and Gal is performed by DEAE-cellulose ion-exchange chromatography. Fractions of 1 ml in volume are collected and assayed for Hex activity using the fluorescent substrates MUG, which is hydrolyzed by both α- and β-subunits forming Hex isoenzymes, and MUGS, which is hydrolyzed only by the α-subunit-containing isoform. Gal activity is determined by using MUGal substrate. The same amount of proteins for each sample was loaded.

activity assays with both MUG and MUGS substrates provides useful information about the subunit composition of the Hex isoforms expressed in the cells. At the experimental conditions employed to perform DEAE-cellulose chromatographic analysis, Hex B (ββ-dimer) is unretained by the column and eluted with the void volume, whereas Hex A (αβ-dimer) and other Hex isoforms were eluted by a linear saline gradient (Tancini et al., 2012) (Fig. 2.8).

1. Resuspend DEAE-cellulose resin in 0.2 M Na/P buffer, pH 6.0.
2. Equilibrate a 0.5 ml column with 20 volumes of 10 mM Na/P buffer, pH 6.0.
3. Set the flow rate at 1 ml/min.
4. Dilute 0.3 mg of total proteins with 10 mM Na/P buffer, pH 6.0, up to a final volume of 0.5 ml and apply sample to the column.
5. Wash the column with 10 ml of 10 mM Na/P buffer, pH 6.0.
6. Elute the protein retained by the column using a linear 0.0–0.4 M NaCl gradient in 40 ml of 10 mM Na/P buffer, pH 6.0.
7. Collect fractions (1 ml) and assay the activity of Hex using MUG and MUGS substrates and Gal using MUGal substrate, as detailed earlier.

6.4. Cholesterol depletion and isolation of lipid microdomains

The effect of membrane cholesterol perturbation on lipid microdomain structure is assessed by treatment of cells with methyl-β-cyclodextrin (MbCD), which causes cholesterol depletion. pEGFP-N1-HEXA transfected cells are incubated with MbCD prior to the TX-100 treatment.

1. Incubate 1×10^7 cells/ml in serum-free medium for 30 min at 37 °C in presence of 15 mM MbCD (Sigma-Aldrich).
2. Isolate lipid microdomains from normal (HEXA) or MbCD-treated pEGFP-N1-HEXA cells (HEXA-MbCD), pEGFP-N1 cells (Mock) and HEK293 as control (CTRL), as described above with some modifications. Incubate 2×10^7 cells with 0.5 ml TNE containing 1% TX-100 and 20 μl of protease inhibitor cocktail for 30 min on ice.
3. After the protein extraction, dilute 1 mg of total proteins with 0.5 ml of TNE buffer containing 1% TX-100 and upload to the discontinuous sucrose-density gradient as reported in Section 2.1.
4. After the ultracentrifugation, recover 11 fractions of equal volume (450 μl) from the top to the bottom of the gradient for enzyme activity assays, Western and dot blot analysis (Fig. 2.9).
5. Incubate flot-2-positive fraction 3 for 10 min at 37 °C to disaggregate the lipid microdomain complexes.
6. Load in DEAE column (0.5 ml) for ion-exchange chromatography, as above reported (Fig. 2.9).

6.5. Fluorescence microscopy analysis

1. Plate HEXA cells onto glass coverslip and grow the cells for 24 h in a humidified incubator under 5% CO_2 at 37 °C.

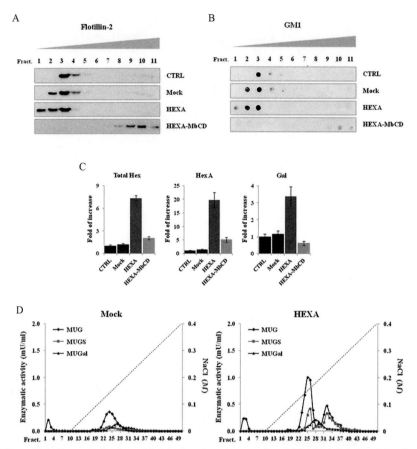

Figure 2.9 Characterization of lipid microdomain-associated Hex and Gal in HEXA cells. (A) Immunoblot analysis of gradient fractions by using anti-flotillin-2 antibody. (B) Dot blot analysis of gradient fractions to reveal the ganglioside GM1 by using cholera toxin B. (C) Hex and Gal enzyme activities in flot-2 positive fraction 3. The fold of increase is calculated with respect to the CTRL. (D) Chromatographic pattern of lipid microdomain-associated Hex and Gal enzymes. Control cells, CTRL. pEGFP-N1 transfected cells, Mock. pEGFP-N1-HEXA transfected cells, HEXA. pEGFP-N1-HEXA transfected cells treated with MbCD, HEXA-MbCD. Fraction number, Fract. Fractions are collected from the top to the bottom of the sucrose-density gradient. Gray triangle indicates the sucrose concentration increase.

2. Wash the cells three times with D-PBS and fix with 4% paraformaldehyde/PBS for 30 min at RT.
3. Rinse the slides three times with D-PBS and mount on glass coverslip using Vectashield with DAPI (Vector Laboratories).

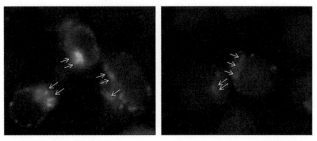

Figure 2.10 Intracellular distribution of recombinant Hex A-GFP. Fluorescent microscopy images of pEGFP-N1-HEXA cells. Nucleus is evidenced by DAPI staining. Magnification, 100×. (See color plate.)

4. Carry out fluorescence microscopy analysis (Fig. 2.10) using a 100 × oil immersion objective and process the images using an appropriate software.

6.6. Colocalization studies by confocal microscopy analysis

1. Plate HEXA cells onto glass coverslip and keep in a humidified incubator under 5% CO_2 at 37 °C for 24 h.
2. Wash the cells for three times with D-PBS.
3. Add 5 μM lipophilic membrane dye DiQ (Molecular Probes), previously diluted in D-PBS containing 2% BSA, and incubate for 30 min at RT.
4. Wash the cells three times with D-PBS and fix with 4% paraformaldehyde/PBS for 30 min at RT.
5. Mount on glass coverslip using Mowiol without DAPI and keep 2 h at RT.
6. Perform confocal analysis using an Ar/Kr laser. Detect the fluorescence of GFP and membrane probe DiQ at the excitation wavelengths of 488 and 568, respectively. Elaborate images by a SGI Octane workstation (SGI, Mountain View, CA) with the Imaris software (Bitplane, Zurich, CH) using the "Shadow projection" and "Isosurface" modules for three-dimensional reconstruction (Figs. 2.11 and 2.12).

Figure 2.11 Association of Hex A-GFP with lipid microdomains. Immunoblot analysis of gradient fractions by using anti-GFP antibody. Mock, pEGFP-N1 transfected cells; HEXA, pEGFP-N1-HEXA transfected cells; Fract, fraction number. Gray triangle indicates the sucrose concentration increase.

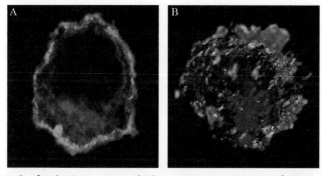

Figure 2.12 Confocal microscopy and 3D-reconstruction images of HEXA cells. Membranes are stained by lipophilic membrane dye DiQ (red). (A) Confocal microscopy image of pEGFP-N1-HEXA cells. Colocalization is reported in yellow. (B) Three-dimensional reconstruction of confocal microscopy image reported in (A). Colocalization is reported in blue. (See color plate.)

7. SUMMARY

The lysosomal glycohydrolases associated to the plasma membrane are fully processed, thus indicating the existence of a lysosome-to-plasma membrane delivery mechanism mediating the translocation of lysosomal enzymes to the cell surface. In this review we describe different methods to purify total cell lipid microdomains and separate cell surface lipid microdomain-associated glycohydrolases. Moreover, we report an experimental strategy based on the production of Hex A-GFP-overexpressing cells to analyze intracellular trafficking and membrane association of Hex by confocal microscopy analysis.

The presence of glycohydrolases associated to the external side of the plasma membrane is widely documented and a role on the modification of the cell surface glycosphingolipids participating to the modulation of cell functions has been suggested.

ACKNOWLEDGMENT

This work was supported by ELA Foundation (Agreement No. 2011-037C1B) and AIRC 5 per mille Special program 2011, Pr. 12214.

REFERENCES

Aureli, M., Masilamani, A. P., Illuzzi, G., Loberto, N., Scandroglio, F., Prinetti, A., et al. (2009). Activity of plasma membrane beta-galactosidase and beta-glucosidase. *FEBS Letters*, *583*, 2469–2473.

Bossi, G., & Griffiths, G. M. (2005). CTL secretory lysosomes: Biogenesis and secretion of a harmful organelle. *Seminars in Immunology*, *17*, 87–94.

Bradford, M. M. (1976). A rapid and sensitive method for the quantitation of microgram quantities of protein utilizing the principle of protein-dye binding. *Analytical Biochemistry*, *72*, 248–254.

Chevallet, M., Santoni, V., Poinas, A., Rouquié, D., Fuchs, A., Kieffer, S., et al. (1998). New zwitterionic detergents improve the analysis of membrane proteins by two-dimensional electrophoresis. *Electrophoresis*, *19*, 1901–1909.

Emiliani, C., Martino, S., Stirling, J. L., & Orlacchio, A. (1995). Influence of cell differentiation and protein kinase C activation on sub-cellular distribution of beta-N-acetylhexosaminidases of HL60 cells. *Physiological Chemistry and Physics and Medical NMR*, *27*, 369–376.

Emiliani, C., Urbanelli, L., Racanicchi, L., Orlacchio, An, Pelicci, G., Sorbi, S., et al. (2003). Up regulation of glycohydrolasee expression in fibroblasts from patients with Alzheimer's disease correlate with ras activation. *Journal of Biological Chemistry*, *278*, 38453–38460.

Liang, F., Seyrantepe, V., Landry, K., Ahmad, R., Ahmad, A., Stamatos, N. M., et al. (2006). Monocyte differentiation up-regulates the expression of the lysosomal sialidase, Neu1, and triggers its targeting to the plasma membrane via major histocompatibility complex class II-positive compartments. *Journal of Biological Chemistry*, *281*, 27526–27538.

Magini, A., Mencarelli, S., Tancini, B., Ciccarone, V., Urbanelli, L., Hasilik, A., et al. (2008). Identification and characterization of mature beta-hexosaminidases associated with human placenta lysosomal membrane. *Bioscience Reports*, *28*, 229–237.

Magini, A., Polchi, A., Tancini, B., Urbanelli, L., Hasilik, A., & Emiliani, C. (2012). Glycohydrolases β-hexosaminidase and β-galactosidase are associated with lipid microdomains of Jurkat T-lymphocytes. *Biochimie*, *94*, 684–694.

Mahuran, D. J. (1999). Biochemical consequences of mutations causing the GM2 gangliosidoses. *Biochimica et Biophysica Acta*, *1455*, 105–138.

Marchese, A., Paing, M. M., Temple, B. R., & Trejo, J. (2008). G protein coupled receptor sorting to endosomes and lysosomes. *Annual Review of Pharmacology and Toxicology*, *48*, 601–629.

Medina, D. L., Fraldi, A., Bouche, V., Annunziata, F., Mansueto, G., Spampanato, C., et al. (2011). Transcriptional activation of lysosomal exocytosis promotes cellular clearance. *Developmental Cell*, *21*, 421–430.

Mencarelli, S., Cavalieri, C., Magini, A., Tancini, B., Basso, L., Lemansky, P., et al. (2005). Identification of plasma membrane associated mature beta-hexosaminidase A, active towards GM2 ganglioside, in human fibroblasts. *FEBS Letters*, *579*, 5501–5506.

Okada, S., & O'Brien, J. S. (1968). Generalized gangliosidosis: Beta-galactosidase deficiency. *Science*, *160*, 1002–1004.

Orlacchio, A., Martino, S., Sarchielli, P., Gallai, V., & Emiliani, C. (1998). Beta-N-acetylhexosaminidase in peripheral blood lymphocytes and monocytes in the different forms and stages of multiplesclerosis. *Journal of Neurochemistry*, *71*, 1168–1176.

Reddy, A., Caler, E. V., & Andrews, N. W. (2001). Plasma membrane repair is mediated by Ca^{2+}-regulated exocytosis of lysosomes. *Cell*, *106*, 157–169.

Rodgers, W., & Rose, J. K. (1996). Exclusion of CD45 inhibits activity of p56lck associated with glycolipid-enriched membrane domains. *Journal of Cell Biology*, *135*, 1515–1523.

Rosati, E., Mencarelli, S., Magini, A., Sabatini, R., Tassi, C., Orlacchio, A., et al. (2007). Enhancement of lysosomal glycohydrolase activity in human primary B lymphocytes during spontaneous apoptosis. *International Journal of Immunopathology and Pharmacology*, *20*, 279–287.

Schuck, S., Honsho, M., Ekroos, K., Shevchenko, A., & Simons, K. (2003). Resistance of cell membranes to different detergents. *Proceedings of the National Academy of Sciences of the United States of America*, *100*, 5795–5800.

Scriver, C. R., Beaudet, A. L., Sly, W. S., & Valle, D. (Eds.), (1995). *The metabolic and molecular bases of inherited disease* (pp. 2787–2823). New York: McGraw-Hill Publishing Co.

Settembre, C., & Ballabio, A. (2011). TFEB regulates autophagy: An integrated coordination of cellular degradation and recycling processes. *Autophagy*, *7*, 1379–1381.

Simons, K., & Toomre, D. (2000). Lipid rafts and signal transduction. *Nature Reviews. Molecular Cell Biology*, *1*, 31–39.

Sonnino, S., Aureli, M., Loberto, N., Chigorno, V., & Prinetti, A. (2010). Fine tuning of cell functions through remodeling of glycosphingolipids by plasma membrane-associated glycohydrolases. *FEBS Letters*, *584*, 1914–1922.

Tancini, B., Emiliani, C., Mencarelli, S., Cavalieri, C., Stirling, J. L., & Orlacchio, A. (2000). Evidence for the regulation of β-N-acetylhexosaminidase expression during pregnancy in the rat. *Biochimica et Biophysica Acta*, *1475*, 184–190.

Tancini, B., Magini, A., Bortot, B., Polchi, A., Urbanelli, L., Sonnino, S., et al. (2012). β-Hexosaminidase over-expression affects lysosomal glycohydrolases expression and glycosphingolipid metabolism in mammalian cells. *Molecular and Cellular Biochemistry*, *363*, 109–118.

Tancini, B., Magini, A., Latterini, L., Urbanelli, L., Ciccarone, V., Elisei, F., et al. (2010). Occurrence of an anomalous endocytic compartment in fibroblasts from Sandhoff disease patients. *Molecular and Cellular Biochemistry*, *335*, 273–282.

> CHAPTER THREE

Visualizing of Signaling Proteins on Endosomes Utilizing Knockdown and Reconstitution Approach

Myoungkun Jeoung, Emilia Galperin[1]
Department of Molecular and Cellular Biochemistry, University of Kentucky, Lexington, Kentucky, USA
[1]Corresponding author: e-mail address: emilia.galperin@uky.edu

Contents

1. Introduction	48
2. Description of Methods	49
2.1 MEK2 KDAR	49
2.2 SHOC2 KDAR	54
3. Conclusions	61
Acknowledgments	62
References	62

Abstract

Spatial distribution of intracellular signaling molecules and assembly of signaling complexes are yet to be fully understood. Studies of signaling events in time or space present a particular challenge due to the adverse effects that overexpression of signaling proteins may have on their functions and localization. To follow the distribution of signaling proteins in living cells we developed a methodology named knockdown and reconstitution (KDAR) that allows one to visualize proteins at levels of expression that are close to physiological. This methodology provides a stable expression of "endogenous" shRNA for long-term silencing of the targeted gene and simultaneous expression of a DNA cassette coding for a fluorescently labeled protein, which is insensitive to the targeting shRNA. In this chapter we discuss the needed reagents and outline two experimental approaches to generate KDAR stable cell lines. First, we demonstrate how the plasmid-mediated KDAR approach is successfully utilized to visualize spatial distribution of the GFP-labeled MEK2 in living cells. We then show how the lentivirus-mediated KDAR approach is used to reconstitute and visualize expression of the ERK1/2 scaffold protein Shoc2.

1. INTRODUCTION

Binding of epidermal growth factor to the EGF receptor (EGFR) at the cell surface induces rapid internalization of ligand–receptor complexes by mechanisms of endocytosis. It also induces activation of several signal transduction cascades, including the activation of the mitogen-activated protein kinase (MAPK1/2)/extracellular signal-regulated kinase (ERK1/2) signaling cascade. Linear events of the canonical ERK1/2 cascade have been studied extensively and the effects of activating the ERK1/2 pathway on a broad array of cellular functions including proliferation, survival, motility, transcription, and metabolism are well documented (Katz, Amit, & Yarden, 2007). Proteins of the ERK1/2 pathway are located on various cellular compartments such as the plasma membrane, nucleus, and endosomal membranes (Sorkin & Goh, 2009; Sorkin & Von Zastrow, 2002). However, little is known about the mechanisms regulating the spatial distribution of these signaling molecules and the subcellular targeting of ERK1/2 k signals.

Advanced optical approaches such as FRET and single particle tracking together with use of GFP-like (XFP) fluorescent protein tags have proven to be very effective tools for live-cell imaging and have been used to map spatial organization of signaling events (Galperin, Verkhusha, & Sorkin, 2004; Padilla-Parra & Tramier, 2012). Localization of signaling proteins of the ERK1/2 pathway tagged with XFPs and interactions of these proteins with EGFR by live-cell microscopy have been previously reported (Sorkin, 2001; Sorkin, McClure, Huang, & Carter, 2000). However, most of these studies analyzed localization of overexpressed XFP-fusion proteins in the presence of an endogenous counterpart, which may lead to aberrant localization and characterized activities of signaling proteins. For instance, overexpression of small GTPase H-Ras results in activation of a large pool of these molecules due to changes in the stoichiometric ratio of its effectors and GTPase-activating proteins (GAPs), leading to changes in H-Ras subcellular localization in an EGFR-independent manner (Jiang & Sorkin, 2002).

Adaptors or scaffolds are another example of proteins for which function is highly dependent on their concentration (Bhattacharyya, Remenyi, Yeh, & Lim, 2006). Overexpression of scaffolds may direct formation of nonspecific complexes, sequestration of signaling partners, and reduced number of functional complexes, resulting in outcomes that are opposite to the proposed functions of these proteins (Cacace et al., 1999; Morrison, 2001). Certainly, the notion that overexpression of signaling proteins may affect protein functions and cellular localization always makes difficult the interpretation of the experimental outcomes.

In this chapter, we describe a knockdown and reconstitution (KDAR) methodology that prevents potential complications of protein overexpression and allows cellular imaging of signaling proteins expressed at physiological levels. We provide two strategies for utilizing the KDAR method. Both strategies involve simultaneous depletion and replacement of endogenous protein with its exogenous fluorescent protein-fused counterpart allowing for a more biologically relevant view of signaling events.

2. DESCRIPTION OF METHODS
2.1. MEK2 KDAR

MAPK kinases, MEK kinase 1 and 2 (MEK1 and MEK2), are direct activators for ERK1/2 kinases (English & Cobb, 2002). Functions of these two proteins have been studied extensively, and the significance of their localization to the plasma membrane and to the nucleus has been determined by a variety of methods, including use of the XFP protein tags. In addition to the localization at the plasma membrane and nucleus, phosphorylated/activated MEK1 has been recently reported to localize to endosomal membranes. This endosomal membrane localization was suggested to be a part of the mechanism that controls duration and amplitude of ERK1/2 activity (Teis, Wunderlich, & Huber, 2002). However, the methods utilized to study MEK1/2 localization are based on either indirect immunofluorescence or an overexpression of its GFP-tagged counterpart. This section summarizes an experimental approach developed to visualize MEK2 in living cells at near endogenous levels of expression (experimental flow is outlined in Fig. 3.1).

2.1.1 Expression constructs

To visualize MEK2 compartmentalization in living cells, a fluorescently labeled version of MEK2 was prepared. Full-length cDNA sequence encoding wild-type human MEK2 was obtained from Dr. L Heasley (University of Colorado, CO) and was transferred from pcDNA3.1-MEK2 vector to pEGFP-N2 vector (Clontech Mountain View, USA). To generate the GFP-tagged version of MEK2 a forward primer containing an *Xho*I site (5′-CCGCTCGAGATGCCCAAGAAGAAGCCGAC-3′) and reverse primer

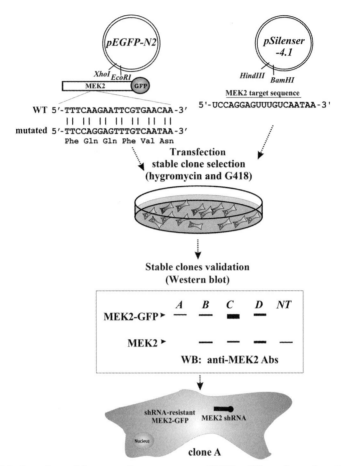

Figure 3.1 Overview of the procedure to generate KDAR cell lines. For a description see text. Sequence in MEK2–GFP shows silent mutations that have been introduced into MEK2–GFP sequence to generate siRNA-resistant cDNA. MEK2 mRNA sequence that has been targeted by siRNA is presented as MEK2 target sequence.

containing an EcoRI site (5′-CTGCAGAATTCCACGGCGGT-GCGCGTGGG-3′) after the stop codon were used to amplify the MEK2 sequence by polymerase chain reaction (PCR). PCR product was digested using XhoI and EcoRI restriction sites and ligated into pEGFP-N2 digested with XhoI and EcoRI enzymes (Fig. 3.1). Pfu polymerase was purchased from Stratagene (La Jolla, CA). Fusion to the C-terminus of MEK2 did not affect its biological kinase activity (Galperin & Sorkin, 2008). Activity of MEK2–GFP was tested using commercial antibodies that detect the activated phosphorylated forms of both MEK1 and MEK2 kinases as well as commercial antibodies that detect the activated phosphorylated forms of both MEK1 and MEK2 isoforms (Cell Signaling, Danvers, MA).

To generate a vector stably expressing shRNA, pSilencer-4.1-CMVhygro plasmid (Ambion, Inc. Grand Island, NY, USA) was used. A set of oligonucleotides corresponding to the MEK2-specific sequence of siRNA duplex #3 (5′-UCCAGGAGUUUGUCAAUAA-3′) containing *Hin*dIII and *Bam*HI restriction sites was synthesized. The following primers were used: 5′-GATCCGGAAGCTGATCCACCTTGATTCAAGAGA TCAAGGTGGATCAGCTTCCTTA-3′ and 5′-AGCTTAAGGAAGCT GATCCACCTTGATCTCTTGAATCAAGGTGGATCAGCTTCCG-3′. Oligonucleotides were annealed in annealing buffer containing 100 mM potassium acetate, 30 mM HEPES, pH 7.4, 2 mM magnesium acetate according to the manufacturer's recommendations and ligated into pSilencer-4.1-CMVhygro using *Bam*HI and *Hin*dIII restriction sites.

To generate a vector stably expressing MEK2–GFP that is resistant to shRNA knockdown, silent mutations changing the DNA sequence of MEK2 but not the amino acid sequence were introduced using a QuickChange site-directed mutagenesis kit according to the manufacturer's directions (Stratagene, La Jolla, CA, USA). The following primers were used: 5′-GTTCACCCCCGACTTTCAAGAATTCGTGAACAAATGCCTCA TCAAG-3′, and 5′-CTTGATGAGGCATTTGTTCTCGAATTCTTGAA AGTCGGGGGTGAAC-3′ to introduce silent mutations. All construct were verified by dideoxynucleotide sequencing.

2.1.2 Choice of cell type, transfections, clone selection, and imaging

The localization of fluorescently fused proteins was examined in human cervical carcinoma HeLa cells. HeLa cells were chosen as the preferred expression system due to robust ERK1/2 activation and high levels of EGFR in these cells. HeLa cells were grown in high glucose Dulbecco modified Eagle's medium (DMEM) (Invitrogen, NY, USA) containing 10% fetal bovine serum (FBS) (HyClone ThermoFisher Scientific, MA, USA) and supplemented with penicillin, streptomycin, and glutamine.

To silence protein expression by RNA interference, we first obtained "SMARTpool" and four individual duplexes to human MEK2 from Dharmacon (ThermoFisher Scientific, MA, USA) and used these for transient transfection to evaluate efficiency of each siRNA duplex. HeLa cells were seeded in 12-well plates (50–60% confluent; 1 ml of DMEM/10% FBS antibiotic free media per well) at least 20 h before transfection. siRNA transfections were performed at 24–36 h intervals according to the manufacturer's recommendations (25 nM final concentration of siRNA), using Dharmafect reagent 2 (Dharmacon, ThermoFisher Scientific, MA, USA). The efficiency of the siRNA knockdown was validated by Western blotting using MEK2-specific antibodies (Cell Signaling, Danvers, MA).

To establish cell lines that stably express both MEK2–GFP and MEK2 shRNA, cells were selected by growing the transfected cells in the presence of G418 (0.4 mg/ml) and Hygromycin (0.2 mg/ml). Individual clones were selected and validated for MEK2–GFP expression levels and expression of the endogenous MEK2 by Western blot analysis using polyclonal antibodies to MEK2. Clones in which endogenous MEK2 was efficiently silenced and MEK2–GFP was expressed at levels no more than twofold higher than endogenous levels were considered for future experiments (Fig. 3.1). All clones were monitored for possible alterations in growth characteristics, and variation in levels of the ERK1/2 cascade proteins (Raf-1, B-Raf, ERK1/2, and EGFR) and other control proteins (GAPDH, actin) (Galperin & Sorkin, 2008).

For fluorescent microscopy, cells were transfected with DNA constructs using Effectene reagent (Qiagen, Hilden, Germany) in six-well plates and then replated 1 day after transfection onto glass-bottom dishes (MatTek, MA, USA). The cells were then incubated in serum-free and phenol red-free medium containing 0.2% bovine serum albumin (BSA) for 20 h prior to the microscopy experiments. Cells were examined using a Marianas™ imaging workstation (Intelligent Imaging Innovation, Denver, CO, USA) which is based on an inverted Zeiss Axio Observer Z1 microscope equipped with 100 plan-apo/1.4NA objective, 175W Xenon illumination source (Sutter Instruments Company, CA, USA), CoolSNAP HQ CCD camera (Roper Scientific, Germany), z-step motor, independently controlled excitation and emission filter wheels (Sutter Instruments Company, CA, USA), and controlled by SlideBook software (Intelligent Imaging Innovation, Denver, CO, USA). The final arrangement of images was performed using adobe Photoshop (Adobe Systems, Mountain View, CA, USA).

2.1.3 Visualization of endosomal MEK2–GFP

MEK2–GFP-tagged protein was visualized in two cell systems: HeLa cells expressing endogenous MEK1 and MEK2, and KDAR HeLa/MEK2–GFP cells (Fig. 3.1, clone A) depleted of endogenous MEK1/2. MEK2–GFP transiently expressed in HeLa cells displayed a cytosolic distribution in serum-starved cells (not shown). In cells treated with EGF (10 ng/ml), transiently expressed MEK2–GFP was also predominantly localized in cytosol, occasionally detected at the plasma membrane, and no translocation of MEK2–GFP to the intracellular membranes was observed. We interpreted the absence of the expected membrane localization as the result of high total levels of MEKs, given that MEK2–GFP was expressed in the presence of endogenous MEK1/2.

Live-cell fluorescence microscopy of KDAR HeLa/MEK2–GFP cells revealed that MEK2–GFP was diffusely distributed in the cytosol in serum-starved cells as well (Fig. 3.2A). However, upon stimulation cells with EGFR at 37 °C for 5–10 min, MEK2–GFP was rapidly translocated to the plasma membrane. Moreover, after 10 min of EGF treatment we observed that MEK2–GFP accumulated in the intracellular vesicular structures (Fig. 3.2A). We were able to observe MEK2–GFP accumulated in vesicles for an additional 20–30 min. MEK2–GFP-containing vesicles were located predominantly in the perinuclear region. Most of these vesicles appeared as fluorescent dots of approximately similar intensity, although

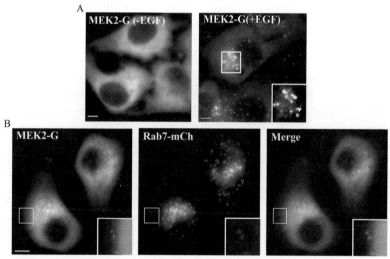

Figure 3.2 Localization of MEK2–GFP and characterization of MEK2–GFP-containing compartments. (A) Serum-starved KDAR HeLa/MEK2–GFP cells were treated with EGF (10 ng/ml) for 10 min at 37 °C. The living cells were imaged live before and after EGF treatment. MEK2–GFP is located in the cytosol. Insets show high-magnification images of the regions of the cell indicated by white rectangles. Scale bar, 10 μm. (B) KDAR HeLa/MEK2–GFP cells were transfected with mCherry-Rab7. Serum-starved cells were treated with EGF (10 ng/ml) for 10 min at 37 °C. Living cells were imaged through fluorescein isothiocyanate and Cy3 filter channels. Insets show high-magnification images of the regions of the cell indicated by white rectangles. Scale bars, 10 μm. (For color version of this figure, the reader is referred to the online version of this chapter.)

donut-like profiles of MEK2–GFP-decorating structures with visible limiting membrane could be occasionally observed (Fig. 3.2A, inset). The majority of the vesicles showed rapid lateral and directed movement over long distances, while others were relatively static. Often, several MEK2–GFP-containing vesicles were sequentially moving along a single track within a short time, characteristic of microtubule-dependent motility of endosomes. To better define MEK2–GFP-containing endosomes, we examined whether MEK2–GFP-containing endosomes colocalized with markers of early endosomes (Rab5/EEA.1) (data not shown), late endosomes, and lysosomes (Rab7 and LysoTracker™) in living cells. HeLa/MEK2–GFP cells were transiently transfected with mRFP-Rab5 or mCherry-Rab7, or preloaded with LysoTracker™. MEK2–GFP compartments colocalized with a small subset of early and late endosomes labeled with mRFP-Rab5 (not shown) and mCherry-Rab7, respectively (Fig. 3.2B). Similar results were obtained in experiments with several clones of HeLa/MEK2–GFP cells (data not shown).

Using a KDAR experimental approach we demonstrated that upon EGFR activation, MEK2 is activated at the plasma membrane and also recruited to a subset of early and late endosomal compartments. These observations also suggested that expression of MEK2–GFP in HeLa cells in the presence of endogenous MEK2 results in a competition of MEK2–GFP with the endogenous counterpart. This competition lead to our inability to detect overexpressed MEK2–GFP on endosomes. Our results emphasize a distinct advantage of KDAR methodology for studies of signaling proteins.

2.2. SHOC2 KDAR

Organization of the components of the ERK1/2 signaling cascade into scaffold complexes plays an important role in the signal transduction and the regulation of intracellular signaling in time, space, and/or signal strength (Alexa, Varga, & Remenyi, 2010; Bhattacharyya et al., 2006; Brown & Sacks, 2008; von Zastrow & Sorkin, 2007; Zeke, Lukacs, Lim, & Remenyi, 2009). Changes in concentration of scaffold proteins or stoichiometric ratios in relation to their interacting partners may result in the sequestration of the interaction partners, thus disturbing the function of the scaffolds and the consequent signaling regulation (Dhanasekaran, Kashef, Lee, Xu, & Reddy, 2007; Morrison & Davis, 2003). In this section, we describe an optimized KDAR approach to reconstitute and visualize the leucine-rich repeats protein Shoc2 in living cells (experimental design is outlined in Fig. 3.3). Shoc2 is a scaffold protein of the ERK1/2 pathway that positively regulates Ras-mediated signaling by interacting with Ras and Raf-1 (Li, Han, & Guan, 2000; Sieburth, Sun, & Han, 1998). Changes in Shoc2 protein levels lead to a dramatic change in ERK1/2 activation (Galperin, Abdelmoti, & Sorkin, 2012). The optimized KDAR approach presented here is particularly beneficial for experiments that need to be performed in difficult to transfect cells (e.g., cancer cells, primary cells).

2.2.1 Expression constructs

To visualize Shoc2 compartmentalization in living cells, a fluorescently labeled version of Shoc2 was prepared. Human Shoc2 full-length cDNA was obtained from Dr. Rodriguez-Viciana (UCL Cancer Institute, UK) and transferred from the pcDNA3.1-Shoc2 vector to the tagRFP-N1 vector (red fluorescent protein) (courtesy of Dr. V.V. Verkhusha, Albert Einstein College of Medicine). To generate the tagRFP-tagged version of Shoc2, a forward primer containing *Xho*I site (5′-CCGCTCGAGATGAGTA-GTAGTTTAGG-3′) and reverse primer containing a *Bam*HI site after the stop codon (5′-CGCGATCCCGTCAGACCATGGCACGATAT-GGACCC-3′) were used to amplify the human Shoc2 sequence by PCR using Pfu polymerase (Stratagene, La Jolla, CA). PCR product was digested using *Xho*I and *Bam*HI restriction enzymes and ligated into tagRFP-N1 digested with *Xho*I and *Bam*HI enzymes (Fig. 3.3). The construct was verified by dideoxynucleotide sequencing. Fusion to the C-terminus of Shoc2 did not affect capacity of Shoc2–tRFP to bind its partners, which was tested using coimmunoprecipitation analysis with HA-tagged H-Ras (Galperin et al., 2012).

To generate a lentiviral vector (LV) that stably expresses Shoc2-specific shRNA, pLVTHM vector was used (Addgene plasmid 12247). A set of oligonucleotides corresponding to the Shoc2-specific sequence of siRNA duplex #1 (5′-GAAGAGAAUUCAAUGCGUU-3′) was synthesized. To design the shRNA we generated oligonucleotides that contained *Mlu*I and *Cla*I sites at the ends of the hairpin. Forward primer has 5′ prefix: CGCGTCCCC and 3′ suffix: TTTTTGGAAAT. Reverse primer has 5′ prefix: CGATTTCCAAAAA and 3′ suffix: GGGGA (Fig. 3.3). The following primers were used: 5′-CGCGTCCCCGAAGAGAATTCAATGCGTT-TTCAAGAGAAACGCATTGAATTCTCTTCTTTTTGGAAT-3′ and 5′-CGATTTCCAAAAAGAAGAGAATTCAATGCGTTTCTCTTGA-AAACGCATTGAATTCTCTTCGGGGA-3′. Oligonucleotides (forward

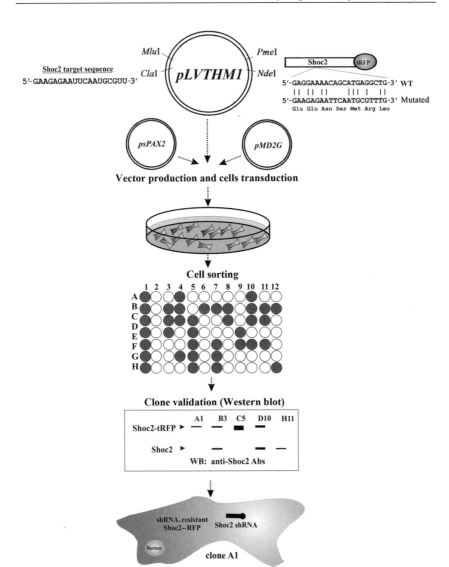

Figure 3.3 Overview of the procedure to generate KDAR cell lines using lentiviruses. For a description see text. Sequence in Shoc2–tRFP shows silent mutations that have been introduced into Shoc2–tRFP sequence to generate siRNA-resistant cDNA. Shoc2 mRNA sequence that has been targeted by siRNA is presented as Shoc2 target sequence. (For color version of this figure, the reader is referred to the online version of this chapter.)

and reverse) were dissolved in water to a final concentration of 1 µM and annealed in a buffer containing 100 mM potassium acetate, 30 mM HEPES, pH 7.4, and 2 mM magnesium acetate. Annealed oligos were phosphorylated by T4 polynucleotide kinase (PNK) (NEB, MA, USA) in the presence of T4 ligase buffer containing 1 mM ATP for 30 min at 37 °C. PNK was heat inactivated for 10 min at 70 °C. The pLVTHM vector was then digested with *Mlu*I for 2 h at 37 °C. Digested pLVTHM DNA was then purified to remove the salts using the QIAquick PCR Purification kit and digested with *Cla*I overnight at 37 °C. The *Mlu*I–*Cla*I-digested vector was purified again using the QIAquick Gel Extraction kit and concentration was determined. Next, phosphorylated oligos were ligated into the *Mlu*I–*Cla*I-digested pLVTHM vector by T4 ligase in the presence of T4 ligase buffer containing 1 mM ATP overnight. The pLVTHM-Shoc2 shRNA construct was verified by enzymatic digest with *Eco*RI and *Cla*I and dideoxynucleotide sequencing. After *Eco*RI and *Cla*I digest, the positive clones produced DNA fragments of 10,849 and 286 bp, while negative clones produced 10,849 and 236 bp.

To generate an LV simultaneously expressing Shoc2 shRNA and the fluorescently labeled Shoc2 protein, vector expressing Shoc2–tRFP resistant to shRNA knockdown was generated. Silent mutations in Shoc2–tRFP that changed the DNA sequence but not the amino acid sequence made the constructs resistant to shRNA knockdown. The following primers were used: 5′-GAGCTCAACAAATGCCGGGAGGAAAACAGCATGAGGCTG-GACTTATCCAAGAGAT-3′, and 5′-ATCTCTTGGATAAGTCCAG-CCTCATGCTGTTTTCCTCCCGGCATTTGTTGAGCTC-3′. Next, shRNA-resistant Shoc2–tRFP cDNA was amplified by PCR using Shoc2–tRFP as a template, a forward primer containing a *Pme*I site (5′-AGCTTTGTTTAAACATGAGTAGTAGTTTAGG-3′) and a reverse primer containing an *Nde*I site after the stop codon (5′-GGAATTCC-ATATGTCAATTAAGTTTGTGCCCCAGTTTGC-3′) and Taq polymerase (Invitrogen, NY, USA). PCR product was purified using the QIAquick Gel Extraction kit and ligated using the pcDNA™3.3-TOPO TA cloning kit (Invitrogen, NY, USA). pcDNA™3.3-TOPO–Shoc2–tRFP DNA was purified, digested with *Pme*I and *Nde*I, and ligated into the *Pme*I–*Nde*I digested pLVTHM-Shoc2 shRNA vector, thus replacing GFP using T4 ligase in the presence of T4 ligase buffer containing 1 mM ATP overnight (Fig. 3.3). All construct was verified by enzymatic digest and dideoxynucleotide sequencing.

2.2.2 DNA transfections and lentivirus production

To produce lentivirus expressing Shoc2 shRNA, LVs packaging plasmids psPAX2 (Addgene plasmid 12260) and the envelope plasmids pMD2G (Addgene plasmid 12259) were obtained from Addgene (http://www.adgene.org). Virus production was done in 293FT cells (Invitrogen, NY, USA). The 293FT cell line is a fast-growing, highly transfectable clonal isolate that is derived from human embryonal kidney cells transformed with the SV40 large T antigen, which produces high viral titers and allows very high levels of protein to be expressed from vectors containing the SV40 origin. Levels of expressed protein in 293FT cells are higher than other 293 cell lines or the T antigen-transformed Cos1 cells classically used (http://products.invitrogen.com/ivgn/product/R70007). Cells were maintained according to Invitrogen specifications in DMEM supplemented with 10% FBS, MEM Non-Essential Amino Acids, penicillin, streptomycin, L-Glutamine, sodium pyruvate, and 500 µg/ml G418. Two 10-cm plates were used ($2-2.5 \times 10^6$ of 293FT cells per 10-cm plate) per experiment. The transfections of DNA constructs were performed using Lipofectamine 2000 (Invitrogene, NY, USA). For 2×10 cm plates, the following transfection mix was prepared in 15 ml falcon tubes: 20 µg vector plasmid (pLVTHM-Shoc2), 8 µg envelops plasmid (pMD2G), and 14.6 µg of packaging plasmid (psPAX2). Media was removed around 14–16 h posttransfection (typically early in the morning) and replaced with 14 ml/dish of fresh preheated media. Supernatant was collected for the first time in the evening, and two to three times thereafter, every 12 h. Collected media was kept at 4 °C during the collecting period. The collected supernatants then were pooled, centrifuged for 5 min at 1500 rpm to remove cell debris, and filtrated on a 0.22 µm filter and concentrated by ultracentrifugation at $47,000 \times g$ for 2 h at 16 °C in a swinging bucket rotor (19,500 rpm in a Beckman SW32 Ti rotor). The pellet was resuspended in 100–200 µl of PBS total and stored at −80 °C.

Note: Additional information on cloning into the pLVHM1 vector and lentivirus preparation can be found at Lentiweb.com.

2.2.3 Choice of cells, cell transduction, clone selection, and imaging

The localization of tRFP-tagged Shoc2 proteins was examined in Cos1 fibroblast-like cells derived from African green monkey kidney tissue. Cos1 cells were chosen as the expression system in these experiments due to their adherent morphology and the ease of detecting the EGFR signaling complexes on the endocytic compartment in cells stimulated with EGF (Galperin & Sorkin, 2005). These cells have low auto-fluorescence background and a flattened cell shape that is convenient for epifluorescence microscopy. Moreover, Cos1 cells that express the large T antigen allow tRFP/CFP/YFP vectors, containing the SV40 origin of replication, to induce high levels of fluorescent protein expression and, therefore, better visualization of the chromophores with low brightness (e.g., CFP). Cos1 cells were grown in high glucose DMEM (Invitrogen, NY, USA) containing 10% FBS (HyClone, NY, USA), and supplemented with antibiotics and glutamine.

The experimental strategy for preparation of Shoc2 KDAR cells is presented here as three sequential steps. The first step was to generate lentiviruses that carry only Shoc2-specific shRNA sequences, followed by validation of the shRNA knockdown efficiency upon virus incorporation. To silence protein expression by RNA interference, we first obtained four individual duplexes to human Shoc2 (Dharmacon, ThermoFisher Scientific, MA, USA) and evaluate the efficiency of each siRNA duplex after transient transfection into Cos1 cells. Cos1 cells were seeded in 12-well plates (50–60% confluent; 1 ml of DMEM/FBS per well) at least 20 h before transfection. Two siRNA transfections were performed at 24–36 h intervals according to the manufacturer's recommendations, using Dharmafect reagent 2 (Dharmacon, ThermoFisher, MA, USA). The efficiency of the siRNA knockdown was validated by Western blotting using Shoc2-specific antibodies. Sequences of two of the most efficient siRNA duplexes were then introduced into pLVTHM for stable expression. To validate efficiency of shRNA knockdown introduced by viral DNA incorporation, viruses caring two different shRNA duplexes were produced as described earlier and used for further cell transduction. The efficiency of the shRNA knockdown was validated by Western blotting using Shoc2-specific antibodies (Abcam, OR, USA).

Note: In our experience the most efficient siRNA oligos will not necessarily lead to an efficient shRNA silencing upon integration into the cell genome. Therefore, at least two siRNA sequences need to be tested for future virus-transduced shRNA silencing.

The next step in the generation of lentivirus-mediated KDAR cells is the construction of pLVTHM vectors that carry both shRNA and Shoc2-tRFP cDNA. Generation of the expressing viral vectors and the preparation of the virus that carries shRNA and Shoc2–tRFP cDNA are described in Sections 2.2.1 and 2.2.2. For lentivirus transduction, Cos1 cells were plated in 12-well dishes at 50–60% confluence. Shoc2–tRFP fluorescence was detected 36–48 h after lentivirus transduction. The efficiency of shRNA knockdown and cDNA expression was validated using Shoc2- and tRFP-specific antibodies.

Note: It is important to choose a fluorescent protein tag with the highest brightness possible. *A goal of KDAR methodology is to generate stable cells in which endogenous protein has been efficiently depleted and its exogenous counterpart is being expressed at levels that are close to endogenous.* Thus, a fluorescent protein tag with low brightness may lead to difficulties in protein visualization due to the limitations of the fluorescence microscope.

The final step in generation of KDAR cells was selection and validation of individual clones. The pLVTHM vector does not carry a marker for antibiotic selection. Therefore, to achieve a homogenous population of cells with desired levels of Shoc2–tRFP expression, the transduced cells were sorted by FACS into 96-well cell culture plates. Cell sorting proved as a fast way to get the desired KDAR populations (Fig. 3.3). Individual clones were then validated for Shoc2–tRFP expression and Shoc2 depletion by Western blot analysis using polyclonal antibodies to Shoc2 (Abcam, OR, USA). Clones in which endogenous Shoc2 was efficiently silenced and Shoc2–tRFP was expressed at the levels no more than twofold higher than endogenous levels were considered for future experiments (Fig. 3.4A). The clones were monitored for possible growth alterations and variation in levels of the ERK1/2 cascade proteins (Raf-1, B-Raf, MEK1/2, ERK1/2, and EGFR) and other control proteins (GAPDH, actin).

2.2.4 Visualization of endosomal Shoc2–tRFP

For fluorescent microscopy, cells were plated on glass-bottom dishes (MatTek, MA, USA) and then incubated in serum-free and phenol red-free medium containing 0.2% BSA for 16 h prior to the microscopy experiments. The Marianas™ imaging workstation (Intelligent Imaging Innovation, Denver, Co) identical to the one used for MEK2–GFP visualization was used in these experiments.

KDAR cells expressing Shoc2–tRFP and Shoc2-specific shRNA (i.e., Cos-SR) displayed a diffused cytosolic/nuclear pattern. Upon stimulating cells with EGF (2 ng/ml) at 37 °C for 10–20 min, we observed a small pool of Shoc2–tRFP accumulate on intracellular vesicular structures. The majority of these vesicles appeared as fluorescent dots of approximately similar intensity (Fig. 3.4B).

To define Shoc2–tRFP-containing vesicles, we examined whether these vesicles colocalized with markers of early endosomes (Rab5/EEA.1), or late endosomes and lysosomes (Rab7 and LysoTracker™) in living cells.

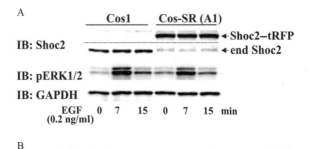

Figure 3.4 Validation of Cos-SR cells and localization of Shoc2–tRFP. (A) Parental Cos1 and Cos-SR (cells stably expressing Shoc2-shRNA and Shoc2-tagRFP, clone A1) were serum starved and treated with 0.2 ng/ml EGF for indicated times at 37 °C. The lysates were probed Shoc2, activated ERK1/2 (pERK1/2), and GAPDH. Dotted line shows area of the blot that was cropped. (B) Shoc2–tRFP was imaged in living Cos-SR cells through Cy3 filter channel. Scale bars, 10 μm.

Cos-SR cells were transiently transfected with CFP-Rab5 or CFP-Rab7. Shoc2–tRFP compartments were found to partially colocalize with a small subset of late endosomes labeled with CFP-Rab7 (not shown). Similar results were obtained in experiments with other clones of Cos-SR cells (data not shown). Cumulatively, our data demonstrated that upon EGFR activation, Shoc2–tRFP is recruited to a subset of late endosomes.

3. CONCLUSIONS

Results presented in this chapter demonstrate that the use of KDAR methodology allowed observations that were not achievable in experiments with overexpressed proteins. It also suggests that KDAR methodology provides a great advantage for studies of mechanisms regulating spatial distribution of signaling processes particularly in living cells. The method described in this chapter is straightforward and cost effective and, when possible, permits avoiding costly development of knockout or knockin animals. The main advantage of the KDAR methodology is that proteins are studied at expression levels close to endogenous which allows for a more accurate analysis of protein localization during signal transduction. Use of protein tags different than fluorescent protein tags will allow for KDAR methodology to be applied to other needs that still require proteins to be expressed at their physiological levels (e.g., cell surface biotinylation). Additionally, this methodology will be of great advantage when used together with the set of methods of single cell fluorescence microscopy (FRET, FRAP) to study spatiotemporal regulation of signal transduction in human cells and simultaneous measurements of interactions of three signaling proteins within the scaffold complexes.

ACKNOWLEDGMENTS

This work was supported by grants from the National Cancer Institute (R00CA126161 to E. G.) and NIH Grant P20GM103486 from the National Center for Research Resources. Its contents are solely the responsibility of the authors and do not necessarily represent the official views of the NIH or the NIGMS. We thank Dr. Mathew Gentry and Stacy Smith for critical reading of the manuscript and the Viral Production Core at the Department of Molecular and Cellular Biochemistry (University of Kentucky) for assistance with production of lentiviruses. The authors thank Maciej Wiznerowicz for valuable advice.

REFERENCES

Alexa, A., Varga, J., & Remenyi, A. (2010). Scaffolds are 'active' regulators of signaling modules. *The FEBS Journal, 277*(21), 4376–4382.

Bhattacharyya, R. P., Remenyi, A., Yeh, B. J., & Lim, W. A. (2006). Domains, motifs, and scaffolds: The role of modular interactions in the evolution and wiring of cell signaling circuits. *Annual Review of Biochemistry, 75*, 655–680.

Brown, M. D., & Sacks, D. B. (2008). Compartmentalised MAPK pathways. *Protein-Protein Interactions as New Drug Targets.* In: E. Klussmann and J. Scott, Springer vol. 186, (pp. 205–235).

Cacace, A. M., Michaud, N. R., Therrien, M., Mathes, K., Copeland, T., Rubin, G. M., et al. (1999). Identification of constitutive and ras-inducible phosphorylation sites of KSR: Implications for 14-3-3 binding, mitogen-activated protein kinase binding, and KSR overexpression. *Molecular and Cellular Biology, 19*(1), 229–240.

Dhanasekaran, D. N., Kashef, K., Lee, C. M., Xu, H., & Reddy, E. P. (2007). Scaffold proteins of MAP-kinase modules. *Oncogene, 26*(22), 3185–3202.

English, J. M., & Cobb, M. H. (2002). Pharmacological inhibitors of MAPK pathways. *Trends in Pharmacological Sciences, 23*(1), 40–45.

Galperin, E., Abdelmoti, L., & Sorkin, A. (2012). Shoc2 is targeted to late endosomes and required for Erk1/2 activation in EGF-stimulated cells. *PLoS One, 7*(5), e36469.

Galperin, E., & Sorkin, A. (2005). Visualization of Rab5 activity in living cells using FRET microscopy. *Methods in Enzymology, 403*, 119–134.

Galperin, E., & Sorkin, A. (2008). Endosomal targeting of MEK2 requires RAF, MEK kinase activity and clathrin-dependent endocytosis. *Traffic, 9*(10), 1776–1790.

Galperin, E., Verkhusha, V. V., & Sorkin, A. (2004). Three-chromophore FRET microscopy to analyze multiprotein interactions in living cells. *Nature Methods, 1*(3), 209–217.

Jiang, X., & Sorkin, A. (2002). Coordinated traffic of Grb2 and Ras during epidermal growth factor receptor endocytosis visualized in living cells. *Molecular Biology of the Cell, 13*(5), 1522–1535.

Katz, M., Amit, I., & Yarden, Y. (2007). Regulation of MAPKs by growth factors and receptor tyrosine kinases. *Biochimica et Biophysica Acta, 1773*(8), 1161–1176.

Li, W., Han, M., & Guan, K. L. (2000). The leucine-rich repeat protein SUR-8 enhances MAP kinase activation and forms a complex with Ras and Raf. *Genes & Development, 14*(8), 895–900.

Morrison, D. K. (2001). KSR: A MAPK scaffold of the Ras pathway? *Journal of Cell Science, 114*(Pt 9), 1609–1612.

Morrison, D. K., & Davis, R. J. (2003). Regulation of MAP kinase signaling modules by scaffold proteins in mammals. *Annual Review of Cell and Developmental Biology, 19*, 91–118.

Padilla-Parra, S., & Tramier, M. (2012). FRET microscopy in the living cell: Different approaches, strengths and weaknesses. *Bioessays, 34*(5), 369–376.

Sieburth, D. S., Sun, Q., & Han, M. (1998). SUR-8, a conserved Ras-binding protein with leucine-rich repeats, positively regulates Ras-mediated signaling in C. elegans. *Cell, 94*(1), 119–130.

Sorkin, A. (2001). Internalization of the epidermal growth factor receptor: Role in signalling. *Biochemical Society Transactions, 29*(Pt 4), 480–484.

Sorkin, A., & Goh, L. K. (2009). Endocytosis and intracellular trafficking of ErbBs. *Experimental Cell Research, 315*(4), 683–696.

Sorkin, A., McClure, M., Huang, F., & Carter, R. (2000). Interaction of EGF receptor and grb2 in living cells visualized by fluorescence resonance energy transfer (FRET) microscopy. *Current Biology: CB, 10*(21), 1395–1398.

Sorkin, A., & Von Zastrow, M. (2002). Signal transduction and endocytosis: Close encounters of many kinds. *Nature Reviews Molecular Cell Biology, 3*(8), 600–614.

Teis, D., Wunderlich, W., & Huber, L. A. (2002). Localization of the MP1-MAPK scaffold complex to endosomes is mediated by p14 and required for signal transduction. *Developmental Cell, 3*(6), 803–814.

von Zastrow, M., & Sorkin, A. (2007). Signaling on the endocytic pathway. *Current Opinion in Cell Biology, 19*(4), 436–445.

Zeke, A., Lukacs, M., Lim, W. A., & Remenyi, A. (2009). Scaffolds: Interaction platforms for cellular signalling circuits. *Trends in Cell Biology, 19*(8), 364–374.

CHAPTER FOUR

Virus-Induced Signaling Influences Endosome Trafficking, Virus Entry, and Replication

David J. Marchant[*], Leanne Bilawchuk[*], Gordon Nish[*], Richard G. Hegele[†,1]

[*]Li Ka Shing Institute of Virology, Department of Medical Microbiology and Immunology, University of Alberta, Edmonton, Alberta, Canada
[†]Department of Laboratory Medicine and Pathobiology, University of Toronto, Toronto, Ontario, Canada
[1]Corresponding author: e-mail address: richard.hegele@utoronto.ca

Contents

1. Introduction	66
2. Isolation of Fibroblasts from Wild-Type or MyD88$^{-/-}$ Mice and Purification of Virus	66
2.1 Isolation of primary pulmonary fibroblasts	67
2.2 Virus purification	67
3. Screening Assays for the Determination of Principal Host Kinases Involved During Virus Replication	68
3.1 Kinase inhibitor screen of GFP-virus-infected cells	68
4. Virus–Host Cell Receptor Interactions and Kinase Activation	70
4.1 Determination of MAPK phosphorylation by receptor neutralization, phospho-ELISA, and Western blot Assay	71
4.2 Inhibition of toll-like receptor 4 signaling using a neutralization antibody	71
4.3 Receptor neutralization of ERK signaling using monoclonal antibodies against decay acceleration factor and CAR	72
5. Detection of Virus-Induced Cell Signaling and Virus Entry into Endosomes Using Immunofluorescence Confocal Microscopy	73
5.1 Preparation of specimens for immunofluorescence confocal microscopy	73
5.2 Image acquisition, three-dimensional modeling, and rendering	74
6. Summary	75
References	75

Abstract

Viruses are obligate intracellular pathogens that interact with host cell machinery for enabling entry, replication, and spread. This chapter describes methods for studying the interaction of viruses with host cell signaling pathways and surface receptors during cellular infection, with an emphasis on protein kinases. We also describe how use of immunofluorescence confocal microscopy for imaging virus–host interactions provides a powerful approach for obtaining structural correlations that extend results of immunological and biochemical assays.

1. INTRODUCTION

Viruses utilize a broad range of endocytosis pathways to enter host cells (Mercer, Schelhaas, & Helenius, 2010), and they use an even broader array of signaling networks and strategies to activate endocytosis of virus particles into host cells (Pelkmans, 2005; Pelkmans, et al., 2005). The activation of p38 MAP kinase through the TLR4–MyD88 signaling pathway, though it leads to recognition of pathogens by the immune system, has also been usurped by viruses to trigger endocytosis into the host cell (Marchant et al., 2010). The activation of p38 MAP kinase signaling enhances endocytosis of extracellular constituents, including viruses. Internalization of viruses by endocytosis, while leading to internalization of foreign antigen for expression of viral antigen via major histocompatibility complex type 1, also leads to entry of internalized virus and productive infection. Therefore, activation of p38 MAP kinase via pattern recognition receptors is evolutionarily favorable to viruses.

Entry into some endocytosis pathways, like Arf6, is not amenable to infection, but acts as a decoy pathway that leads to a virus replication dead-end (Marchant et al., 2009). Arf6-mediated endocytosis of coxsackievirus is activated by ERK MAP kinase signaling, triggered by virus–cell receptor interaction. The majority of virus particles are internalized via clathrin endocytosis as a result of ERK MAP kinase activation, leading to productive infection. However, some virus becomes trapped in the Arf6-mediated endocytosis pathway during entry that either does not support virus fusion, or directs the entering virus particles away from a site in the cell that is conducive to virus replication.

2. ISOLATION OF FIBROBLASTS FROM WILD-TYPE OR MyD88$^{-/-}$ MICE AND PURIFICATION OF VIRUS

Genetic models, both *in vivo* and *ex vivo* were used to investigate the roles of the MyD88 signaling axis in activation of p38 MAP kinase during virus entry (Marchant et al., 2010).

2.1. Isolation of primary pulmonary fibroblasts

Treat all animals humanely, in accordance with institutional and federal regulations and guidelines.

1. C57 background mice: wild-type or MyD88 knockout (MyD88$^{-/-}$).
2. Anesthetize mice by isofluorane and then euthanize by cervical dislocation.
3. Remove lungs, mince into several pieces on tissue culture surface (10 cm, six-well dish), and allow pieces to dry (~5 min) so that they adhere to the tissue culture surface.

    Note: Tissue must remain adhered to surface after addition of media. Fibroblasts will only adhere to the tissue culture surface directly below the site of tissue adhesion.

4. Drop-wise, overlay minced tissue with DMEM, 10% FBS, Penicillin–Streptomycin and incubate overnight at 37 °C, 5% CO_2.
5. Remove minced tissue after overnight incubation, wash once in growth media, leave in growth media and replate cells 1 week later.

    Note: Tissue must be removed the following day to prevent outgrowth of penicillin static/resistant commensal bacteria remaining in the tissue pieces.

6. Maintain cells in DMEM, 10% FBS, Penicillin–Streptomycin at 37 °C, 5% CO_2. Change media daily to remove contaminating erythrocytes.
7. Use 0.05% Trypsin–0.53 mM EDTA to detach cells from culture surface and if fibroblasts are healthy they will need to be expanded into larger growth flasks.
8. Plate into suitable multiwell tissue culture trays for experimentation.

2.2. Virus purification

It is ideal (sometimes mandatory, as demanded by reviewers!) to purify virus stocks for measuring the signaling response in target cells. Unpurified virus supernatant contains cytokines, chemokines, and other products of cell lysis/necrosis brought on by replication of virus during virus stock preparation. Enveloped viruses are most often purified on sucrose gradients, typically recovered on a sucrose cushion at a density of 1.15–1.2 g/mL buffered sucrose and nonenveloped viruses can be purified in CsCl gradients as pellets. Commercial kits (e.g., Adeno-X™ Maxi Purification kit (Clontech, Mountain View, CA), ViraBind™ Adenovirus Purification kit (Cell Biolabs, Inc., San Diego, CA)) can be used for the concentration and purification of adenovirus, saving time and money in eliminating trial and error associated with protocol optimization. The purified adenovirus product derived from kits is generally of high quality and high titer.

2.2.1 Sucrose purification of respiratory syncytial virus

There are many good protocols in the literature available for sucrose purification of respiratory syncytial virus (RSV) so we will outline the following protocol only briefly, a variation of the method reported by Ueba (Ueba, 1978).

1. HEp-2 cells in a large flask are infected at high multiplicity of infection (MOI = number of infectious units of virus per cell = ~3).
2. Between 24 and 48 h later RSV supernatant from inoculated HEp-2 epithelial cells is harvested and clarified by centrifugation at $5000 \times g$ for 20 min.
3. The supernatant is precipitated in polyethylene glycol 6000 and precipitate is resuspended in 20% sucrose-NT buffer (0.15 M NaCl and 0.05 M TRIS).
4. The resuspended RSV is layered onto a 35%, 45%, and 60% sucrose density gradient and centrifuged at $165,000 \times g$ for 180 min.
5. The virus band is collected at the 35–45% interface (1.15–1.2 g/mL).
6. Virus is stored in liquid nitrogen in 100 μL aliquots.

3. SCREENING ASSAYS FOR THE DETERMINATION OF PRINCIPAL HOST KINASES INVOLVED DURING VIRUS REPLICATION

3.1. Kinase inhibitor screen of GFP-virus-infected cells

One approach to study interactions of viruses with host cell kinases is to examine the effects of pretreatment of cells with pharmacological kinase inhibitors on subsequent virus replication. A panel of kinase inhibitors (Table 4.1) was used to determine which host cell kinases can be implicated in the replication of a variety of human respiratory viruses (Marchant et al., 2010).

Anisomycin was purchased from Sigma-Aldrich (St. Louis, MO) and used at a concentration of 10 μM.

1. HeLa, 1HAEo-, and A549 cells were seeded to confluence in 96-well optical-bottom plates in 10% fetal calf serum and 10% DMEM.
2. Kinase inhibitors were added at the concentrations indicated in Table 4.1 for 1 h prior to infection.
3. Cells were infected with recombinant viruses that express GFP upon successful entry and replication: adenovirus (AdV)-GFP, RSV-GFP, human parainfluenza virus type 3 (hPIV3)-GFP, and coxsackievirus type B3 (CVB3)-GFP.
    a. Infect at MOI of 1 and 10.

Table 4.1 Drug inhibitors[a] for common kinases

Inhibitor (target kinase)	Final concentration (μM)
API-2 (Akt/protein kinase B [PKB])	1
BAY11-7085 (NF-κB)	10
LY294002 (PI3K)	25
MG132 (proteasome)	10
PP2 (Src family kinases)	5
SB203580 (p38 MAPK)	5
SB216763 (GSK3β)	10
SP600125 (JNK)	20
U0126 (MEK1/2)	20

[a]Inhibitors purchased from Tocris Pharmaceuticals (Ellisville, MO) were used at the final concentrations indicated (target kinase is in parentheses).

4. Fix cells in methanol–acetone (3:1 v/v)[1] 24 h postinfection, remove and replace with PBS.
   a. CVB3-GFP: Fix and harvest cells, as above, 8–10 h postinfection.
5. Detect and quantify GFP fluorescence with a Tecan GENios fluorescence plate reader (excitation [E_x] 485, emission [E_m] 535, gain 60; MTX Lab Systems).
6. Treat cells with Hoechst 33342 (1/10,000 in ddH$_2$0 for 10 min) to stain nuclei then detect fluorescence (E_x 360, E_m 465, gain 60) as a cell input control.

3.1.1 Influenza infectivity via flow cytometry

Here, we focus on influenza virus as an example of an approach for quantification of cellular infectivity by virus after pretreatment of cells with kinase inhibitors.

1. Pretreat cells with inhibitors as described in Table 4.1.
2. Infect cells at influenza virus MOIs of 1 and 10.

---

[1] This fixation method permeabilizes cells and causes leaching of GFP into the PBS used to rehydrate fixed cells. The GFP detected in the wells is correlated to the amount of infection in the culture well. More sensitive plate readers now exist that accurately detect and quantify intracellular GFP, imaging and recording the fluorescent channel of the well, using morphometry software to accurately partition the region of interest in the cell, more accurately quantifying GFP concentration.

At this point in the assay one may detect influenza virus in a drug screen by a variety of means, such as a series of immunofluorescence staining steps that involves staining influenza with primary anti-influenza antibodies, followed by washing and then incubation with a fluorescence-conjugated antibody. Fluorescent influenza-infected cells are enumerated by flow cytometry. Alternatively, one may use an immune β-galactosidase staining method followed by colorimetric detection rather than fluorescence-conjugated secondary antibodies.

3. At 16 h postinfection (to detect the first round of influenza virus replication) fix cells with paraformaldehyde for 10 min.
   a. Centrifuge cells at $500 \times g$ for 5 min and reconstitute in PBS, 0.3% Triton X-100, and 1% bovine serum albumin (BSA) for 10 min at room temperature (RT).
   b. Add goat anti-influenza H1N1 (GenWay Biotech, San Diego, CA, catalogue no. 18-783-77828-1) at a 1/100 dilution in PBS/Triton/BSA buffer for 30 min at RT.
   c. Wash cells three times using PBS and subsequent centrifugation at $500 \times g$ for 3 min. Allow 5 min between each washing step.
   d. Add secondary Alexa 594 donkey anti-goat antibody (Invitrogen, cat. no. A11058) at a 1/200 dilution in 0.3% Triton X-100 in PBS for 30 min at RT.
   e. Wash cells three times as above and enumerate with a Beckman Coulter Epics XL-MCS flow cytometer.

For these experiments, it is imperative to include uninfected cells stained in parallel with infection experiments. One determines infectivity of test cells by gating from the stained but uninfected cells.

The antibody staining of influenza method described earlier is considerably slower and more labor intensive than using recombinant GFP viruses as infectivity indicators in Section 3.1. The centrifugation steps required between each staining and washing step bring extra effort and time. *In situ* staining with influenza-specific antibodies conjugated to fluorophores or β-galatosidase for X-Gal staining can be imaged using imaging plate readers like the Perkin Elmer Operetta high-content imaging system, allowing application of high-throughput segmentation algorithms postacquisition.

4. VIRUS–HOST CELL RECEPTOR INTERACTIONS AND KINASE ACTIVATION

Viruses can interact with molecules expressed on the cell surface, including pattern recognition receptors. In this section, we describe methods for determining the level of kinase activation in the setting of virus interaction with a pattern recognition receptor (toll-like-receptor 4 for RSV and influenza virus infections) and virus-entry receptors (decay accelerating factor and coxsackie-adenovirus receptor (CAR) for coxsackievirus B3 infection).

4.1. Determination of MAPK phosphorylation by receptor neutralization, phospho-ELISA, and Western blot Assay

1. ELISA kits for ERK, JNK, ATF2, and p38 MAPK phosphorylation were purchased from Biosource (Invitrogen, Carlsbad, CA).
2. Seed cells into 12- or 24-well trays and infect with virus or cell-conditioned medium in 10% FBS-DMEM.
3. Harvest cells on ice using protein-lysis buffer (10 mM HEPES [pH 7.4], 50 mM Na$_4$P$_2$O$_7$, 50 mM NaF, 50 mM NaCl, 5 mM EDTA, 5 mM EGTA, 1 mM Na$_3$VO$_4$, 0.5% Triton X-100, 10 µg/mL leupeptin, and 1 mM phenylmethylsulfonyl fluoride).[2]
4. Centrifuge sample at approximately $16,000 \times g$ and retrieve postnuclear supernatant.
5. Determine protein concentration of sample.
6. Load 10 µg of sample per well of each ELISA mixture and assay samples according to manufacturer's instructions.
7. Subtract the basal levels of MAPK activation (conditioned medium) from the values obtained from the virus-infected cells at each time point.

4.1.1 Western blot assay

8. Harvest cells as described earlier in protein-lysis buffer and centrifuge at $14,000 \times g$ for 10 min at 4 °C to isolate postnuclear supernatant.
9. Boil samples and resolve on a 12% polyacrylamide gel at 120 V for 1 h.
10. Transfer to a nitrocellulose membrane, then probe for RSV, CVB3, p42/p44 total ERK MAPK, MyD88, and total and phospho-p38 MAPK.

4.2. Inhibition of toll-like receptor 4 signaling using a neutralization antibody

1. Pretreat 1HAEo- and HeLa cells for 1 h with a dilution series of rabbit IgG control (ab46540) and rabbit polyclonal antihuman TLR4 antibody from AbCam (ab13556, Cambridge, MA) in serum-free DMEM.

---

[2] The ELISA kits for p38, ERK, and JNK MAP kinases call for different denaturing conditions requiring separate sample aliquots for each ELISA. As per manufacturer's instructions.

2. Infect with a dilution series of RSV-A2 or influenza virus.
3. Enumerate infected cells the following day using flow cytometry (described above) and harvest lysates for Western blot analysis (described earlier, to monitor successful inhibition of phospho-p38 MAP kinase or phospho-ATF-2).
4. Enumerate influenza virus-infected cells by flow cytometry after antibody staining (described earlier).

In each experiment, the activation (phosphorylation) of p38 MAP kinase and ERK MAP kinase was monitored during virus entry to confirm that TLR4 was inhibiting MAP kinase signaling and not sterically blocking virus from interacting with target host cells. As such, other forms of experimentation are required to support the results of antibody neutralization.

4.3. Receptor neutralization of ERK signaling using monoclonal antibodies against decay acceleration factor and CAR

The MAP kinases induce massive signaling cascades that mediate processes that range from apoptosis, cell division, and endocytosis. ERK MAP kinase activates Arf6-mediated endocytosis (Robertson et al., 2006), and vice versa (Tushir & D'Souza-Schorey, 2007). CVB3 binds its receptor, CAR, which activates ERK signaling. ERK signaling in turn triggers Arf6-mediated endocytosis and uptake of virus particles into the Arf6-mediated endocytosis pathway. Viruses that enter into the Arf6 pathway do not infect the cell as this is a dead-end pathway of entry for CVB3 (Marchant et al., 2009).

1. Experiments were performed at 37 °C. HeLa cells were treated with an anti-DAF (decay acceleration factor) or anti-CAR mAb at a 1:1000 dilution (200 ng/mL) for 30 min prior to and during CVB3 infection at an MOI = 10.
2. A higher concentration (1:50; 4 μg/mL) of antibody was added to cells, alone, to activate signaling via antibody–receptor binding.
3. Cells were harvested with protein-lysis buffer [10 mM HEPES (pH 7.4), 50 mM Na$_4$P$_2$O$_7$, 50 mM NaF, 50 mM NaCl, 5 mM EDTA, 5 mM EGTA, 1 mM Na$_3$VO$_4$, 0.5% Triton X-100, 10 μg/mL leupeptin and 1 mM PMSF] on ice at 15 min postinfection and lysates were probed by Western blotting for phospho-(pTh202/pTyr204) ERK MAP kinase and total ERK.

5. DETECTION OF VIRUS-INDUCED CELL SIGNALING AND VIRUS ENTRY INTO ENDOSOMES USING IMMUNOFLUORESCENCE CONFOCAL MICROSCOPY

Immunofluorescence confocal microscopy of CVB3 during cell entry has been described previously (Marchant et al., 2009). An AOBS Leica confocal TCS SP2 microscope (63×/numerical aperture [NA], 1.2) (Heidelberg, Germany) was used (Fig. 4.1).

5.1. Preparation of specimens for immunofluorescence confocal microscopy

1. Add virus at an MOI 3–5 to cells on ice for 1 h then wash cells once with PBS and proceed with infection at 37 °C, 5% CO_2 until the appropriate time points.

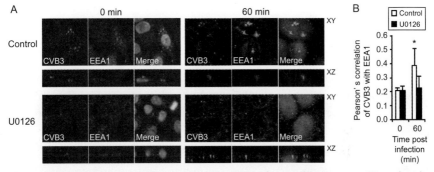

Figure 4.1 Confocal immunofluorescence imaging of coxsackievirus type B3 and early endosomes can be quantified postacquisition using Pearson's correlation of pixel overlap. (A) HeLa cells were infected with coxsackievirus (CVB3) on ice to synchronize virus entry. The 0-min time point was fixed prior to transfer of cultures to 37 °C. Infected cells were transferred to a humidified 37 °C incubator and fixed 60 min postwarming to obtain the 60-min time point. Cells were stained for early endosomes using a monoclonal antibody for early endosome antigen-1 (EEA1) and CVB3 was stained with a monoclonal antibody against viral protein 1. Endosomes (EEA1) and CVB3 (VP1) were detected with secondary antibodies conjugated to Alexafluors 594 and 488, respectively. Blue nuclei were stained with DAPI. Note the colocalization of green CVB3 in red EEA1 endosomes 60-min postinfection, that is not apparent 0 min or during treatment with ERK MAP kinase inhibitor U0126. (B) Pearson's correlation was conducted on Z-stacks of red (EEA1) and green (CVB3) images using Volocity software. Pearson's correlation quantifies the proportion of overlapping red and green voxels (pixel in three dimensions), that is, the proportion of CVB3 that has entered into early endosomes 60-min postinfection. *$p < 0.05$. Data derived from Marchant et al. (2009).

2. Fix cells by adding room temperature methanol–acetone (3:1 vol/vol) for 2 min then block with 1% BSA–PBS or block and permeabilize using 0.3% Triton X-100, 1% BSA, and PBS (PERM buffer) for 10 min at RT.

5.1.1 Cell staining

3. Add rabbit anti-phospho-ATF2 (9221; Cell Signaling Technologies, Danvers, MA) and goat anti-TLR4 (ab53629; AbCam, Cambridge, MA) primary antibodies at a 1:100 dilution in PERM (or non-PERM control) buffer at 4 °C overnight. Wash with PBS three times, 5 min each, between each antibody treatment.
   a. Add secondary Alexa 594 donkey anti-goat or Alexa 488 goat anti-rabbit were at a 1/400 dilution in PERM (or non-PERM control buffer) sequentially, with an additional wash step, to avoid secondary antibody cross-reactivity.

5.1.2 Virus staining

4. Add primary monoclonal anti-RSV (RSV3; a mix of three monoclonal RSV antibodies: glycoprotein F, protein P, and protein N—Novocastra, Buffalo Grove, IL) as above.
   a. Add secondary Alexa 405 or Alexa 488 antibodies (Invitrogen, Grand Island, NY) as above, minding possible cross-reactivity.

5.1.3 Nuclear staining

5. Stain nuclei using DAPI (4′,6-diamidino-2-phenylindole) contained within the mounting medium, Vectashield (Vector Laboratories, Burlingame, CA).

5.2. Image acquisition, three-dimensional modeling, and rendering

The AOBS SP2 system allows for user-defined photo multiplier tube (PMT) filters of custom bandwidth for specific detection of each fluorophore by a separate PMT detector. This allows for the exclusion of bleed-through of signal into the emission profile of fluorophores with longer wavelengths from fluorophores with emission at shorter wavelengths.

1. Perform lambda scans throughout the experiments to confirm the specificity of Alexa Fluor dyes, as well as sequential scans to prevent the detection of artifactual colocalization.

2. Collate image stacks (100–150 × 512 × 512 pixels; six-frame average) using Improvision Volocity v5 software (Coventry, United Kingdom), with minor brightness and contrast adjustment (not greater than 20% of original values).
3. Filter out minor background noise for all fluorescence confocal images using the fine noise filter in Volocity.
4. Count virus particles automatically in a defined region of interest (ROI) at $t=0$ min for RSV and influenza virus using the measurement function, detected by intensity greater than 30% and larger than 0.05 μm.
5. Draw the ROI around each DAPI-stained nucleus with an extra 1 μm buffer to isolate each individual cell for determination of the number of virus particles per cell.

6. SUMMARY

It has been well established that viruses induce cellular signaling throughout the entire virus life cycle (Coyne, Shen, Turner, & Bergelson, 2007; Garmaroudi et al., 2010; Janes & Yaffe, 2006; Jensen et al., 2013; Lou et al., 2002; Marchant et al., 2009, 2010; Pelkmans, 2005; Si et al., 2005). Particularly, viruses can induce signaling within the first 10 min of infection that persists up to the point at which virus buds from the host cell or is released from infected cells by cell lysis (Coyne & Bergelson, 2006; Coyne et al., 2007; Garmaroudi et al., 2010; Marchant et al., 2009, 2010). The entry phase of the virus life cycle is activated by virus (virion) interaction with the virus entry receptor (Coyne & Bergelson, 2006; Coyne et al., 2007; Marchant et al., 2009) and even pattern recognition receptors that have evolved to trigger a host response to an incoming pathogen.

REFERENCES

Coyne, C. B., & Bergelson, J. M. (2006). Virus-induced Abl and Fyn kinase signals permit coxsackievirus entry though epithelial tight junctions. *Cell*, *124*, 119–131.

Coyne, C. B., Shen, L., Turner, J. R., & Bergelson, J. M. (2007). Coxsackievirus entry across epithelial tight junctions requires occludin and the small GTPases Rab34 and Rab5. *Cell Host & Microbe*, *2*, 181–192.

Garmaroudi, F. S., Marchant, D., Si, X., Khalili, A., Bashashati, A., Wong, B. W., et al. (2010). Pairwise network mechanisms in the host signaling response to coxsackievirus B3 infection. *Proceedings of the National Academy of Sciences of the United States of America*, *107*, 17053–17058.

Janes, K. A., & Yaffe, M. B. (2006). Data-driven modelling of signal-transduction networks. *Nature Reviews Molecular Cell Biology*, *7*, 820–828.

Jensen, K. J., Garmaroudi, F. S., Zhang, J., Lin, J., Boroomand, S., Zhang, M., et al. (2013). An ERK-p38 subnetwork coordinates host cell apoptosis and necrosis during coxsackievirus B3 infection. *Cell Host & Microbe, 13*, 67–76.

Lou, H., Yanagawa, B., Zhang, J., Luo, Z., Zhang, M., Esfandiarei, M., et al. (2002). Coxsackievirus B3 replication is reduced by inhibition of the extracellular signal-regulated kinase (ERK) signaling pathway. *Journal of Virology, 76*, 3365–3373.

Marchant, D., Sall, A., Si, X., Abraham, T., Wu, W., Luo, Z., et al. (2009). ERK MAP kinase-activated Arf6 trafficking directs coxsackievirus type B3 into an unproductive compartment during virus host-cell entry. *Journal of General Virology, 90*, 854–862.

Marchant, D., Singhera, G. K., Utokaparch, S., Hackett, T. L., Boyd, J. H., Luo, Z., et al. (2010). Toll-like receptor 4-mediated activation of p38 mitogen-activated protein kinase is a determinant of respiratory virus entry and tropism. *Journal of Virology, 84*, 11359–11373.

Mercer, J., Schelhaas, M., & Helenius, A. (2010). Virus entry by endocytosis. *Annual Review of Biochemistry, 79*, 803–833.

Pelkmans, L. (2005). Viruses as probes for systems analysis of cellular signalling, cytoskeleton reorganization and endocytosis. *Current Opinion in Microbiology, 8*, 331–337.

Pelkmans, L., Fava, E., Grabner, H., Hannus, M., Habermann, B., Krausz, E., et al. (2005). Genome-wide analysis of human kinases in clathin- and caveolae/raft-mediated endocytosis. *Nature, 436*, 78–86.

Robertson, S. E., Setty, S. R., Sitaram, A., Marks, M. S., Lewis, R. E., & Chou, M. M. (2006). Extracellular signal-regulated kinase regulates clathrin-independent endosomal trafficking. *Molecular Biology of the Cell, 17*, 645–657.

Si, X., Luo, H., Morgan, A., Zhang, J., Wong, J., Yuan, J., et al. (2005). Stress-activated protein kinases are involved in coxsackievirus B3 viral progeny release. *Journal of Virology, 79*, 13875–13881.

Tushir, J. S., & D'Souza-Schorey, C. (2007). ARF6-dependent activation of ERK and Rac1 modulates epithelial tubule development. *EMBO Journal, 26*, 1806–1819.

Ueba, O. (1978). Respiratory syncytial virus. I. Concentration and purification of the infectious virus. *Acta Medica Okayama, 32*, 265–272.

CHAPTER FIVE

Methods to Evaluate Zinc Transport into and out of the Secretory and Endosomal–Lysosomal Compartments in DT40 Cells

Taiho Kambe[1]
Graduate School of Biostudies, Kyoto University, Kyoto, Japan
[1]Corresponding author: e-mail address: kambe1@kais.kyoto-u.ac.jp

Contents

1. Introduction   78
2. Intracellular Zinc Transporters Localized to the Secretory and Endosomal–Lysosomal Compartments Play Crucial Roles   79
3. Establishment of DT40 Cells Deficient in Zinc Transporters Genes   82
    3.1 Basic characteristics of DT40 cells   82
    3.2 Construction and preparation of knockout vectors   82
    3.3 Electroporation to introduce knockout vectors and establish DT40 cell lines deficient in one gene   82
    3.4 Preparation of genomic DNA from the clones   83
    3.5 Establishment of DT40 cells deficient in multiple genes using the estrogen receptor chimeric Cre recombinase, CreER   84
4. Experimental Procedures Used in Studies of DT40 Cells Deficient in Zinc Transport   84
    4.1 Confirmation of phenotypes of DT40 mutants   84
    4.2 Reexpression of human orthologs of ZIP and ZnT transporters in DT40 mutants   85
    4.3 Analyses of complex formation of Zinc transporters by immunoprecipitation   85
    4.4 Transient transfection in DT40 mutants   86
5. Functional Analysis of Zinc Mobilization into or out of the Secretory and Endosomal–Lysosomal Compartments   86
    5.1 Immunofluorescence analysis of zinc mobilization in DT40 mutants deficient in ZIP and ZnT transporters   86
    5.2 Monitoring cytosolic zinc levels and zinc-related homeostasis in DT40 cells   87
    5.3 Evaluation of zinc status in subcellular compartments using zinc-dependent enzymes such as alkaline phosphatase   87
    5.4 Measurement of ALP activity   87
6. Concluding Remarks   89
Acknowledgment   89
References   89

Abstract

Zinc plays crucial roles in diverse biological processes. Recently, in addition to zinc mobilization into and out of the cell, zinc mobilization into and out of intracellular organelles, including the secretory and endosomal–lysosomal compartments, has received growing interest. In vertebrate cells, the Zrt/Irt-like proteins (ZIPs) and Zn transporters (ZnTs) are the two major families of zinc transport proteins involved in zinc mobilization across cellular membranes. Importantly, nearly half of them are localized to subcellular compartments. Thus, to elucidate the numerous zinc-related cellular events, understanding those ZIP and ZnT functions is critical. This chapter describes advanced methods used in our laboratory to examine zinc mobilization by them. Specifically, genetic and molecular approaches using chicken DT40 cells deficient in multiple ZIPs and ZnTs are described. Moreover, procedures to evaluate zinc-related phenotypes caused by the impairment of zinc mobilization into and out of the secretory and endosomal–lysosomal compartments are also described. These methods should be useful in characterizing the roles of zinc in diverse cellular events including endosomal signaling.

1. INTRODUCTION

Zinc is an essential trace element required for all living organisms. Analysis of the proteome reveals that approximately 4–10% of proteins have potential zinc-binding motifs, which confirm its importance (Andreini, Banci, Bertini, & Rosato, 2006). Recent studies have highlighted the various dynamic functions of zinc including structural, catalytic, and regulatory roles in protein function. For example, zinc is essential for the proper regulation of protein kinase signaling, apoptosis signaling, the unfolded protein response (UPR), trafficking of transcription factors, and channel or transporter modulation (Aras & Aizenman, 2011; Colvin, Holmes, Fontaine, & Maret, 2010; Fukada & Kambe, 2011; Hershfinkel, Aizenman, Andrews, & Sekler, 2010). Moreover, roles of zinc as a second messenger for transducing extracellular signals have been revealed in some types of immune cells (Haase & Rink, 2009; Hirano et al., 2008). Thus, growing evidence emphasizes the indispensable nature of zinc for cellular integrity and homeostasis.

Recently, the molecular mechanisms behind how zinc performs its crucial functions are starting to be elucidated, and it has been revealed that zinc mobilized into and out of the secretory and endosomal–lysosomal compartments has key cellular roles. Hence, exploring the molecular functions of zinc transport proteins involved in these processes will be beneficial. In this chapter, the methods to perform basic experiments using vertebrate cells to study zinc transport are described. Specifically, methods to establish knockout cells lacking multiple zinc transport proteins and assays for examining their zinc-related phenotypes are described. In the secretory and endosomal–lysosomal compartments, numerous zinc transport proteins are functional, and therefore determining each one's function at the molecular level will be important. Thus, systems that enable multiple gene disruptions, like those described here, are useful.

2. INTRACELLULAR ZINC TRANSPORTERS LOCALIZED TO THE SECRETORY AND ENDOSOMAL–LYSOSOMAL COMPARTMENTS PLAY CRUCIAL ROLES

In vertebrate cells, two zinc transport protein families, SLC39A/ZIP and SLC30A/ZnT, are primarily responsible for zinc transport and homeostasis. Zrt/Irt-like protein (ZIP) transporters import zinc into the cytosol from the extracellular or intracellular compartments, while Zn transporter (ZnT) proteins export cytosolic zinc to extracellular or intracellular compartments (Fukada & Kambe, 2011; Gaither & Eide, 2001; Kambe, Weaver, & Andrews, 2008; Lichten & Cousins, 2009) (Fig. 5.1). Both protein families are expressed in a cell- and tissue-specific manners, are developmentally regulated, and are localized to various subcellular compartments including the cytoplasmic membrane (Fukada & Kambe, 2011; Kambe, 2013; Lemaire, Chimienti, & Schuit, 2012). In mammals, 23 ZIP/ZnT proteins are operative (Kambe, Suzuki, Nagao, & Yamaguchi-Iwai, 2006), nearly half of which are localized to the secretory and endosomal–lysosomal compartments (Hennigar & Kelleher, 2012; Kambe, 2011) (Table 5.1). The secretory and endosomal–lysosomal compartments are thought to be important zinc storage sites (Falcon-Perez & Dell'Angelica, 2007; Hogstrand, Kille, Nicholson, & Taylor, 2009; McCormick, Velasquez, Finney, Vogt, & Kelleher, 2010), and growing evidence has revealed that zinc mobilization into and out of these compartments by ZIP and ZnT transporters is involved in a number of biological processes (Fukada & Kambe, 2011; Fukada, Yamasaki, Nishida, Murakami, & Hirano, 2011; Kambe, 2011).

Evidence for the importance of the role of ZIPs in subcellular zinc efflux is growing. For example, ZIP7 participates in cytosolic signaling by releasing zinc from the endoplasmic reticulum (ER) when phosphorylated by CK2 protein kinase (Taylor, Hiscox, Nicholson, Hogstrand, & Kille, 2012). ZIP8 mediates release of zinc from the lysosomes, and plays regulatory roles in T-cell activation, likely through inhibiting calcineurin and thus sustaining CREB activity (Aydemir, Liuzzi, McClellan, & Cousins, 2009). ZIP9 is required for the regulation of Akt and Erk activation in the B cell receptor

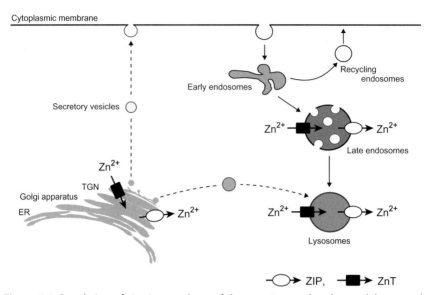

Figure 5.1 Regulation of zinc into and out of the secretory and endosomal–lysosomal compartments by ZIP and ZnT transporters. ZIP and ZnT transporters localized to the secretory and endosomal–lysosomal compartments regulate zinc homeostasis in both the cytosol and the lumen of the subcellular compartments. ER, endoplasmic reticulum; TGN, *trans*-Golgi network.

Table 5.1 Expression of zinc transporters located in the subcellular compartments in DT40 cells

Gene	Reported subcellular localization	Expression in DT40 cells
ZnT1	Mostly localized to the plasma membrane, ER	+
ZnT2	Endosomes/lysosomes, vesicles	−[a]
ZnT4	Endosomes/lysosomes, TGN, vesicles	+
ZnT5	Early secretory pathway, vesicles	+
ZnT6	Early secretory pathway, vesicles	+
ZnT7	Early secretory pathway, TGN	+
ZnT10	Golgi apparatus, recycling endosomes	−
ZIP7	ER, Golgi apparatus	−
ZIP8	Plasma membrane, endosomes/lysosomes	+
ZIP9	Golgi apparatus	+
ZIP13	Golgi apparatus, vesicles	+

[a] The cDNA fragment was detected by RT-PCR, but the functional ORF was not. This table omits the zinc transporters that show highly tissue- or cell-specific expression patterns.

signaling pathway (Taniguchi et al., 2013), and thought to function in secretory pathway homeostasis (Matsuura et al., 2009). ZIP13 is involved in the nuclear translocation of Smad proteins, and thereby functions in the BMP/TGF-β signaling pathway by transporting zinc from the Golgi (Fukada et al., 2008), and ZIP13 also maintains ER homeostasis by regulating zinc release from vesicular stores (Jeong et al., 2012).

Important biological functions associated with ZnTs are also abundant. ZnT1 was shown to affect intracellular zinc distribution and to downregulate transcription factors stimulated by MTF-1, c-Jun, and Elk, when localized to the ER by forming a heterocomplex with EVER1 and EVER2 (Lazarczyk et al., 2008). ZnT4, ZnT5, ZnT6, and ZnT7 are all suggested to maintain secretory pathway homeostasis and regulate zinc-requiring enzyme activities (Fukunaka et al., 2011; McCormick & Kelleher, 2012). ZnT5 and ZnT7 are also engaged in regulation of cytosolic signaling cascades involving PKC/NF-kappaB and PI3K/Akt (Nishida, Hasegawa, Nakae, Oboki, et al., 2009; Zhang, Liang, Guo, Deng, et al., 2013). ZnT10 localizes to the secretory and endosomal compartments (recycling endosomes) and functions in the regulation of Erk signaling by controlling cellular zinc levels (Bosomworth, Thornton, Coneyworth, Ford, & Valentine, 2012; Patrushev, Seidel-Rogol, & Salazar, 2012). Zinc influx into the endosomal–lysosomal compartments by ZnT transporters has also been suggested to influence cholesterol metabolism in experiments of ZnT2 overexpression in mutant cells lacking both metallothionein and ZnT1 (Kobayashi et al., 1999). Finally, in yeast, drosophila, and plants, the homologs of the ZIP and ZnT transporters mentioned earlier have been shown to be important in various biological processes including ER-associated degradation, Golgi membrane trafficking, regulation of catecholamine production, and auxin metabolism and response (Ellis et al., 2004; Fang et al., 2008; Lasswell, Rogg, Nelson, Rongey, & Bartel, 2000; Stathakis et al., 1999). Together, these data strongly suggest that zinc plays important roles in endosomal signaling, despite the lack of direct evidence.

To further explore roles for zinc in endosomal signaling, approaches using DT40 cells will be helpful because of their high frequency of homologous recombination and their expression of a number of ZIP and ZnT transporters (Table 5.1). In addition to ZIP and ZnT transporters, several other membrane proteins are thought to function in subcellular zinc transport. The L-type calcium channel is employed for zinc release from the ER after extracellular stimulation (this phenomenon is called "zinc wave") (Yamasaki et al., 2012). TRPMLs are thought to be involved in zinc mobilization in the endosomal–lysosomal compartments (Eichelsdoerfer, Evans, Slaugenhaupt, & Cuajungco, 2010). Importantly, these related proteins are also expressed in DT40 cells.

3. ESTABLISHMENT OF DT40 CELLS DEFICIENT IN ZINC TRANSPORTERS GENES

3.1. Basic characteristics of DT40 cells

The chicken B lymphocyte line DT40 is a well-known model used for gene knockout studies in vertebrate cells, as described in the excellent book *Reviews and Protocols in DT40 Research* (Buerstedde & Takeda, 2006). The utility of the cells has been shown in a number of functional studies of cellular processes such as DNA damage repair, recombination, and chromosome segregation, as well as B lymphocyte-specific functions such as B cell receptor signaling. Several strategies for establishing DT40 cells lacking genes of interest are described elsewhere (Buerstedde & Takeda, 2006). Here, however, the methods used in our laboratory to establish DT40 cells deficient in multiple ZIP and ZnT transporter genes are described. Seven drug-resistant genes are available for gene disruption and gene reexpression in DT40 cells, making cells deficient in multiple genes feasible. Alternatively, drug-resistant genes flanked by loxP sequences for excision by Cre recombinase are available if needed.

3.2. Construction and preparation of knockout vectors

1. Obtain DNA fragments that will serve as the right and left arms of the knockout vector. The total length of both arms is usually 5 kb or longer for increased recombination efficiency, but shorter lengths will work. Genetic information for choosing sequences for the arms and designing primers can be found online at http://www.ncbi.nlm.nih.gov/projects/genome/guide/chicken/.
2. Ligate the arms to either side of a drug-resistant cassette containing a promoter, drug-resistant gene, and a polyA addition site. If necessary, the cassette can be flanked by loxP sequences.
3. Linearize 25–30 μg of the constructed plasmid DNA using an appropriate restriction enzyme. The restriction site should be located outside of the homology region.
4. Precipitate the DNA with ethanol and wash with 70% ethanol, air dry for 10 min, and resuspend in sterile TE to 0.5–1.0 μg/μl.

3.3. Electroporation to introduce knockout vectors and establish DT40 cell lines deficient in one gene

DT40 cells are cultured similarly to mammalian cells, except that chicken serum is required. Specifically, DT40 cells are cultured in RPMI 1640

supplemented with 10% heat-inactivated fetal calf serum, 1% chicken serum, and 50 μM 2-mercaptoethanol (2-ME) at 39.5 °C. DT40 cells can be cultured at 37 °C, but they grow slowly.
1. Spin down 1×10^7 cells and wash with phosphate-buffered saline (PBS) once.
2. Resuspend the cells in 0.5 ml PBS and transfer to an electroporation cuvette (0.4-cm electrode gap).
3. Add 25–30 μg of the linearized DNA and incubate on ice for 10 min.
4. Electroporate under conditions of 550 V, 25 μF, and incubate on ice for 10 min.
5. Transfer the cells to a 10-cm dish containing 20 ml of fresh media without selection, and culture the cells overnight at 39.5 °C, 5% CO_2 to allow recovery.
6. Resuspend the electroporated cells in 60–80 ml of fresh media containing selection. The final concentrations of drugs used are 2 mg/ml for Neomycin, 0.5 mg/ml for Puromycin, 0.15–0.25 mg/ml for Hygromycin B, 0.3 mg/ml for Zeocin, 25 μg/ml for Blasticidin S, 1 mg/ml for Histidinol, and 25 μg/ml for Mycophenolic acid.
7. Seed 200 μl (about 3×10^4 cells) into each well of three or four 96-well plates.
8. Select the cells at 39.5 °C, 5% CO_2. Stable clones will be visible in about a week. When the clones grow to about 2 mm in diameter, transfer them to 24-well plates with 1 ml of the same selection medium. The drug does not have to be added to the culture medium once stable clones are established in 24-well plates.

3.4. Preparation of genomic DNA from the clones

We recover genomic DNA as described previously (Laird et al., 1991) with slight modification. The recovered DNA prepared by this method is sufficient to perform genomic polymerase chain reaction (PCR) amplification. Scale up if a large amount of DNA is needed for methods such as Southern blotting.
1. Collect the cells from one well of a 24-well plate into a snap cap microcentrifuge tube, and lyse the cells in 100 μl of lysis solution (200 mM NaCl, 20 mM EDTA, 40 mM Tris–HCl, pH 8.0, 0.5% SDS, 0.5% 2-ME). Add 4 μl of Proteinase K (20 mg/ml) and incubate 2 h or overnight at 55 °C.
2. Add 1 μl of RNase A (10 mg/ml) and then incubate for an additional 10–15 min at room temperature.

3. Add 70 μl of saturated NaCl, shake well for 15 s and let stand for 15 min on ice. Centrifuge at 12,000 × g for 15–30 min at 4 °C.
4. Transfer the cleared DNA solution into a clean tube and precipitate it with 250 μl of 100% ethanol. Shake well and centrifuge at 12,000 × g for 2 min at 4 °C.
5. Wash the DNA precipitate with 500 μl of 70% ethanol. Centrifuge at 12,000 × g for 1 min at 4 °C, and air dry for a few minutes.
6. Dissolve in 15 μl of TE, incubate at 55 °C for 30 min to 1 h to get into solution.

3.5. Establishment of DT40 cells deficient in multiple genes using the estrogen receptor chimeric Cre recombinase, CreER

1. Establish a drug-resistant cell line expressing the estrogen receptor-Cre recombinase fusion protein using the same procedure described in Section 3.3. The CreER protein is retained within the cytoplasm until driven to the nucleus by a ligand such as 4-hydroxy-tamoxifen. Only then can the enzyme excise DNA flanked by loxP sites.
2. Treat ~5×10^4 cells in a 6-cm dish containing 5 ml of media with 200 nM 4-hydroxy-tamoxifen for 24–48 h at 39.5 °C, 5% CO_2.
3. Resuspend the cells in fresh medium without the drug, and plate 200 μl containing one cell in each well of three 96-well plates.
4. In about a week, transfer the single clones to 24-well plates containing 1 ml medium without 4-hydroxy-tamoxifen.
5. Confirm excision of the drug-resistant cassette by genomic PCR and check for drug sensitivity of the clones.
6. Electroporate the cells with the next knockout vector using the same procedures described in Section 3.3 to establish double transfectants.
7. Repeat the same steps described in Section 3.4 to isolate the DNA and identify DT40 cells with recombination.
8. Repeat steps 2–6 to establish DT40 cells deficient in multiple genes.

4. EXPERIMENTAL PROCEDURES USED IN STUDIES OF DT40 CELLS DEFICIENT IN ZINC TRANSPORT

4.1. Confirmation of phenotypes of DT40 mutants

Disruption of ZIP and ZnT transporter genes in DT40 cells causes a number of cellular defects (Kambe, 2011). It is essential to confirm that the defects in the DT40 mutants are complemented by reexpression of the disrupted gene.

Exogenous reexpression of the human orthologs is one option, and it is also helpful in determining human gene function.

4.2. Reexpression of human orthologs of ZIP and ZnT transporters in DT40 mutants

In our laboratory, plasmids harboring an appropriate drug-resistant gene, where the cDNA of interest is under the control of the chicken β actin promoter, are usually used for reexpression; however, most mammalian expression systems can be used (Buerstedde & Takeda, 2006). Transfection for reexpression in DT40 mutants is performed in the same way described in Section 3.3.

4.3. Analyses of complex formation of Zinc transporters by immunoprecipitation

Zinc transporters function by forming dimers (Bin et al., 2011; Kambe, 2012). Homo or hetero complex formation is easily confirmed by immunoprecipitation (Fukunaka et al., 2009; Suzuki, Ishihara, Migaki, Nagao, et al., 2005).

1. Establish cell lines expressing the zinc transporter of interest tagged with an antigenic epitope using the electroporation method described in Section 3.3.
2. After washing cells once in PBS, lyse the cells ($\sim 2 \times 10^7$ cells) with 1 ml of Nonidet P-40 lysis buffer (50 mM HEPES–HCl, pH 7.4, 100 mM NaCl, 1.5 mM MgCl$_2$, 1% (v/v) Nonidet P-40) containing protease inhibitors. Rotate the cell lysates for 2 h at 4 °C, and adjust the concentration to 1 μg/μl with Nonidet P-40 lysis buffer[1].
3. To precipitate the expressed zinc transporter, add antibodies against the epitope to the lysate and rotate it for 1 h. Add 25 μl of Protein A- or Protein G-sepharose beads (GE Healthcare, Waukesha, WI) and rotate the mixture for an additional 2 h at 4 °C.
4. Centrifuge the mixture at $400 \times g$ for 5 min at 4 °C and wash the pelleted beads containing the immunoprecipitate three times with Nonidet P-40 lysis buffer. Elute the precipitated proteins in $5\times$ Ling's solubilizing buffer (150 mM sucrose, 50 mM Tris–HCl, pH 8.0, 20 mM DTT, 10% SDS, 5 mM EDTA).
5. Add an equal volume of $2\times$ Urea buffer (8 M urea, 30 mM sucrose, 10 mM Tris–HCl, pH 8.0, 4 mM DTT, 2% SDS, 1 mM EDTA) and incubate at 37 °C for 30 min before electrophoresis.
6. For the input fraction of the immunoprecipitation, precipitate an aliquot of the original cell lysate with three volumes of acetone for 2 h at -20 °C

---

[1] Note that Nonidet P-40 is also referred to as Igepal.

and centrifuge at 12,000 × g for 15 min at 4 °C. Lyse the pellet with 5 × Ling's solubilizing buffer and treat with 2 × Urea buffer as described above before electrophoresis. This procedure is optional in the case that protein concentrations need to be increased.

4.4. Transient transfection in DT40 mutants

Transient transfection is also possible in DT40 cells (Buerstedde & Takeda, 2006) and is useful for analyses using reporter genes to evaluate cytosolic zinc levels or monitor zinc homeostasis in subcellular compartments. Electroporation is usually used, but lipofection is also possible.

1. Spin down 5×10^6 cells and wash with K–PBS (30.8 mM NaCl, 120.7 mM KCl, 8.1 mM Na$_2$HPO$_4$, 1.46 mM KH$_2$PO$_4$, and 5 mM MgCl$_2$).
2. Resuspend the cells in 0.3 ml K-PBS in a snap cap microcentrifuge tube.
3. Promptly add 10–15 μg of plasmid DNA diluted in 0.2 ml K-PBS to the cells. After mixing gently, transfer to an electroporation cuvette (0.4-cm electrode gap). Incubate on ice for 10 min.
4. Electroporate under conditions of 600 V, 25 μF, and incubate on ice for 10 min.
5. Transfer the cells to a 10-cm dish containing 20 ml of fresh media and culture at 39.5 °C, 5% CO$_2$. Move the cells into multiwell plates after a several hour recovery period, if the cells will need to be treated.

5. FUNCTIONAL ANALYSIS OF ZINC MOBILIZATION INTO OR OUT OF THE SECRETORY AND ENDOSOMAL–LYSOSOMAL COMPARTMENTS

5.1. Immunofluorescence analysis of zinc mobilization in DT40 mutants deficient in ZIP and ZnT transporters

There are a number of fluorescent zinc probes with different affinities to zinc (Huang & Lippard, 2012) that can be used as reporters. Upon zinc binding, these probes elicit a bright fluorescent signal, which can be used to measure zinc levels or follow subcellular zinc release in live cells.

1. Wash approximately 5×10^5 cells with cold Hanks' balanced salt solution (HBSS) twice and then resuspend them in HBSS containing the zinc fluorescent probe (typically μM order). Treat the cells with ZnSO$_4$ if more zinc is needed in the cells and subcellular compartments. TPEN, a zinc chelator, is also added to ensure that the fluorescence signal results from zinc binding to the probe.

2. Incubate the stained cells on coverslips coated with poly-L-lysine in a humidified chamber for 30 min at 37 °C.
3. Fix the stained cells on the coverslips with 4% paraformaldehyde. Cytospin (Thermo Scientific) is also available to fix the cells on cover glass (Matsuura et al., 2009).
4. Observe the stained cells under a fluorescence microscope.

5.2. Monitoring cytosolic zinc levels and zinc-related homeostasis in DT40 cells

Metallothionein (*MT*) mRNA has been used to elucidate cytosolic zinc levels in mammalian cells because of its rapid induction in response to zinc. Similar procedures are possible in DT40 cells because chicken *MT* mRNA is regulated in the same way as it is in mammalian cells (Matsuura et al., 2009). Other zinc-regulated responses such as the UPR are also conserved in DT40 cells (Ishihara et al., 2006). Assessment of these events can be used to evaluate intracellular zinc. In fact, most mammalian reporter plasmids are functional in DT40 cells (Buerstedde & Takeda, 2006). For example, the MT-luciferase reporter used to evaluate cytosolic zinc (Kimura et al., 2008) and the 5xATF6 luciferase used to monitor the UPR (Wang et al., 2000) are both effective in DT40 cells and can be used to assess changes in zinc transport (Ishihara et al., 2006; Jeong et al., 2012).

5.3. Evaluation of zinc status in subcellular compartments using zinc-dependent enzymes such as alkaline phosphatase

The activity of alkaline phosphatase (ALP) is also zinc dependent, and thus is useful to assess the zinc status in the secretory compartments of vertebrates (Jeong et al., 2012; Suzuki, Ishihara, Migaki, Matsuura, et al., 2005) or in the endosomal–lysosomal compartments (vacuoles) in yeast (Qiao, Ellis, Steffen, Wu, & Eide, 2009). For example, activation of the endogenous nonspecific ALP (TNAP) or the exogenously expressed placenta ALP (PLAP) is impaired in DT40 cells deficient in *ZnT5, ZnT6,* and *ZnT7* genes (Fig. 5.2).

5.4. Measurement of ALP activity

1. Lyse approximately $0.5-1 \times 10^6$ cells in 100 μl ALP lysis buffer (10 mM Tris–HCl, pH 7.5, 0.5 mM MgCl$_2$ and 0.1% Triton X-100).
2. Incubate the lysate at 65 °C for 30 min to inactivate TNAP if measuring PLAP (placenta-derived ALP) activity (Fig. 5.2). TNAP is thermosensitive, while PLAP is thermoresistant.

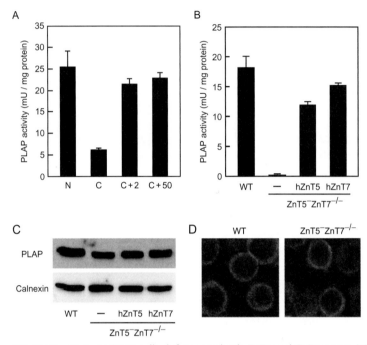

Figure 5.2 PLAP activity in DT40 cells deficient in both *ZnT5* and *ZnT7* genes. (A) Zinc-dependent PLAP activity in DT40 cells. N, normal medium; C, zinc-deficient medium; C+2 and C+50, zinc-deficient medium supplemented with 2 and 50 μM ZnSO$_4$, respectively. (B) PLAP activity in DT40 cells deficient in both *ZnT5* and *ZnT7* genes (*ZnT5$^-$ZnT7$^{-/-}$* cells). The decreased activity was restored by expression of either hZnT5 or hZnT7. (C) and (D) PLAP protein is located in the cytoplasmic membrane in *ZnT5$^-$ZnT7$^{-/-}$* cells, despite having little activity. *Modified from Suzuki, Ishihara, Migaki, Matsuura, et al. (2005).* (For color version of this figure, the reader is referred to the online version of this chapter.)

3. Incubate 1–5 μg of the lysate in triplicate in a 96-well plate for 10 min at room temperature. In addition, prepare dilutions of purified ALP for a standard curve.
4. Add 100 μl of substrate solution (2 mg/ml *p*-nitrophenyl phosphate in 1 M diethanolamine buffer, pH 9.8, containing 0.5 mM MgCl$_2$) using a multichannel pipette, mix well using microplate mixer, and incubate for 5–10 min at room temperature before measuring the activity.
5. Measure the *p*-nitrophenol released by TNAP or PLAP using the 405 nm absorbance (yellow color).

6. CONCLUDING REMARKS

Recent findings indicate that zinc mobilization across the membranes of subcellular compartments by both ZIP and ZnT transporters acts as a fundamental regulator of a variety of biological functions (Fukada & Kambe, 2011; Fukada et al., 2011; Haase & Rink, 2009; Kambe, 2011). Indeed, zinc mobilization into and out of the endosomes is thought to play crucial roles in endosomal signaling. However, many details still need to be determined. The methods described in this chapter will be helpful in the clarification of the functions of zinc in endosomal signaling. They are focused on the use of DT40 cells, but could also be widely applied to studies using mammalian cells deficient in ZIP or ZnT transporters.

ACKNOWLEDGMENT

I thank Drs. Shunichi Takeda and Minoru Takata for providing us the DT40 cells and the protocol for gene disruption experiments. This work was supported by Grants-in-Aid from the Ministry of Education, Culture, Sports, Science and Technology of Japan.

REFERENCES

Andreini, C., Banci, L., Bertini, I., & Rosato, A. (2006). Zinc through the three domains of life. *Journal of Proteome Research, 5*, 3173–3178.

Aras, M. A., & Aizenman, E. (2011). Redox regulation of intracellular zinc: Molecular signaling in the life and death of neurons. *Antioxidants & Redox Signaling, 15*, 2249–2263.

Aydemir, T. B., Liuzzi, J. P., McClellan, S., & Cousins, R. J. (2009). Zinc transporter ZIP8 (SLC39A8) and zinc influence IFN-gamma expression in activated human T cells. *Journal of Leukocyte Biology, 86*, 337–348.

Bin, B. H., Fukada, T., Hosaka, T., Yamasaki, S., Ohashi, W., Hojyo, S., et al. (2011). Biochemical characterization of human ZIP13 protein: A homo-dimerized zinc transporter involved in the spondylocheiro dysplastic Ehlers-Danlos syndrome. *Journal of Biological Chemistry, 286*, 40255–40265.

Bosomworth, H. J., Thornton, J. K., Coneyworth, L. J., Ford, D., & Valentine, R. A. (2012). Efflux function, tissue-specific expression and intracellular trafficking of the Zn transporter ZnT10 indicate roles in adult Zn homeostasis. *Metallomics, 4*, 771–779.

Buerstedde, J. M., & Takeda, S (Eds.), (2006). *Reviews and protocols in DT40 research. Subcellular biochemistry*, Vol. 40. New York: Springer.

Colvin, R. A., Holmes, W. R., Fontaine, C. P., & Maret, W. (2010). Cytosolic zinc buffering and muffling: Their role in intracellular zinc homeostasis. *Metallomics, 2*, 306–317.

Eichelsdoerfer, J. L., Evans, J. A., Slaugenhaupt, S. A., & Cuajungco, M. P. (2010). Zinc dyshomeostasis is linked with the loss of mucolipidosis IV-associated TRPML1 ion channel. *Journal of Biological Chemistry, 285*, 34304–34308.

Ellis, C. D., Wang, F., MacDiarmid, C. W., Clark, S., Lyons, T., & Eide, D. J. (2004). Zinc and the Msc2 zinc transporter protein are required for endoplasmic reticulum function. *Journal of Cell Biology, 166*, 325–335.

Falcon-Perez, J. M., & Dell'Angelica, E. C. (2007). Zinc transporter 2 (SLC30A2) can suppress the vesicular zinc defect of adaptor protein 3-depleted fibroblasts by promoting zinc accumulation in lysosomes. *Experimental Cell Research, 313*, 1473–1483.

Fang, Y., Sugiura, R., Ma, Y., Yada-Matsushima, T., Umeno, H., & Kuno, T. (2008). Cation diffusion facilitator Cis4 is implicated in Golgi membrane trafficking via regulating zinc homeostasis in fission yeast. *Molecular Biology of the Cell, 19,* 1295–1303.

Fukada, T., Civic, N., Furuichi, T., Shimoda, S., Mishima, K., Higashiyama, H., et al. (2008). The zinc transporter SLC39A13/ZIP13 is required for connective tissue development; its involvement in BMP/TGF-beta signaling pathways. *PLoS One, 3,* e3642.

Fukada, T., & Kambe, T. (2011). Molecular and genetic features of zinc transporters in physiology and pathogenesis. *Metallomics, 3,* 662–674.

Fukada, T., Yamasaki, S., Nishida, K., Murakami, M., & Hirano, T. (2011). Zinc homeostasis and signaling in health and diseases: Zinc signaling. *Journal of Biological Inorganic Chemistry, 16,* 1123–1134.

Fukunaka, A., Kurokawa, Y., Teranishi, F., Sekler, I., Oda, K., Ackland, M. L., et al. (2011). Tissue nonspecific alkaline phosphatase is activated via a two-step mechanism by zinc transport complexes in the early secretory pathway. *Journal of Biological Chemistry, 286,* 16363–16373.

Fukunaka, A., Suzuki, T., Kurokawa, Y., Yamazaki, T., Fujiwara, N., Ishihara, K., et al. (2009). Demonstration and characterization of the heterodimerization of ZnT5 and ZnT6 in the early secretory pathway. *Journal of Biological Chemistry, 284,* 30798–30806.

Gaither, L. A., & Eide, D. J. (2001). Eukaryotic zinc transporters and their regulation. *Biometals, 14,* 251–270.

Haase, H., & Rink, L. (2009). Functional significance of zinc-related signaling pathways in immune cells. *Annual Review of Nutrition, 29,* 133–152.

Hennigar, S. R., & Kelleher, S. L. (2012). Zinc networks: The cell-specific compartmentalization of zinc for specialized functions. *Biological Chemistry, 393,* 565–578.

Hershfinkel, M., Aizenman, E., Andrews, G., & Sekler, I. (2010). Zinc bells rang in Jerusalem!. *Science Signaling, 3,* mr2.

Hirano, T., Murakami, M., Fukada, T., Nishida, K., Yamasaki, S., & Suzuki, T. (2008). Roles of zinc and zinc signaling in immunity: Zinc as an intracellular signaling molecule. *Advances in Immunology, 97,* 149–176.

Hogstrand, C., Kille, P., Nicholson, R. I., & Taylor, K. M. (2009). Zinc transporters and cancer: A potential role for ZIP7 as a hub for tyrosine kinase activation. *Trends in Molecular Medicine, 15,* 101–111.

Huang, Z., & Lippard, S. J. (2012). Illuminating mobile zinc with fluorescence from cuvettes to live cells and tissues. *Methods in Enzymology, 505,* 445–468.

Ishihara, K., Yamazaki, T., Ishida, Y., Suzuki, T., Oda, K., Nagao, M., et al. (2006). Zinc transport complexes contribute to the homeostatic maintenance of secretory pathway function in vertebrate cells. *Journal of Biological Chemistry, 281,* 17743–17750.

Jeong, J., Walker, J. M., Wang, F., Park, J. G., Palmer, A. E., Giunta, C., et al. (2012). Promotion of vesicular zinc efflux by ZIP13 and its implications for spondylocheiro dysplastic Ehlers-Danlos syndrome. *Proceedings of the National Academy of Sciences of the United States of America, 109,* E3530–E3538.

Kambe, T. (2011). An overview of a wide range of functions of ZnT and Zip zinc transporters in the secretory pathway. *Bioscience, Biotechnology, and Biochemistry, 75,* 1036–1043.

Kambe, T. (2012). Molecular architecture and function of ZnT transporters. *Current Topics in Membranes, 69,* 199–220.

Kambe, T. (2013). Regulation of zinc transport. In R. A. Scott & V. Culotta (Eds.), *Encyclopedia of inorganic and bioinorganic chemistry, metals in cells* (pp. 301–309). Hoboken, NJ: John Wiley and Sons, Inc.

Kambe, T., Suzuki, T., Nagao, M., & Yamaguchi-Iwai, Y. (2006). Sequence similarity and functional relationship among eukaryotic ZIP and CDF transporters. *Genomics, Proteomics & Bioinformatics, 4,* 1–9.

Kambe, T., Weaver, B. P., & Andrews, G. K. (2008). The genetics of essential metal homeostasis during development. *Genesis, 46*, 214–228.

Kimura, T., Li, Y., Okumura, F., Itoh, N., Nakanishi, T., Sone, T., et al. (2008). Chromium(VI) inhibits mouse metallothionein-I gene transcription by preventing the zinc-dependent formation of an MTF-1-p300 complex. *Biochemical Journal, 415*, 477–482.

Kobayashi, T., Beuchat, M. H., Lindsay, M., Frias, S., Palmiter, R. D., Sakuraba, H., et al. (1999). Late endosomal membranes rich in lysobisphosphatidic acid regulate cholesterol transport. *Nature Cell Biology, 1*, 113–118.

Laird, P. W., Zijderveld, A., Linders, K., Rudnicki, M. A., Jaenisch, R., & Berns, A. (1991). Simplified mammalian DNA isolation procedure. *Nucleic Acids Research, 19*, 4293.

Lasswell, J., Rogg, L. E., Nelson, D. C., Rongey, C., & Bartel, B. (2000). Cloning and characterization of IAR1, a gene required for auxin conjugate sensitivity in Arabidopsis. *Plant Cell, 12*, 2395–2408.

Lazarczyk, M., Pons, C., Mendoza, J. A., Cassonnet, P., Jacob, Y., & Favre, M. (2008). Regulation of cellular zinc balance as a potential mechanism of EVER-mediated protection against pathogenesis by cutaneous oncogenic human papillomaviruses. *Journal of Experimental Medicine, 205*, 35–42.

Lemaire, K., Chimienti, F., & Schuit, F. (2012). Zinc transporters and their role in the pancreatic β-cell. *Journal of Diabetes Investigation, 3*, 202–211.

Lichten, L. A., & Cousins, R. J. (2009). Mammalian zinc transporters: Nutritional and physiologic regulation. *Annual Review of Nutrition, 29*, 153–176.

Matsuura, W., Yamazaki, T., Yamaguchi-Iwai, Y., Masuda, S., Nagao, M., Andrews, G. K., et al. (2009). SLC39A9 (ZIP9) regulates zinc homeostasis in the secretory pathway: Characterization of the ZIP subfamily I protein in vertebrate cells. *Bioscience, Biotechnology, and Biochemistry, 73*, 1142–1148.

McCormick, N. H., & Kelleher, S. L. (2012). ZnT4 provides zinc to zinc-dependent proteins in the trans-Golgi network critical for cell function and Zn export in mammary epithelial cells. *American Journal of Physiology Cell Physiology, 303*, C291–C297.

McCormick, N., Velasquez, V., Finney, L., Vogt, S., & Kelleher, S. L. (2010). X-ray fluorescence microscopy reveals accumulation and secretion of discrete intracellular zinc pools in the lactating mouse mammary gland. *PLoS One, 5*, e11078.

Nishida, K., Hasegawa, A., Nakae, S., Oboki, K., Saito, H., Yamasaki, S., et al. (2009). Zinc transporter Znt5/Slc30a5 is required for the mast cell-mediated delayed-type allergic reaction but not the immediate-type reaction. *Journal of Experimental Medicine, 206*, 1351–1364.

Patrushev, N., Seidel-Rogol, B., & Salazar, G. (2012). Angiotensin II requires zinc and downregulation of the zinc transporters ZnT3 and ZnT10 to induce senescence of vascular smooth muscle cells. *PLoS One, 7*, e33211.

Qiao, W., Ellis, C., Steffen, J., Wu, C. Y., & Eide, D. J. (2009). Zinc status and vacuolar zinc transporters control alkaline phosphatase accumulation and activity in Saccharomyces cerevisiae. *Molecular Microbiology, 72*, 320–334.

Stathakis, D. G., Burton, D. Y., McIvor, W. E., Krishnakumar, S., Wright, T. R., & O'Donnell, J. M. (1999). The catecholamines up (Catsup) protein of Drosophila melanogaster functions as a negative regulator of tyrosine hydroxylase activity. *Genetics, 153*, 361–382.

Suzuki, T., Ishihara, K., Migaki, H., Matsuura, W., Kohda, A., Okumura, K., et al. (2005). Zinc transporters, ZnT5 and ZnT7, are required for the activation of alkaline phosphatases, zinc-requiring enzymes that are glycosylphosphatidylinositol-anchored to the cytoplasmic membrane. *Journal of Biological Chemistry, 280*, 637–643.

Suzuki, T., Ishihara, K., Migaki, H., Nagao, M., Yamaguchi-Iwai, Y., & Kambe, T. (2005). Two different zinc transport complexes of cation diffusion facilitator proteins localized in the secretory pathway operate to activate alkaline phosphatases in vertebrate cells. *Journal of Biological Chemistry*, *280*, 30956–30962.

Taniguchi, M., Fukunaka, A., Hagihara, M., Watanabe, K., Kamino, S., Kambe, T., et al. (2013). Essential role of the zinc transporter ZIP9/SLC39A9 in regulating the activations of Akt and Erk in B-cell receptor signaling pathway in DT40 cells. *PLoS ONE*, *8*, e58022.

Taylor, K. M., Hiscox, S., Nicholson, R. I., Hogstrand, C., & Kille, P. (2012). Protein kinase CK2 triggers cytosolic zinc signaling pathways by phosphorylation of zinc channel ZIP7. *Science Signaling*, *5*, ra11.

Wang, Y., Shen, J., Arenzana, N., Tirasophon, W., Kaufman, R. J., & Prywes, R. (2000). Activation of ATF6 and an ATF6 DNA binding site by the endoplasmic reticulum stress response. *Journal of Biological Chemistry*, *275*, 27013–27020.

Yamasaki, S., Hasegawa, A., Hojyo, S., Ohashi, W., Fukada, T., Nishida, K., et al. (2012). A novel role of the L-type calcium channel alpha(1D) subunit as a gatekeeper for intracellular zinc signaling: Zinc wave. *PLoS One*, *7*, e39654.

Zhang, X., Liang, D., Guo, B., Deng, W., Chi, Z. H., Cai, Y., et al. (2013). Zinc transporter 5 and zinc transporter 7 induced by high glucose protects peritoneal mesothelial cells from undergoing apoptosis. *Cellular Signalling*, *25*, 999–1010.

CHAPTER SIX

Interactions Between Endosomal Maturation and Autophagy: Analysis of ESCRT Machinery During *Caenorhabditis elegans* Development

Marion Manil-Ségalen, Emmanuel Culetto, Renaud Legouis[1], Christophe Lefebvre[1]

Centre de Génétique Moléculaire, CNRS UPR3404, Université Paris-Sud, 91198 Gif-sur-Yvette Cedex, France
[1]Corresponding authors: e-mail address: legouis@cgm.cnrs-gif.fr; lefebvre@cgm.cnrs-gif.fr

Contents

1. Introduction	94
2. Strains and Reagents	95
2.1 Strains	95
2.2 RNAi clones	96
2.3 Antibodies	96
2.4 Solutions	97
3. Fluorescent-Tagged Protein Construction and Transgenesis	98
3.1 Constructs	98
3.2 Transformation procedures	98
4. Analysis of Developmental Phenotypes in ESCRT Mutants	100
4.1 Genetics	100
4.2 RNAi experiments	101
4.3 Observations of developmental phenotypes	103
5. Analysis of Vesicular Compartments	104
5.1 Immunolocalization of endosomal and autophagy proteins	104
5.2 Electron microscopy	108
5.3 Vital dyes and fluorescent constructs	109
6. Analysis of Autophagy	110
6.1 Counting LGG-1 and LGG-2 dots	111
6.2 Autophagic flux	111
6.3 Pharmacological modulation of autophagy	112
7. Methods to Visualize Amphisome, the Fusion Organelle Between Endosomes and Autophagosomes	114
8. Conclusions	116

Acknowledgments 116
References 116

Abstract

Endocytosis and autophagy are key vesicular pathways involved in degradation and recycling of cellular material. Both degradative pathways finally fuse with lysosome but are indeed interconnected at several levels. In particular, the fusion between endosomes and autophagosomes can generate intermediate vesicles named amphisomes. We analyzed the physiological and developmental roles of the ESCRT machinery in a model organism, the nematode *Caenorhabditis elegans* and showed that the blockage of the endosomal maturation triggers the induction of autophagic activity.

This chapter describes several methods for studying endocytosis, autophagy, and their interconnection in *C. elegans*. A series of genetic, biochemical, and microscopy analyses has been used to study at the cellular and developmental levels, the cross talks between autophagy and endocytosis.

1. INTRODUCTION

Eukaryotes are characterized by a very complex endomembrane network with highly dynamic properties. This network is based on multiple transport vesicle types that usually undergo many maturation steps to select and deliver cargos to specific cellular compartments. Endosomes are vesicles involved in material transport from plasma membrane toward degradative lysosome compartment. During endocytosis, invagination of plasma membrane domains allows the internalization and recycling of lipids and membrane proteins, by *de novo* formation of vesicles. Subsequently, these vesicles fuse to form early endosomes. Endosomes go through a series of maturation steps during which they continuously exchange material with the secretory pathway. Early endosomes can form either recycling endosomes that will ultimately fuse with plasma membrane or lead to late endosomes also named multivesicular bodies (MVBs) (Huotari & Helenius, 2011). MVB internalize membrane protein cargos specified for degradation in intraluminal vesicles (ILVs) of 40–100 nm diameter. The ILVs are formed by the invagination of the limiting membrane of the MVB which requires four endosomal sorting complex required for transport (ESCRT-0 à -III), composed of Class E vacuolar protein sorting (VPS-E). Finally, MVB fuse with lysosomes to allow the degradation of cargos and recycling of amino acids and lipids (Raiborg & Stenmark, 2009).

Autophagy, which usually refers to macroautophagy, is another cellular mechanism required for the degradation of cell constituents, specifically involved in recycling cytoplasmic long-lived proteins and organelles (Reggiori & Klionsky, 2002). This recycling mechanism is dependent on dedicated double membrane vesicles, named autophagosomes, that engulf cytoplasmic material either by selective or nonselective sequestration and deliver it to the lysosomes.

Endocytosis and autophagy are indeed two interdependent recycling processes. In mammalian cells the final maturation of autophagosome involves a fusion with endosome to generate a hybrid organelle named amphisome (Gordon & Seglen, 1988), which subsequently fuse with lysosomes. Our group is studying the role of endocytosis and its connections with the autophagic pathway during *Caenorhabditis elegans* development. Our initial observations pointed out that mutations in ESCRT genes lead to various developmental defects such as embryonic or larval arrest and molting defects. Additionally, at the cellular level, ESCRT mutants show an increase in the number of autophagosomes (Roudier, Lefebvre, & Legouis, 2005). Similar observations have also been reported in mammals and drosophila (for review, see Manil-Segalen, Lefebvre, Culetto, & Legouis, 2012).Two main hypotheses can explain the accumulation of autophagosomes: a blockage of the terminal maturation and fusion with the lysosome or an induction of the autophagic flux. Interestingly, depending on the organism studied, the two situations have been demonstrated. In mammals and drosophila, the alteration of ESCRT function is responsible for a blockage of fusion between autophagosomes and the endolysosomal system, while in *C. elegans* it triggers, by a yet unknown mechanism, the increase of the autophagic flux. Our studies have shown that amphisomes do exist in *C. elegans* but are rare or very transient. We postulated that the increase of autophagy observed upon defective endosome maturation is a specific adaptive mechanism to maintain cellular homeostasis (Djeddi et al., 2012). In this chapter we describe a series of specific protocols that we have used to analyze the relations between endosomes and autophagosomes in *C. elegans*.

2. STRAINS AND REAGENTS
2.1. Strains

The list of strains that we have used for this work is provided in Table 6.1.

Table 6.1 Genotype and name of strains used in this study

Strain number	Genotype
RD189	vps-32(ok1355)/dpy-10(e128) unc-4(e120) III; Ex [vps-27::gfp; rol-6(su1006)]
RD122	vps-27(ok579)/unc-24(e138) dpy-20(e1282)IV
RD106	N2; Ex[vps-27::gfp; rol-6(su1006)]
RD110	N2; Ex[vps-36::gfp; rol-6(su1006)]
RD163	N2; Ex[vps-22::gfp; rol-6(su1006)]
RT327	unc-119(ed3); Is[Pvha6::gfp::rab-5; unc-119(+)]
DA2123	N2; adIs2122[gfp::lgg-1; rol-6(su1006)]
RD108	N2; Ex[gfp::lgg-2; rol-6(su1006)]
RD197	lgg-1(tm3489)/dpy-10(e128) unc-4(e120) III
RD210	unc-119(ed3); Is[Ppie-1::gfp::cherry::lgg-1; unc-119(+)]
KR3875	let-363(h111) dpy-5(e61)I; sDp(I;f)
RD204	unc-119(ed3); Is[Ppie-1::gfp::lgg-1; unc-119(+)]

2.2. RNAi clones

Bacteria (HT115) clones expressing double-stranded RNA (dsRNA) against *C. elegans* genes were obtained either from the Ahringer library (Kamath & Ahringer, 2003) or from Open Biosystems.

1. *ESCRT genes*: ESCRT-0: *vps-27/hgrs*-1 (C07G1.5), ESCRT-I: *vps-23/tsg-101* (C09G12.9), *vps-28* (Y87G2A.10), *vps-37* (CD4.4), ESCRT-II: *vps-22* (C27F2.5), *vps-25* (W02A11.2), ESCRT-III: *vps-20* (Y65B4A.3), *vps-24* (T27F7.1), *vps-32* (C56C10.3), ESCRT-III associated: *vps-4* (Y34D9A.10).
2. *Autophagy genes*: *tor/let-363* (B0261.2), *atg-7* (M7.5), *lgg-1* (C32D5.9), *lgg-2* (ZK593.6), *rab-7* (W03C9.3).

2.3. Antibodies

2.3.1 Primary antibodies

Mouse monoclonal antibodies: anti-GFP (Roche, 1814460), antitubulin (Sigma, T9026).
Rat polyclonal antibodies: anti-LGG-1 (Tian et al., 2010).

Rabbit polyclonal antibodies: anti-VPS-23 (Michelet et al., 2009), anti-VPS-27 (Roudier et al., 2005), anti-VPS-32 (Michelet et al., 2009), anti-LGG-2 (Al Rawi et al., 2011), anti-GFP (Invitrogen, A11122).

2.3.2 Secondary antibodies
Alexa Fluor-488-, Alexa Fluor-568-, Alexa Fluor-633-conjugated antibodies (Molecular Probes).
Horse Radish peroxidase-conjugated antibodies (Promega).

2.4. Solutions
1. M9: 6 g Na_2HPO_4, 3 g KH_2PO_4, 5 g NaCl add water to 1 l. Aliquot per 200 ml. Autoclave. Cool at room temperature and add 200 µl sterile 1 M $MgSO_4$.
2. S Basal: 5.9 g NaCl, 50 ml 1 M potassium phosphate, pH 6, 1 ml cholesterol (5 mg/ml in EtOH) add water to 1 l. Autoclave.
3. 1 M potassium phosphate, pH 6: 136 g KH_2PO_4, add water to 900 ml. Adjust pH to 6 with KOH. Add water to 1 l. Autoclave.
4. S Medium: 1 l S Basal, add 10 ml 1 M potassium citrate, pH 6, 10 ml trace metals solution, 3 ml 1 M $CaCl_2$ and 3 ml 1 M $MgSO_4$.
5. 1 M potassium citrate, pH 6: 293.5 g of tripotassium citrate monohydrate, 20 g of citric acid monohydrate, water to 900 ml. Adjust pH to 6 with KOH. Add water to 1 l. Autoclave.
6. Trace metals solution: 1.86 g Na_2EDTA, 0.69 g $FeSO_4 \cdot 7H_2O$, 0.29 g $ZnSO_4 \cdot 7H_2O$, 0.016 g $CuSO_4$, add water to 1 l. Autoclave. Store in the dark.
7. 10× PBS: 40 g NaCl, 1 g KCl, 14.34 g $Na_2HPO_4 \cdot 12H_2O$, 1 g KH_2PO_4, add water to 300 ml. Adjust pH to 7.4. Autoclave.
8. 1× bleach solution: 0.5 ml 10 N NaOH, 1.5 ml 10% HClO, add water to 10 ml. Store at 4 °C up to 1 week.
9. NGM agar stock: 3 g NaCl, 2.5 g bactopeptone, 1 ml cholesterol (5 mg/ml in EtOH), water to 1 l. Aliquot per 500 ml and add 10 g of agar per bottle. Autoclave.
10. NGM agar plate: melt down 500 ml of NGM agar in microwave, cool to 55 °C, add 500 µl 1 M $CaCl_2$, 500 µl 1 M $MgSO_4$, 10 ml 1 M KH_2PO_4, 1.650 ml 1 M K_2HPO_4.
11. Double-stranded RNA interference (RNAi) agar plate: melt down 500 ml of NGM agar in microwave, cool to 55 °C, add 500 µl 1 M $CaCl_2$, 500 µl 1 M $MgSO_4$, 10 ml 1 M KH_2PO_4, 1.650 ml 1 M

K$_2$HPO4, 125 μl carbenicillin (100 mg/ml) and 625 μl 0.8 M isopropyl β-D-1-thiogalactopyranoside (IPTG).

3. FLUORESCENT-TAGGED PROTEIN CONSTRUCTION AND TRANSGENESIS

Because most of the antibodies against mammal proteins do not cross-react with *C. elegans* homologs, we have generated a series of transgenic strains expressing fluorescent (GFP, Cherry, etc.) tagged proteins involved in endocytosis and autophagy pathways (see Table 6.1).

3.1. Constructs

Our constructs were generated by either classical molecular cloning, PCR fusion method (Hobert, 2002), or by Gateway cloning procedure (Invitrogen). The Gateway *C. elegans* dedicated plasmids have been constructed by Seydoux's lab which Website provides most of the required information regarding vector sequences and detail of cloning procedures (http://www.bs.jhmi.edu/MBG/SeydouxLab/vectors/index.html).

3.2. Transformation procedures

Initially, *C. elegans* transgenesis was carried out by microinjection in the germ line of DNA construct of interest associated with phenotypic markers such as *rol-6* or GFP (Mello & Fire, 1995). The main limitation of this approach is the formation of an extra chromosome containing a high copy number of the transgene. However, the biolistic bombardment of gold particles coupled to DNA allows the direct integration of transgene into the genome. Due to the low frequency of integration, a positive selection using *unc119* mutant worms is necessary to select transgenic animals.

3.2.1 Transformation by microinjection

Microinjection in *C. elegans* germ line is a rapid and easy way to obtain transgenic animals (Mello, Kramer, Stinchcomb, & Ambros, 1991) and has been described in detail (Mello & Fire, 1995). Briefly, the DNA construct (plasmid, cosmid, fosmid, and PCR product) is injected in the distal part of a young adult gonad (http://www.dnatube.com/video/687/Celegans-microinjection). This part of the gonad is a syncytium that will give the future oocytes; thereby all transgenic embryos coming from one injected adult are independent. Injected DNA concatenates to form an extrachromosomal structure that is transmitted to the progeny with a variable transmission rate.

3.2.2 Biolistic transformation
3.2.2.1 Worms preparation
1. Day 1: bleach 500 *unc-119* gravid worms and incubate embryo in M9 overnight on a rotating wheel at 20 °C in 15 ml tubes.
2. Day 2: suspend L1 larvae into 50 ml of S-medium buffer, 5 µg/ml nystatin, 50 µg/ml streptomycin and add frozen pellet of 200 ml HB101 bacteria culture.
3. Allow worms to grow during 3 days at 20 °C. Repeat steps 1–3 to amplify worms.
4. Day 8: Most of the worms are young adults and are ready for bombardment (100,000 worms per bombardment are used).

3.2.2.2 Gold particles preparation and DNA coating
1. Weigh 60 mg of 1 µm gold beads (Biorad) into a 2-ml tube.
2. Add 2 ml 70% EtOH. Vortex 5 min. Incubate 15 min on a rotating wheel.
3. Pellet the beads by spinning 15 s in a microcentrifuge.
4. Discard the supernatant.
5. Repeat the following steps three times: add 1 ml sterile water, vortex for 1 min, pellet the beads by spinning 15 s in a microfuge, discard the supernatant.
6. Add 1 ml sterile 50% glycerol and keep the tube at −20 °C until needed.
7. Resuspend the gold beads by vortexing for 5 min to disrupt agglomerated particles.
8. For each bombardment, place 40 µl gold beads in a 1.5-ml tube.
9. Vortex vigorously and then add in order: 50 µl linearized DNA (100–200 ng/µl), 50 µl 2.5 M $CaCl_2$, 20 µl 0.1 M spermidine.
10. Homogenize by pipetting. Do not vortex.
11. Incubate for 30 min on ice and homogenize by pipetting every 5 min.
12. Spin for 15 s and discard the supernatant.
13. Add 600 µl 70% EtOH.
14. Spin for 15 s and discard the supernatant.
15. Add 1 ml 100% EtOH.
16. Spin for 15 s and discard the supernatant.

17. Add 15 μl 100% EtOH and gently resuspend the pellet by flicking the tube several times and store on ice until bombardment.

3.2.2.3 Bombardment
1. Just before the bombardment: spin the worms for 2 min at 1000 rpm and discard the supernatant.
2. Using a short Pasteur pipette transfer the pellet worms (100,000 young adults) to the center of a plate (10 cm).
3. Let the plate without lid to allow the evaporation of the liquid.
4. Perform the bombardment using the Bio-Rad Biolistic PDS-1000/He particle delivery system (follow the instruction manual).

3.2.2.4 Plating worms
One hour after the shoot, wash the bombarded worms off the plate using M9 buffer. Transfer them into five new fresh NGM plates seeded with OP50. Place plates into a 20 °C incubator for 10–15 days.

3.2.2.5 Screening
Plates are screened for WT animals. From one to many WT worms should be found per plate. Clone each WT animal to an individual plate, incubate at 20 °C for 4 days and analyze the progeny.

4. ANALYSIS OF DEVELOPMENTAL PHENOTYPES IN ESCRT MUTANTS

C. elegans allows to explore *in vivo* the roles of endocytosis and autophagy, at the levels of the cell, the tissue and the whole organism. We have focused on the developmental aspects by using both Mendelian genetics and RNAi. In particular, we have combined genetic approaches to modulate autophagy and endocytosis, to perform epistatic analyses, and to study subcellular localization of key markers in different genetic contexts.

4.1. Genetics

We first identified the *C. elegans* homologs of genes involved in endocytosis and autophagy and look for mutants. Most of the mutants have been obtained from the two main resources dedicated to generate deletion mutant for each *C. elegans* gene (http://celeganskoconsortium.omrf.org/; http://www.shigen.nig.ac.jp/c.elegans).

Next we used classical genetic crosses to combine mutant alleles and/or markers to allow functional *in vivo* analyses. For instance, we combined ESCRT mutants and GFP fusion transgenes to perform time-lapse studies of the endosomal maturation. A detailed example of such a genetic approach allowing us to generate *vps-32* mutant expressing VPS-27::GFP is shown in Fig. 6.1. This type of experiment revealed that VPS-32 and VPS-27 are enriched in distinct subdomains at the endosomal membrane and that the formation of VPS-27 subdomains is not affected by the depletion of ESCRT-II or -III components (Michelet et al., 2009).

4.2. RNAi experiments

When mutant alleles of a gene are not available, RNAi is a good alternative. In our study, we used RNAi against various ESCRT components (see Table 6.1) to analyze their role on autophagy. Firstly described in *C. elegans* in 1998, RNAi is a powerful technique to specifically knockdown a gene function (Fire et al., 1998). *C. elegans* is very sensitive to dsRNA introduction both by microinjection and feeding (Timmons & Fire, 1998). Two RNAi libraries are currently available (Kamath & Ahringer, 2003; Rual et al., 2004). They contain 28,268 clones in total and cover 94% of the *C. elegans* genome. In both libraries, specific DNA is cloned between two inverted T7 promoters and the RNAi plasmids are hosted in the HT115 bacterial strain, which contains the IPTG inducible T7 polymerase.

4.2.1 Injection

4.2.1.1 dsRNA transcription (wear gloves during the entire process)
1. PCR on the chosen clone using T7 primers (5′-CGTAATACGAC TCACTATAG-3′).
2. *In vitro* transcription using the Ambion MaxiScript kit:
   – Transcription 1 h at 37 °C (mix by pipetting: 8 µl DNA (from PCR), 2 µl 10× Transcription buffer, 1 µl 10 mM ATP, 1 µl 10 mM CTP, 1 µl 10 mM GT, 1 µl 10 mM UTP, 2 µl T7 enzyme, 4 µl water).
   – Incubate 15 min at RT with 1 µl DNAse 1.
   – Incubate 1 min at 95 °C with 1 µl 0.5 M EDTA.
3. Complementary strands annealing (15 min at 37 °C, then transfer on ice for a few minutes).
4. RNA purification:
   – Precipitate at least 30 min at −20 °C by adding 8.3 µl 3 M ammonium acetate, 200 µl 100% EtOH and 30 µl RNase-free water.

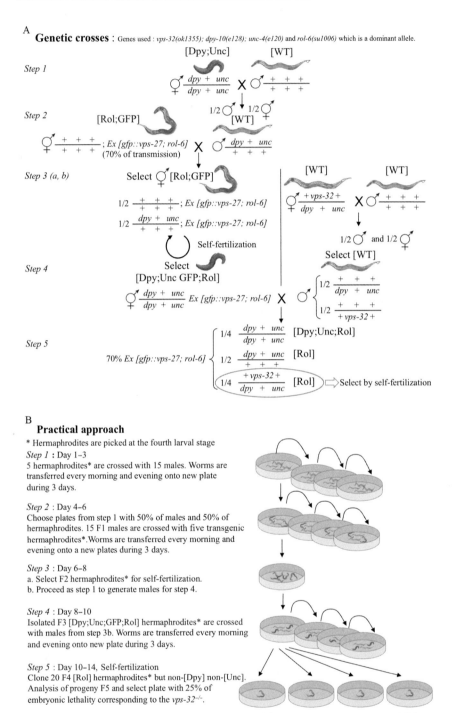

Figure 6.1 Genetic approach to generate *vps-32(ok1355)* mutant expressing VPS-27::GFP. (A) Description of genetic crosses in five steps. (B) Schematic representation and practical description of each step of the crossing. (For color version of this figure, the reader is referred to the online version of this chapter.)

- Centrifuge 15 min at 4 °C at 13,000 rpm; carefully discard the supernatant.
- Wash once in 70% ethanol (centrifuge 15 min at 4 °C at 13,000 rpm; carefully discard the supernatant).
- Resuspend the pellet in 15 μl RNase-free water.
5. The dsRNA can be stored at −80 °C.

4.2.1.2 Injection
Injection of dsRNA is performed as described in Section 3.2 except that dsRNA is injected in the intestine of fourth larval stage (L4) or young adults. The effect of RNAi is observed in the progeny of the injected worms.

4.2.2 Feeding
The entire protocol has been described previously (Kamath & Ahringer, 2003). Briefly, worms are cultivated on specific NGM plates (see Section 2.4). Plates are seeded with 6-h bacteria cultures (RNAi strain in LB, 50 μg/ml ampicillin). L4 animals are put on the RNAi plates either at 15 or 20 °C. After few days, the progeny of the fed worms can then be analyzed. Alternatively, first larval stage (L1) can be fed, using the same protocol. In this case, some phenotypes can be observed directly on the fed worms, instead of the progeny. This technique allows to bypass embryonic lethality phenotype and to score larvae and adult phenotypes. The efficiency of RNAi by feeding can vary a lot from one gene to another.

4.3. Observations of developmental phenotypes
Using RNAi approach and knockout mutants, we inactivated 11 different ESCRT genes and analyzed their cellular and developmental phenotypes. Although, those genes are supposed to be involved in the biogenesis of the MVB, we noticed that the depletion of ESCRT genes led to a heterogeneity in developmental phenotypes ranging from an embryonic or larval lethality to an absence of obvious phenotype (Djeddi et al., 2012). Phenotype analysis of adults, larvae, or embryo was performed using differential interference contrast (DIC) microscope to visualize developmental arrest, molting defects, and organogenesis of epithelial tissues (Fig. 6.2). The protocol for mounting animals for live observation has been well described in Shaham (2006). Briefly animals are placed between slide and coverslip on 2% agarose pad in 4 μl of M9 containing 10–25 mM sodium azide (NaN3) to anesthetize and therefore immobilize animals. Animals are transferred to the pad using a worm pick. Since the worm is transparent, DIC microscopy allows *in vivo* observation of *C. elegans* structures such as nuclei or tissue architecture.

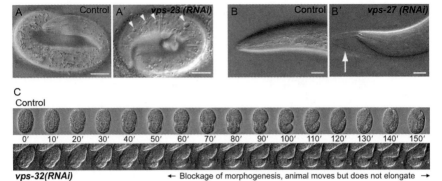

Figure 6.2 Nomarski pictures of control or ESCRT-depleted animals. (A–B′) In *vps-23 (RNAi)* embryos (A′), cytoplasmic vacuoles (arrowheads) are present in the epidermis. During larval development (B′), *vps-27(RNAi)* animals are trapped in their old cuticle (arrow) indicating a molting defect compared to the control (B). Scale bars: 10 μm. (C) Time lapse analyses of control embryo and *vps-32*-depleted embryo, which are unable to make morphogenesis and elongate properly. DIC acquisitions were done every minute, but only one image every 10 min is presented.

5. ANALYSIS OF VESICULAR COMPARTMENTS

5.1. Immunolocalization of endosomal and autophagy proteins

The detection of endogenous or transgenic proteins is done on fixed animals stained with fluorescent antibodies. Classically, immunolocalization protocols contain two main steps: the fixation and permeabilization steps followed by the antibody staining. Worms are then observed with either epifluorescence or confocal microscope. We used this technique to visualize endosomes and autophagosomes at different steps of maturation using specific antibodies (see Section 6.1 for images acquisition).

5.1.1 Immunofluorescence on embryos
Our lab used three protocols with slightly different fixation to perform whole-mount immunolocalization of proteins.

5.1.1.1 Permeabilization by freeze-cracking
1. Prepare poly-L-lysine Teflon coated slide by adding 10 μl poly-L-lysine (Sigma, 0.1%, w/v) per well and let dry. Add 10–15 μl of M9 per well.
2. Pick and transfer to wells synchronized or semisynchronized eggs, laid by well-fed worms.

3. Cover the worms with coverslip and immerse the slide into liquid nitrogen (at least 1 min).
4. Quickly remove the coverslip from the slides using a scalpel blade.

5.1.1.2 Methanol–acetone fixation

1. Transfer the freeze-cracked slide for 30 min into $-20\ ^{\circ}$C methanol bath in a Coplin jar maintained in dry ice. Dehydrated methanol stock is maintained by adding molecular sieves (3 Å, Sigma).
2. Transfer the slide to cold acetone bath (Coplin jar in dry ice) for 10–15 min.
3. Take off the slide from the acetone bath and tap it gently on Kimwipe paper to drain off any residual acetone.
4. Maintain the slide horizontal, add a droplet of $1\times$ PBS to wash off the wells.

   From this point, and for each change of buffer, proceed as follows: remove the liquid over the sample by gently taping one edge of the slide on a Kimwipe paper, drain any residual liquid around the well with a twisted piece of Kimwipe paper, put the slide in horizontal position, add droplet of liquid to cover only the samples within the wells, and incubate for the appropriate time in a humidified chamber (an opaque plastic box with cover, with humidified filter paper at the bottom and slide holder).
5. Incubate in $1\times$ PBS, pH 7.4, 2% BSA, 0.5% Triton X-100 for 30 min. Alternative procedure for fixation for some antibodies: the acetone step could be omitted, also allowing to maintain GFP fluorescence.

5.1.1.3 Paraformaldehyde fixation

Alternatively, after the freeze-crack permeabilization step perform the fixation step in 2% paraformaldehyde.

1. Incubate sample in a drop of 2% paraformaldehyde in FB (60 mM PIPES, 25 mM HEPES, pH 7.05, 10 mM EGTA, 2 mM MgCl$_2$, pH is adjusted to 6.8 with 0.5 M NaOH) at room temperature for 10 min.
2. Add $1\times$ PBS for 5 min. Repeat this washing procedure twice.
3. Incubate in $1\times$ PBS, 50 mM NH$_4$Cl for 15 min (reduction of aldehydes).
4. Add $1\times$ PBS 0.5% Triton X-100 for 5 min.
5. Add $1\times$ PBS 2% BSA, 0.5% Triton X-100 for 1 h.

5.1.1.4 Antibody staining
1. Add primary antibody at the appropriate dilution in $1 \times$ PBS, pH 7.4, 2% BSA, 0.5% Triton X-100 and incubate overnight at 4 °C.
2. Incubate the slide with $1 \times$ PBS, pH 7.4, 2% BSA, 0.5% Triton X-100 for 5–10 min. Repeat this step twice.
3. Apply the secondary antibody at the appropriate dilution in $1 \times$ PBS, pH 7.4, 2% BSA, 0.5% Triton X-100 and incubate in the humidified chamber for 3 h at 37 °C.
4. Incubate the slide with $1 \times$ PBS, pH 7.4, 2% BSA, 0.5% Triton for 10 min. Repeat this procedure twice.
5. Apply mounting medium (DABCO in 70% glycerol) to the embryos and cover with a glass coverslip.

5.1.2 Whole-mount formaldehyde fixed worms immunolocalization protocol

This protocol was originally described in 1990 (Finney & Ruvkun, 1990) for larvae and adult worms and has been modified by L. Pearson and K. Gieseler (University of Lyon). All centrifugations are performed at $200 \times g$ for 1 min.

5.1.2.1 Permeabilization and fixation
1. Wash the plates with 1 ml of cold M9 medium in order to harvest the worms. Transfer them in a 15-ml high-clarity polypropylene conical tube. Centrifuge to settle down the worms. Remove supernatant, add fresh M9, gently shake to disperse the worms and spin down the worms as previously described. Repeat this washing procedure twice to get rid of small debris and bacteria.
2. After the final wash, remove the supernatant and add up to 4 ml of fixative solution. The fixative solution is made by combining 0.25 ml 16% formaldehyde with 2.85 ml 100% methanol and 0.9 ml $4 \times$ fixation buffer (320 mM KCl, 80 mM NaCl, 40 mM EGTA, 60 mM PIPES, pH 7.4, 20 mM spermidine).
3. Mix worms by gently inverting tubes three times. Immerse tubes in liquid nitrogen 1–2 min. At this point, worms can be either stored at −80 °C or proceeded further.
4. Incubate in a cold room with gentle shaking on a slowly rotating wheel for 1–2 h. Alternatively, incubate worms in cold ice for 1–2 h with gentle agitation every 5 min.
5. Centrifuge worms, remove supernatant, add 1 ml of TTB buffer (100 mM Tris–HCl, pH 7.4, 1% Triton X-100, 1 mM EDTA), invert

tubes three times and centrifuge to pellet down worms. Remove the supernatant, add 1 ml TTB, invert tubes, and transfer suspended worms to 1.5 ml Sorenson microcentrifuge (or silicon Eppendorf) tubes. Centrifuge worms and make an additional wash in TTB. From this step, worms are very fragile and are constantly handled with precaution.
6. Resuspend worms in 1 ml TTB 1% β-mercaptoethanol and incubate for 2 h with gentle agitation on a slowly rotating wheel at 37 °C.
7. Centrifuge worms, remove supernatant, add 1 ml 1× BO3 buffer (make fresh from a 50× BO3 stock solution: 3.09 g HBO3, 1 g NaOH, add water to 100 ml. Adjust pH to 9.2 with HCl), mix well by three gentle inversion, spin down, aspirate supernatant. Repeat twice the washing procedure in 1× BO3.
8. Resuspend worms in 1 ml 1× BO3 10% 0.1 M DTT, invert tubes three times, and incubate 15 min with gentle agitation at room temperature.
9. Centrifuge worms, remove supernatant, add 1 ml 1× BO3, invert tubes three times, and pellet down worms by centrifugation. Repeat this procedure once.
10. Remove supernatant, resuspend worms in 1 ml 1× BO3 supplemented with 75 ml 30% H_2O_2 (fresh solution opened for less than 2 months), and incubate for 15 min at room temperature with gentle agitation: invert tubes three times every 5 min.
11. Centrifuge worms, remove the supernatant, add 1.5 ml 1× BO3, invert tubes three times then spin down worms. Repeat this step once.
12. Remove supernatant, resuspend worms in 1 ml AbB (antibody buffer B: 1× PBS, pH 7.4, 0.1% BSA, 0.5% Triton X-100, 1 mM EDTA), and incubate at room temperature on a rotating wheel for 20 min.
13. Pellet down worms by centrifugation, remove supernatant, and resuspend worms in 1 ml AbA (antibody buffer A: 1× PBS, pH 7.4, 1% BSA, 0.5% Triton X-100, 1 mM EDTA).

At this point, worms can be stored at 4 °C for several weeks or proceeded further.

5.1.2.2 Antibody staining
1. Transfer an aliquot of fixed worms (20 µl) to a 1.5-ml Sorenson microcentrifuge tube and add 180 µl of the primary antibody diluted in AbA.
2. Incubate on a rotating wheel for 2 h at room temperature or alternatively overnight in a cold room.

3. Spin down worms, remove supernatant, add 1 ml AbB, and incubate on a rotating wheel for 30 min at room temperature. Repeat this washing procedure three times more.
4. Add 1 ml AbA, invert tubes three times, centrifuge, and remove supernatant.
5. Add 180 µl of the appropriate dilution of secondary antibody in AbA.
6. Incubate on a rotating wheel for 2 h at room temperature in the dark.
7. Spin down worms, remove supernatant, add 1 ml AbB, and incubate on a rotating wheel for 15 min at room temperature in the dark. Repeat this washing procedure three more times.

Worms can be kept at 4 °C for a few days before mounting.

5.1.2.3 Mounting worms
1. Prepare poly-L-lysine Teflon coated slide by adding 10 µl poly-L-lysine (Sigma, 0.1%, w/v) per well and let dry.
2. Transfer an aliquot of worms to the well (5–10 µl), spread the worms with the help of a platinum pick, add 10 µl of mounting medium DABCO.
3. Apply coverslip and gently hit it with tweezers to spread the mounting medium and eliminate any air bubble.

5.2. Electron microscopy

Analysis of the ultrastructure details of endosomes and autophagosomes in wild type and mutant worms could be performed with electronic microscopy (EM) as previously described (Roudier et al., 2005). Briefly, larvae are mounted on low-melting agarose pads containing 4% glutaraldehyde, gathered in groups of four to six animals then cut with a razor blade and recover by a second layer of agarose. Subsequently, larvae are fixed overnight in the agarose sandwich and postfixed with osmium tetroxide and uranyl acetate. Agarose blocks containing groups of larvae are trimmed with razor and orientated in molds either transversely or longitudinally before embedding in the resin. Seventy nanometers thick sections are contrasted with uranyl acetate and lead citrate.

To check whether VPS-27::GFP and GFP::LGG-1 specifically label the endosomal vesicles and the autophagosomes, respectively, we performed immuno-gold electron microscopy on embryos. Animals were transferred to 200-µm-deep flat carriers followed by cryo-immobilization in the EMPACT-2 HPF apparatus (Leica Microsystems) and cryo-substitution as described (Kolotuev, Schwab, & Labouesse, 2010). Ultrathin sections

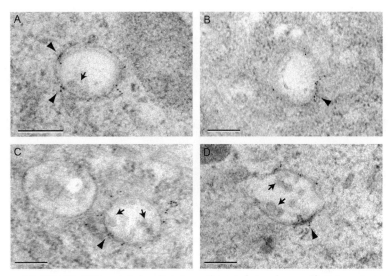

Figure 6.3 VPS-27::GFP protein localizes to endosomal membrane. Electron microscopy images of VPS-27::GFP worms incubated with anti-GFP antibody (strain RD106). High magnification of immuno-gold particles (arrowhead) indicates that VPS-27::GFP protein is enriched in subdomains at the surface of endosomes but not present in the inner vesicle (small arrow) of MVB. Panels A and B show endosomes with almost no internal vesicle compared to C and D. Scale bar is 200 nm.

on formvar/carbon-coated cupper grids are sequentially treated for 10 min with 1% Tween 20, 100 mM glycine, then 1% BSA all in PBS. Samples were then labeled with the primary rabbit anti-GFP antibody (1:400 dilution) and after washes, with the secondary goat anti-rabbit antibody coupled to 10 nm colloidal gold particles (1:20 dilution, BBInternational) both in PBS 1% BSA for 30 min. After extensive washes, sections were contrasted with 2% uranyl acetate for 5 min. Sections were observed with a Jeol 1400 TEM at 120 kV and images are acquired with a Gatan 11 Mpixels SC1000 Orius CCD camera (Fig. 6.3).

5.3. Vital dyes and fluorescent constructs

In order to perform live imaging of endo-lysosomal compartments, we have used either vital dyes or fluorescent construct (as described in Section 3). LysoTracker® (Molecular Probes) allows a detection of acidic compartments such as late endosomes and lysosomes. Briefly, worms are grown on plates containing 2 μM of LysoTracker® (Hersh, Hartwieg, & Horvitz, 2002), kept in darkness. Alternatively, it is possible to inject worms with a 100 μM dilution of Lysotracker® (Boyd, Hajjar, & O'Connell, 2011). In

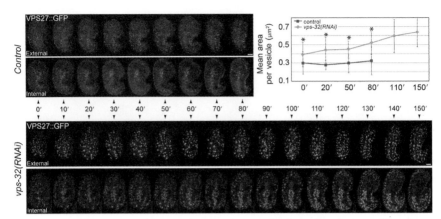

Figure 6.4 VPS-27 accumulates at the membrane of enlarged endosome. Time lapse confocal imaging of VPS-27::GFP in control and vps-32(RNAi) embryos during morphogenesis. Each panel corresponds to a projection of seven confocal sections covering, respectively, the epidermis (external) and digestive tract (internal). Confocal acquisitions were performed every 10 min from the beginning of elongation process until the first muscular movement. As elongation is defective in vps-32(RNAi) embryos, acquisitions could be performed longer. Formation of enlarged endosomes is a slow process during which the cytoplasmic pool of VPS-27::GFP is recruited at the membrane. Quantification of the size of VPS-27::GFP vesicles was performed in six control animals (dark gray) and six vps-32(RNAi) animals (light gray). The mean area per vesicles in vps-32(RNAi) embryos and controls show significantly different distributions (stars indicate $p < 2 \times 10^{-10}$ using a Kolmogorov–Smirnov test). Bar is 5 µm.

both case the progeny will present fluorescent vesicles (excitation wavelength: 543 nm).

In *C. elegans*, fluorescent fusion proteins have been constructed to visualize protein localization and movement *in vivo*. In our studies, we used several GFP fusion proteins to visualize endocytic compartments in living (Fig. 6.4) or fixed animals. GFP::RAB-5 was used to label early endosomes whereas VPS-E::GFP fusions are more specific to the MVBs/late endosomes (see Table 6.1) (Michelet et al., 2009; Roudier et al., 2005).

Lysotracker® and GFP constructs allow *in vivo* visualization of vesicles, using epifluorescence or confocal microscopes. GFP can also be detected by using an anti-GFP antibody (see Section 5.1)

6. ANALYSIS OF AUTOPHAGY

The formation of the autophagic vesicles requires the recruitment of Atg8, an ubiquitin-like protein, to the membrane of the nascent autophagosomes. In *C. elegans* the presence of two Atg8 homologues *lgg-1* and *lgg-2* is an intermediate situation between the unique gene of yeast and the complex situation of mammals. To investigate the autophagic pathway in wild type or ESCRT mutant context, we used either GFP fusion proteins for LGG-1 and LGG-2, or antibodies. Because these proteins are associated with autophagosomal membrane, the first approach used to evaluate the number of autophagosomes is to count the number of dots positive for these proteins using *in vivo* fluorescence or immunofluorescence (see Section 5.1 for immunostaining protocol). We observed a strong increase in the number of autophagosomes in all ESCRT mutants analyzed.

To monitor an increase of autophagic flux *versus* a blockage of autophagosomal maturation we analyzed the fusion proteins GFP::LGG-1 and GFP::LGG-2 by Western blotting (Alberti, Michelet, Djeddi, & Legouis, 2010; Djeddi et al., 2012; Michelet & Legouis, 2012) as shown in Fig. 6.5A. To complete this approach, we have generated a "tandem" LGG-1 protein (fused with both GFP and Cherry) to monitor the acidification of the autophagosomes (Fig. 6.5B).

6.1. Counting LGG-1 and LGG-2 dots

Routinely, fluorescence images were captured on an AxioOberver Z1 microscope (Zeiss) equipped with Evolve EMCCD camera (Roper Scientific). Confocal images were captured on confocal Nikon (TE 2000-U). Z-stacks are acquired from Zeiss or Nikon microscope with 0.5–1 µm step to image the entire embryo. Images analysis was performed with NIH ImageJ Software (http://rsb.info.nih.gov/ij/).

To count LGG-1 or LGG-2 positive dots from pictures obtained by optical or confocal microscope using ImageJ "Cell Counter Analysis plugin," two different approaches were used. For the very early stages of embryo where there is only few autophagosomes, we counted the dots from the entire stack projections while for the later stages, we counted three nonconsecutive single planes per embryo.

6.2. Autophagic flux
6.2.1 Western blot analysis
Entire extraction is performed on ice and with gloves to limit protein degradation. This protocol is used for Western blot on total protein extracts.
1. Cultivate and synchronize as many worms as desired (liquid culture, see Section 3.2.2.1 or culture on NGM/OP50 or RNAi plates see Section 2.4).

2. Wash them three times in M9 buffer (with centrifugations at 4 °C at 200 × g).
3. Evaluate the volume of the worm pellet.
4. Add 1 volume of glass beads (Sigma GP7-72) and 1 volume of PBS 2% triton.
5. Perform two cycles of 60 s at 6000 rpm with Precellys24 instrument (Bertin Technologies) in order to disrupt the worms. Place the tube on ice between the two cycles.
6. Centrifuge 2 min at 4 °C at 12,000 rpm.
7. Mix supernatants with Laemmli buffer (final concentration: 1.3% SDS, 0.08 M Tris, pH 6.8, 0.07% glycerol, 0.1% bromophenol blue).
8. Before electrophoresis, denaturate extracts 5 min at 95 °C.

Protein extracts were separated on a NuPAGE® Novex 4-12% Bis–Tris Gel by electrophoresis (Life technologies). Proteins were transferred to nitrocellulose membranes with i-Blot® system (Life technologies), probed with the affinity-purified mouse anti-GFP antibody (1:100) and revealed using horseradish peroxidase-conjugated antibodies (1:10,000) and the ECL detection system (SuperSignal pico Chemiluminescent Substrate, Pierce, Rockford, IL). Signals were quantified on a Las3000 photoimager (Fuji) using ImageQuant 5.2 software (Molecular Dynamics) (see Fig. 6.5A for blot analysis).

6.2.2 In vivo *analysis of acidification*
To monitor the acidification of autolysosome, we looked at *in vivo* expression of GFP::Cherry::LGG-1 (see Section 6.1). The intraluminal pH of autophagosomes is neutral, both GFP and Cherry proteins emit fluorescence, the resulting merged signal is yellow. Soon after the fusion between autophagosome and endo-lysosome, the internal pH is acidified and the GFP signal, but not the Cherry, is quenched (Fig. 6.5B). Induction of functional autophagic flux is detected by the presence of red vesicles whereas the accumulation of yellow vesicles indicates autophagosomal blockage.

6.3. Pharmacological modulation of autophagy
The autophagic pathways could be modulated by pharmacological treatments. Several drugs can be used to either increase or block the autophagic flux (Fleming, Noda, Yoshimori, & Rubinsztein, 2011). Some of these chemical compounds have been already successfully tested *in vivo* in *C. elegans*. Such approaches could elegantly complement genetic approaches based on elimination of either negative regulator or key activator of

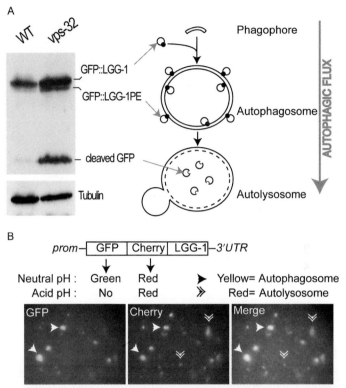

Figure 6.5 Analysis of the autophagic flux in *C. elegans*. (A) The autophagic flux induction is monitored by Western blotting. The quantification of the phosphatidylethanolamine conjugated LGG-1 form and the cleaved GFP fragment is used to analyze the autophagosomes and the autolysosomes, respectively. (A) Strong increase of autophagic flux is observed in *vps-32(RNAi)* animals. (B) Analysis of the acidification of autophagosomes in transgenic worms expressing the tandem GFP::Cherry::LGG-1 protein. Single arrowheads indicate autophagosomes (yellow) and double arrowheads show autolysosomes or amphisomes (red). (For interpretation of the references to color in this figure legend, the reader is referred to the online version of this chapter.)

autophagic pathways. Spermidine has been shown to induce the autophagy process (Djeddi et al., 2012; Morselli et al., 2011) while the type III phosphatidylinositol 3-kinase inhibitors, 3-methyladenine (3-MA) and Wortmannin, inhibit autophagy (Samokhvalov, Scott, & Crowder, 2008).

6.3.1 Spermidine plates setup and C. elegans *exposure*
Spermidine is kept at −20 °C as a 0.1 M solution in water.
1. Spread 150 µl of *E. coli* OP50 overnight culture on freshly poured 10 ml NGM plates and let it air-dry 2 h at room temperature.

2. Expose plates to a UV source for 10 min to kill bacteria.
3. Add 40 μl of 0.1 M spermidine solution in a final volume of 250 μl, spread it over the surface of the plate and let it diffuse overnight. Use plates the following day.
4. Transfer adult gravid worms onto spermidine and control NGM plates for egg laying.
5. Eliminate worms and let the embryos develop in presence of spermidine.
6. Score worms for the investigated phenotype.

6.3.2 The 3-MA and Wortmannin treatments

Wortmannin and 3-MA are conserved frozen in DMSO at a concentration of 1 and 100 mM, respectively. For both compounds treatment is performed in M9 solution at a final concentration of 10 mM 3-MA and 100 μM Wortmannin.

1. Transfer about 100 worms into 1.5-ml tube containing 90 μl of M9 with the appropriate drug at the chosen concentration.
2. Incubate worms for a few hours.
3. Recover worms from the drug treatment by placing them on NGM plates without drug for 24 h.
4. Score worms for the specific phenotypic analysis.
5. Control the drug treatment efficiency by counting the number of LGG-1::GFP foci (significant decreased with either 3-MA or Wortmannin treatment). Control the solvent influence by incubating worms in 10% DMSO in M9.

7. METHODS TO VISUALIZE AMPHISOME, THE FUSION ORGANELLE BETWEEN ENDOSOMES AND AUTOPHAGOSOMES

Endosomal and autophagic compartments can fuse to form amphisomes (Gordon & Seglen, 1988; for review, see Manil-Segalen et al., 2012). To evaluate the presence of amphisomes in *C. elegans*, we quantified the colocalization between LGG-1 and either VPS-27 or VPS-32. This experiment has been done using GFP constructs (GFP::LGG-1; GFP::VPS-27) as well as antibodies (LGG-1, VPS-27, VPS-32). In practice, we counted on confocal images the number of vesicles positive for both autophagic and endosomal markers in each focal plane. In control embryos, amphisomes are difficult to detect but can be observed using either transgenes or endogenous proteins (Fig. 6.6). In absence of *rab-7*, the final step

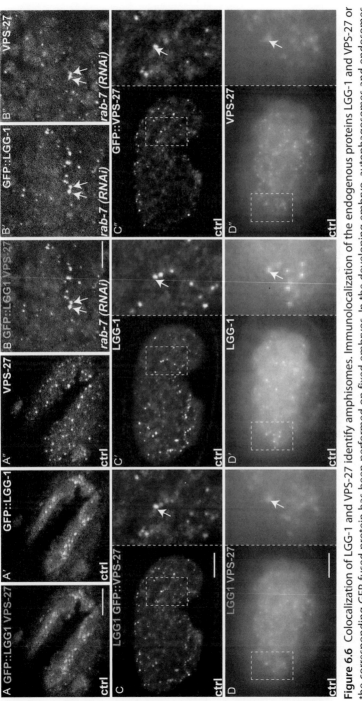

Figure 6.6 Colocalization of LGG-1 and VPS-27 identify amphisomes. Immunolocalization of the endogenous proteins LGG-1 and VPS-27 or the corresponding GFP-fused protein has been performed on fixed embryos. In the developing embryo, autophagosomes and endosomes are labeled by LGG-1 or VPS-27, respectively. Amphisomes are detected by the colocalization of both markers (arrow, B–D). Amphisomes are infrequent in WT situation (A, C, and D) but are easier to detect in *rab-7(RNAi)* embryos (B). Scale bar is 10 μm. (For color version of this figure, the reader is referred to the online version of this chapter.)

of autophagosomal/endosomal maturation is altered and amphisomes accumulate (Fig. 6.6B) (Djeddi et al., 2012). Our data indicate that in *C. elegans* embryo, amphisomes do exist but are very infrequent compared to the total number of autophagosomes (Djeddi et al., 2012). Two main hypotheses could explain this observation:
- Amphisomes can be very transitory compartments and fuse rapidly with the lysosomes.
- Autophagosomes might fuse preferentially with lysosomes, only few of them fusing with endosomes to form amphisomes.

In any case, formation of amphisomes in *C. elegans* appears to be intermediary between the situation in yeast, with no amphisome, and in mammals where amphisomes are frequently observed.

8. CONCLUSIONS

C. elegans has been employed over the past three decades to address fundamental cell biological questions in the context of multicellular organism. In this chapter we have presented the experimental strategies achievable in *C. elegans* to study some specific aspects of two degradative vesicular pathways. Further studies should clarify the relationships between the maturation of endosomes and autophagosomes. In particular, *C. elegans* should be a useful model to analyze the dynamics of these two vesicular pathways *in vivo* with regards to developmental processes.

ACKNOWLEDGMENTS

The Legouis's lab is supported by CNRS, the Fondation ARC pour la Recherche sur le Cancer and Agence Nationale pour la Recherche (ANR) and M. M. S. is a recipient of a fellowship from the Ligue Nationale contre le Cancer.

REFERENCES

Al Rawi, S., Louvet-Vallée, S., Djeddi, A., Sachse, M., Culetto, E., Hajjar, C., et al. (2011). Postfertilization autophagy of sperm organelles prevents paternal mitochondrial DNA transmission. *Science (New York, N.Y.), 334*(6059), 1144–1147. http://dx.doi.org/10.1126/science.1211878.

Alberti, A., Michelet, X., Djeddi, A., & Legouis, R. (2010). The autophagosomal protein LGG-2 acts synergistically with LGG-1 in dauer formation and longevity in C. elegans. *Autophagy, 6*(5), 622–633. http://dx.doi.org/10.4161/auto.6.5.12252.

Boyd, L., Hajjar, C., & O'Connell, K. (2011). Time-lapse microscopy of early embryogenesis in Caenorhabditis elegans. *Journal of Visualized Experiments: JoVE*, (54), e2852. http://dx.doi.org/10.3791/2852.

Djeddi, A., Michelet, X., Culetto, E., Alberti, A., Barois, N., & Legouis, R. (2012). Induction of autophagy in ESCRT mutants is an adaptive response for cell survival in C.

elegans. *Journal of Cell Science*, *125*(Pt. 3), 685–694. http://dx.doi.org/10.1242/jcs.091702.

Finney, M., & Ruvkun, G. (1990). The unc-86 gene product couples cell lineage and cell identity in C. elegans. *Cell*, *63*(5), 895–905.

Fire, A., Xu, S., Montgomery, M. K., Kostas, S. A., Driver, S. E., & Mello, C. C. (1998). Potent and specific genetic interference by double-stranded RNA in Caenorhabditis elegans. *Nature*, *391*(6669), 806–811. http://dx.doi.org/10.1038/35888.

Fleming, A., Noda, T., Yoshimori, T., & Rubinsztein, D. C. (2011). Chemical modulators of autophagy as biological probes and potential therapeutics. *Nature Chemical Biology*, *7*(1), 9–17. http://dx.doi.org/10.1038/nchembio.500.

Gordon, P. B., & Seglen, P. O. (1988). Prelysosomal convergence of autophagic and endocytic pathways. *Biochemical and Biophysical Research Communications*, *151*(1), 40–47.

Hersh, B. M., Hartwieg, E., & Horvitz, H. R. (2002). The Caenorhabditis elegans mucolipin-like gene cup-5 is essential for viability and regulates lysosomes in multiple cell types. *Proceedings of the National Academy of Sciences of the United States of America*, *99*(7), 4355–4360. http://dx.doi.org/10.1073/pnas.062065399.

Hobert, O. (2002). PCR fusion-based approach to create reporter gene constructs for expression analysis in transgenic C. elegans. *BioTechniques*, *32*(4), 728–730.

Huotari, J., & Helenius, A. (2011). Endosome maturation. *The EMBO Journal*, *30*(17), 3481–3500. http://dx.doi.org/10.1038/emboj.2011.286.

Kamath, R. S., & Ahringer, J. (2003). Genome-wide RNAi screening in Caenorhabditis elegans. *Methods (San Diego, California)*, *30*(4), 313–321.

Kolotuev, I., Schwab, Y., & Labouesse, M. (2010). A precise and rapid mapping protocol for correlative light and electron microscopy of small invertebrate organisms. *Biology of the Cell/Under the Auspices of the European Cell Biology Organization*, *102*(2), 121–132. http://dx.doi.org/10.1042/BC20090096.

Manil-Segalen, M., Lefebvre, C., Culetto, E., & Legouis, R. (2012). Need an ESCRT for autophagosomal maturation? *Communicative & Integrative Biology*, *5*(6), 566–571.

Mello, C., & Fire, A. (1995). DNA transformation. *Methods in Cell Biology*, *48*, 451–482.

Mello, C. C., Kramer, J. M., Stinchcomb, D., & Ambros, V. (1991). Efficient gene transfer in C.elegans: Extrachromosomal maintenance and integration of transforming sequences. *The EMBO Journal*, *10*(12), 3959–3970.

Michelet, X., Alberti, A., Benkemoun, L., Roudier, N., Lefebvre, C., & Legouis, R. (2009). The ESCRT-III protein CeVPS-32 is enriched in domains distinct from CeVPS-27 and CeVPS-23 at the endosomal membrane of epithelial cells. *Biology of the Cell/Under the Auspices of the European Cell Biology Organization*, *101*(10), 599–615. http://dx.doi.org/10.1042/BC20090025.

Michelet, X., & Legouis, R. (2012). Autophagy in endosomal mutants: Desperately seeking to survive. *Worm*, *1*(4), 0–4. http://dx.doi.org/10.4161/worm.20848.

Morselli, E., Mariño, G., Bennetzen, M. V., Eisenberg, T., Megalou, E., Schroeder, S., et al. (2011). Spermidine and resveratrol induce autophagy by distinct pathways converging on the acetylproteome. *The Journal of Cell Biology*, *192*(4), 615–629. http://dx.doi.org/10.1083/jcb.201008167.

Raiborg, C., & Stenmark, H. (2009). The ESCRT machinery in endosomal sorting of ubiquitylated membrane proteins. *Nature*, *458*(7237), 445–452. http://dx.doi.org/10.1038/nature07961.

Reggiori, F., & Klionsky, D. J. (2002). Autophagy in the eukaryotic cell. *Eukaryotic Cell*, *1*(1), 11–21.

Roudier, N., Lefebvre, C., & Legouis, R. (2005). CeVPS-27 is an endosomal protein required for the molting and the endocytic trafficking of the low-density lipoprotein receptor-related protein 1 in Caenorhabditis elegans. *Traffic (Copenhagen, Denmark)*, *6*(8), 695–705. http://dx.doi.org/10.1111/j.1600-0854.2005.00309.x.

Rual, J.-F., Ceron, J., Koreth, J., Hao, T., Nicot, A.-S., Hirozane-Kishikawa, T., et al. (2004). Toward improving Caenorhabditis elegans phenome mapping with an ORFeome-based RNAi library. *Genome Research*, *14*(10B), 2162–2168. http://dx.doi.org/10.1101/gr.2505604.

Samokhvalov, V., Scott, B. A., & Crowder, C. M. (2008). Autophagy protects against hypoxic injury in C. elegans. *Autophagy*, *4*(8), 1034–1041.

Shaham, S. (2006). Methods in cell biology. *WormBook*, ed. The C. elegans Research Community, WormBook, doi/10.1895/wormbook.1.49.1, http://www.wormbook.org.

Tian, Y., Li, Z., Hu, W., Ren, H., Tian, E., Zhao, Y., et al. (2010). C. elegans screen identifies autophagy genes specific to multicellular organisms. *Cell*, *141*(6), 1042–1055. http://dx.doi.org/10.1016/j.cell.2010.04.034.

Timmons, L., & Fire, A. (1998). Specific interference by ingested dsRNA. *Nature*, *395*(6705), 854. http://dx.doi.org/10.1038/27579.

CHAPTER SEVEN

Assessment of Cation Trapping by Cellular Acidic Compartments

François Marceau[1], Caroline Roy, Johanne Bouthillier
*Centre de recherche en rhumatologie et immunologie, Centre Hospitalier Universitaire de Québec, Québec, Canada
[1]Corresponding author: e-mail address: francois.marceau@crchul.ulaval.ca

Contents

1. Introduction 120
2. Quinacrine Uptake by Cells 123
    2.1 Cultured adherent cells 124
    2.2 Cells in suspension 125
    2.3 Microscopic, cytofluorometric correlates 127
    2.4 Other techniques 128
3. Macroautophagic Accumulation in Cells That Have Accumulated Cations 128
4. Summary 129
Acknowledgments 130
References 130

Abstract

All nucleated cells, from yeast to animal cells, concentrate cationic chemicals (weak bases with a $pK_a \sim 8\text{--}10$) into acidic cell compartments (low retro-diffusion under a protonated form at low pH = ion trapping). The proton pump vacuolar (V)-ATPase is the driving force of this pseudotransport that concerns acidic organelles (mainly late endosomes and lysosomes). The latter rapidly become swollen (osmotic vacuolization) and macroautophagic. Cation concentration in cells is not proved to involve membrane transporters, but is prevented or reversed by inhibitors of V-ATPase, such as bafilomycin A1. Lipophilicity is a major determinant of the apparent affinity of this pseudotransport because simple diffusion of the uncharged form supports it. Quinacrine is a formerly used antiparasitic drug that is intensely fluorescent, lipophilic, and a tertiary amine. The drug, at micromolar concentrations, is proposed as a superior probe for assessing cation trapping by cellular acidic compartments, being readily quantified using fluorometry in cell extracts and analyzed using microscopy and cytofluorometry (fluorescence settings for fluorescein being applicable). Further, cells respond to micromolar levels of quinacrine by autophagic accumulation (e.g., accumulation of the activated macroautophagic effector LC3 II, immunoblots), an objective and universal response to sequestered amines.

1. INTRODUCTION

Organic amines in aqueous solutions are in equilibrium between a positively charged, protonated form and a neutral one; this is especially relevant for primary, secondary, and tertiary amines, while quaternary amine molecules are predominantly charged at all relevant physiological pH values. Drugs that are weak bases, possessing a pK_a of approximately 8–10, are numerous and distributed among many therapeutic classes (e.g., local anesthetics; antiarrhythmics; antibiotics; antineoplasics; bioamine agonists; and antagonists widely used in psychiatry, autonomic nervous system pharmacology, and allergy). It has been recognized a long time ago that cells concentrate such drugs in their acidic organelles, a form of pseudotransport that is dependent only on the maintenance of a pH difference between the nearly neutral cytosol and the concerned organelles, initially identified as lysosomes (De Duve et al., 1974). Figure 7.1 presents a series of such chemicals that can be considered as substituted triethylamines and that are all concentrated in this manner into cultured cells. In the series, the antiparasitic drug chloroquine is an historical prototype of the "lysosomotropic" drugs. The aromatic "side chains" occasionally possess other protonable functions, but not as determinant as the tertiary amine function for their weak base behavior. The uncharged form of such drugs in the extracellular fluid is postulated to enter the cytosol by simple diffusion (step 1 in Fig. 7.2), and then to diffuse again through the membrane of organelles that express vacuolar (V)-ATPases, a family of proton pumps that consume ATP to acidify the trans-Golgi network and derived organelles (early and late endosomes, lysosomes, secretory granules) (step 2). The central role of simple drug diffusion in the early steps of the phenomenon, at least at the level of intracellular vesicles, implies that the lipophilicity of the organic amine, that can be quantified using the log P scale, is a crucial determinant of this form of transport. Indeed, ion trapping concerns low millimolar concentrations of the hydrophilic compounds lidocaine, triethylamine, and procainamide, the 10–25 μM range for chloroquine and low micromolar levels for quinacrine and amiodarone (apparent affinity correlated to their respective log P; Fig. 7.1). Cation trapping refers to the fact that the amines become predominantly protonated and charged at low pH (6.5 or less in the concerned organelles) and are much less diffusible, especially if H^+ ions continue to be pumped. Current data support that specific V-ATPase inhibitors, such as bafilomycin A1, suppress the concentration of organic cations into cells

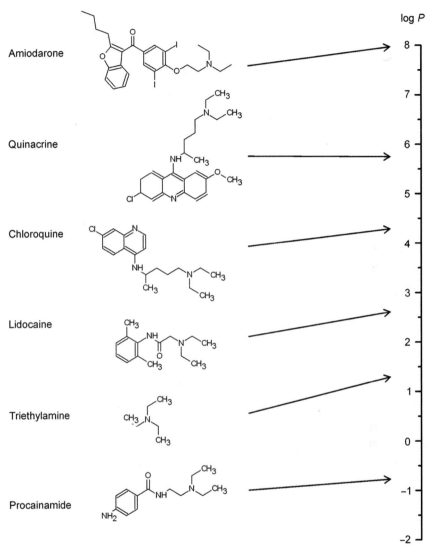

Figure 7.1 Cationic drugs of the triethylamine series that are all documented to be concentrated in cells via V-ATPase-mediated ion trapping. The liposolubility (log P) is a strong predictor of the concentration range for which trapping is observed in cultured cells (see text). The drugs belong to various therapeutic classes, triethylamine being of no medicinal value.

and that the target organelles are more diverse than initially appreciated, while mainly involving the late endosome–lysosome continuum (reviewed by Marceau et al., 2012). Depending on the quantity of amines taken up, endomembrane pH buffering may ensue, with incompetence of lysosomal

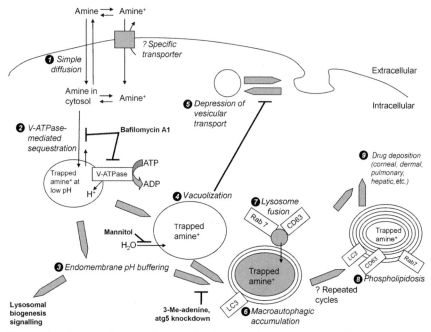

Figure 7.2 Model of successive mechanisms and consequences of the V-ATPase-mediated concentration of cationic drugs into acidic cell compartments. Numbered steps are explained in text.

enzymes and feedback activation of lysosomal biogenesis (step 3; e.g., upregulation of lysosomal proteins LAMP-1 and -2 following cell treatment with chloroquine; Chen, Gombart, & Chen, 2011). Osmotic swelling of the organelles may be observed, a phenotype that may be mitigated by cell dehydration with concentrated mannitol (step 4; Marceau et al., 2012). The vacuolar state of cells depresses various forms of vesicular trafficking in a rather nonspecific manner (e.g., reduced secretion of various proteins, inhibited uptake of transferrin; step 5). It has been only recently appreciated that virtually all enlarged vacuoles induced by treatment with an amine are autophagosomes that have rapidly become positive for the macroautophagic effector microtubule-associated protein light chain 3 (LC3) (Marceau et al., 2012). The autophagic state may derive from both the inhibition of a basal autophagic flux and membrane damage in expanding organelles (Fig. 7.2, step 6). Lysosomes may bring their own markers when fusing with autophagosomes (step 7), perhaps obscuring the nonlysosome origin of some giant vacuoles. Upon prolonged treatment with a cationic drug, the giant vacuoles, initially filled with clear fluid, accumulate debris (typical consequence of lysosome fusion) and concentric multilamellar structures (phospholipidosis, step 8) that have been observed *in vivo* in patients or animals treated with amphiphilic amines (Reasor, Hastings, & Ulrich, 2006). Whether the concentric "onionoid" membrane organization derives from the inhibition of hydrolases (e.g., phospholipases) or from repeated cycles of macroautophagic envelopment is an unresolved issue. The clinical relevance of the amine-induced vacuolar and autophagic cytopathology is open to debate, as the active concentration range of specific drugs is often supratherapeutic. However, the hydrophilic local anesthetics, like lidocaine and procaine, are used as highly concentrated (mM) solutions in confined anatomic areas, and the characteristic cell vacuolization has been observed in tissue sections from treated patients (Vallance, Chaba, Clarke, & Taylor, 2004). The antiarrhythmic drug amiodarone is extensively deposited in various tissues during systemic administration (Fig. 7.2, step 9; Ammoury et al., 2008). More generally, it is suspected that ion trapping of weak bases is a general pharmacokinetic determinant of the very high apparent volume of distribution of most cationic drugs (high ratio of tissue/blood concentrations) and long persistence following cessation (Marceau et al., 2012).

2. QUINACRINE UPTAKE BY CELLS

Various pigmented or fluorescent amines are concentrated in acidic organelles via the V-ATPase-mediated ion trapping mechanisms, as shown by microscopic techniques (Marceau et al., 2012). Acridine orange and the commercial Lysotracker dyes are examples of chemicals used to identify "lysosome-like" organelles. Among the chemicals presented in Fig. 7.1, chloroquine and amiodarone exhibit a weak violet fluorescence under UV excitation, more so for the latter drug. Procainamide has an ultraviolet fluorescence that has been exploited to measure its cell uptake kinetics (Morissette, Lodge, & Marceau, 2008). When searching for a prototype "lysosomotropic" agent, one should consider that chloroquine deenergizes mitochondria in a concentration range that essentially overlaps with the levels that induce ion trapping (Morissette et al., 2008). Quinacrine, an inexpensive and formerly used antiprotozoal drug of low toxicity, stands out as presenting 2 protonable functions (pK_a 10.3, 7.7), high lipophilicity (log P 5.67), and a strong and nearly pH-independent green fluorescence that is essentially compatible with filters for fluorescein in optical instruments (microscopy, cytofluorometry). It is rather metabolically stable, the major *in vivo* metabolite being a des-ethyl secondary amine (Huang et al., 2006) that may essentially be equivalent to the original molecule when considering ion trapping and optical properties. In the rest of this chapter, we will illustrate the use of quinacrine to quantify and characterize cation trapping by cellular acidic compartments. Like other acridine derivatives, quinacrine binds to purified or synthetic DNA (intercalating dye; Doglia, Albinson, Hiort, Nordén, & Grässlund, 1993; Ehsanian, Van Waes, & Feller, 2011) and is the prototype of novel DNA-binding antineoplasic agents (the curaxins, Gasparia et al., 2011). However, nuclear staining is rarely a problem in cultured cells exposed to micromolar levels of quinacrine, and the effect of V-ATPase inhibition proves that DNA is not a significant reservoir in acute experiments (Marceau, Bawolak, Bouthillier, & Morissette, 2009; see also below). The drug is reportedly 80–90% bound to plasma proteins *in vivo* (Ehsanian et al., 2011).

2.1. Cultured adherent cells

The method, based on the fluorometric measurement of quinacrine in a total cell extract, has been applied by Marceau et al. (2009) for analyzing drug uptake by primary cultures of human umbilical artery smooth muscle cells and is suitable for other adherent cell types such as primary murine fibroblasts (Marceau et al., 2012; data not shown).

1. Each experimental point corresponds to the quantity of quinacrine taken up by confluent 25 cm^2 smooth muscle cell flasks containing 3 ml of serum-containing culture medium.
2. Quinacrine dihydrochloride (Sigma-Aldrich) is added from a concentrated stock solution to reach a final concentration of 0–25 μM and the cells are further incubated at 37 °C in a humidified atmosphere containing 5% CO_2 for variable periods (15 min–4 h). Cotreatment with bafilomycin A1 is used to determine the fraction of the cell uptake that is dependent on V-ATPase-mediated ion trapping.
3. After the desired incubation period, each cell flask is rapidly washed three times with 3 ml of phosphate-buffered saline, pH 7.4, at room temperature.
4. Then, the cells are dissolved with 10 ml of 1 N NaOH.
5. Quinacrine is analyzed in the NaOH extract using a luminescence spectrophotometer (such as FluoroLog tau-3, HORIBA JobinYvon Inc., Edison, NJ or SLM-AMINCO-Bowman Series 2) against a standard curve of the authentic drug dissolved in NaOH 1 N (excitation

414 nm, emission 501 nm). The sample volume (10 ml) is taken into account to calculate the cell contents of quinacrine per flask.
6. Control fluorescence from extracts of untreated cells is systematically verified, of small magnitude and subtracted from experimental values.
7. Data analysis is described in Section 2.2.

2.2. Cells in suspension

The techniques described earlier can be adapted to nonadherent suspensions of cells, such as human peripheral blood leukocytes (Roy, Gagné, Fernandes, & Marceau, 2013) or the undifferentiated monoblastoid cell line U937. These cells (Minta & Pambrun, 1985), obtained from the American Type Culture Collection, were propagated in RPMI 1640 medium containing 10% FBS and antibiotics at 37 °C in a 5% CO_2 humidified atmosphere and chemical induction of differentiation was not attempted for the reported experiments (Fig. 7.3, original results).

1. Test drugs (quinacrine 0–30 μM, bafilomycin A1) are added to 1 ml cell suspensions (10^6 cells) based on the serum-containing culture media described earlier according to various schemes and time frames.
2. The cells, protected from light, are incubated under agitation at 37 °C (Thermomixer, Eppendorf), rapidly centrifuged (12,500 rpm, 30 s), washed with 1 ml of phosphate-buffered saline, pH 7.4, at room temperature, recentrifuged, and the final pellet was dissolved in 1 ml of 1 N NaOH.
3. Quinacrine is analyzed in the NaOH extract using a luminescence spectrophotometer against a standard curve of the authentic drug dissolved in NaOH 1 N (excitation 414 nm, emission 501 nm).
4. Control fluorescence from extracts of untreated cells is systematically verified, of small magnitude and subtracted from experimental values.

Considerations on the analysis of transport data:

5. The transport kinetics curves for a fixed 30-min period are fitted by the nonlinear regression method to the Michaelis–Menten equation. The derived V_{max} and K_M and their respective SEM were obtained using Prism (GraphPad Software, San Diego, CA) (Fig. 7.3A, numerical values of the parameters reported in the figure). This analysis is empirically applied to a system that is postulated to be transporter free (Fig. 7.2, step 1) and its apparent saturability may reflect endomembrane pH buffering by quinacrine rather than transporter saturation. However, the analysis conveniently supports comparison between cell types (e.g.,

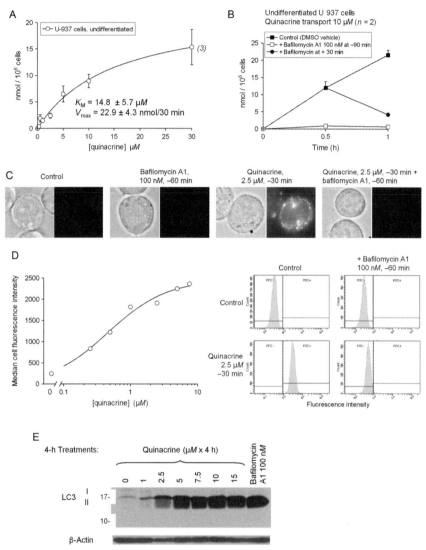

Figure 7.3 Quinacrine uptake into the undifferentiated, nonadherent U937 monoblastoid cells. (A) and (B) Uptake of quinacrine into cell suspensions as measured using extract-associated fluorescence. Each experimental point is derived from a 1-ml-tube containing 10^6 cells. Values are mean±SEM of the number of replicates indicated between parentheses. (A) Effect of drug concentration on uptake during a 30-min period. Data were fitted with a hyperbolic function, from which the apparent K_M and V_{max} values are derived. (B) Time course of quinacrine uptake and effect of the V-ATPase inhibitor bafilomycin A1 (100 nM) on the concentration or retention of the drug. (C) Microscopy of U937 cells treated as indicated. Representative photographs of multiple fields taken during 2 days of experiment. The transmission (left) and green fluorescence

the uptake affinity is of the same order of magnitude in freshly isolated lymphocytes, but much higher in neutrophils, Roy et al., 2013).
6. Pretreatment with the V-ATPase inhibitor bafilomycin A1 (100 nM) virtually abolished quinacrine uptake in U937 cells (Fig. 7.3B), as in other tested cell types. Further, bafilomycin started releasing the drug from the cells if added after a 1-h quinacrine loading period. Cotreatment with other types of drugs, such as inhibitors of organic cation transporters, is also possible and has failed so far to prove that such transporters are significant mediators of quinacrine uptake (Marceau et al., 2009; Roy et al., 2013; Fig. 7.2, step 1). However, there is at least one report of the extrusion of quinacrine by brain endothelial cells via P-glycoprotein (MDR1; Dohgu et al., 2004), and the modulation of quinacrine uptake by mechanisms operating at the plasma membrane remains possible in specific experimental system and may be the basis of cell type/organ specificity in the *in vivo* distribution of the drug (Ehsanian et al., 2011).
7. Considering the cell volume versus that of the culture medium, cell concentrations in the order of ~1000-fold are observed for quinacrine. However, this does not take into account that the drug is not evenly distributed in the cell volume (see below).

2.3. Microscopic, cytofluorometric correlates

1. Epifluorescence microscopy is used to evaluate the cellular uptake and subcellular distribution of quinacrine in adherent cells or cell suspension maintained in their culture medium. Thus, micromolar quinacrine is taken up by U937 cells under the form of fluorescent perinuclear granules of various sizes (Fig. 7.3C). No significant nuclear labeling is seen in most cell types with moderate micromolar concentrations of quinacrine, which is significant because the drug is reported to directly bind DNA at somewhat higher concentrations (Doglia et al., 1993). The effect of

---

(right) of the same square fields (30-μm sides) are represented side by side. (D) Cytofluorometric evaluation of quinacrine uptake by U937 cells during a 30-min incubation period at 37 °C. Left: median green fluorescence intensity as a function of quinacrine concentration. Right: representative distributions of fluorescence in cell populations and effect of bafilomycin A1 pretreatment of the accumulation of quinacrine. (E) Processing of endogenous LC3 in U937 cells subjected to the indicated drug treatments for 4 h and evaluated by immunoblotting for LC3 in the total cell extract. To document equal track loading, samples were immunoblotted for β-actin. *Methods in D and E as in Roy et al. (2013).*

bafilomycin A1 cotreatment supports again that the cell labeling by quinacrine is essentially driven by V-ATPase-mediated ion trapping (Fig. 7.3C).
2. The green fluorescence of quinacrine supports colocalization experiments with probes of other colors using high-power microscopy. In smooth muscle cells, colocalization is nonexistent for the nuclear probe Hoechst 33258 and Mitotracker red, rare for Rab5, but very frequent with Rab7, consistent with the late endosome–lysosome identity of most quinacrine-containing vacuoles (constructions of Rab proteins fused to the Cherry fluorescent protein; Marceau et al., 2009).
3. Suspension of cells may be treated with quinacrine as outlined in Section 2.2, spun, resuspended in Hank's balanced salt solution, and the drug uptake in cell populations assessed using cytofluorometry (using, for instance, the BD SORP LSR II cell analyzer, BD Biociences, Franklin Lakes, NJ; fluorescence settings for FITC are usable). This technique, applied to U937 cells, shows that the uptake of the drug confers a fluorescence well above the autofluorescence level, and that bafilomycin treatment abates this uptake (Fig. 7.3D).

2.4. Other techniques

Few studies report the actual uptake of cationic drugs by organelles that had been isolated by cell fractionation. An elegant series of studies was based on the endocytosis (pinocytosis) of dextran-iron, which transit to lysosomes via the late endosomes as a function of incubation time. Then, the iron-loaded organelles were magnetically isolated after cell disruption and supported transport studies of cationic drugs (Kaufmann & Krise, 2007).

3. MACROAUTOPHAGIC ACCUMULATION IN CELLS THAT HAVE ACCUMULATED CATIONS

Giant vacuoles induced by cationic drugs rapidly become macroautophagic, being labeled with the membrane-bound (lipidated) form of the autophagic effector microtubule-associated protein light chain 3 (LC3 II). This can be microscopically assessed using LC3 fused to GFP for non-fluorescent amines, such as lidocaine (Bawolak, Morissette, & Marceau, 2010), and by immunoblotting LC3 in total cell extracts in all experimental systems, including quinacrine-treated U937 cells (Fig. 7.3E).

Anti-human LC3B antibodies (e.g., rabbit polyclonal from Novus; dilution 1:3000) are used to reveal the cytosolic form LC3 I and processed form LC3 II in total cell extracts (5–15 μg/track) run on 15% SDS-PAGE and transferred to PVDF membranes (β-actin immunoblot to document equal loading). When applied to U937 cells, it was observed that control cells expressed very little endogenous LC3 I and LC3 II (Fig. 7.3E). Quinacrine (4-h treatment) caused the conversion of LC3 at and above 2.5 μM and the LC3 II signal was considerably more intense than LC3 I. Consistent with the hypothetical accumulation of autophagosomes due to lysosomal incompetence, bafilomycin A1 reproduced the accumulation of LC3 II in cells (Fig. 7.3E).

4. SUMMARY

Original studies based on U937 cells and presented in Fig. 7.3 recapitulate findings made using several other cell types: the lipophilic cationic drug quinacrine is avidly taken up by cells in a V-ATPase-dependent manner. Its subcellular location concerns more or less swollen perinuclear granules. The ensuing signaling event that is most noticeable is the accumulation of the autophagic effector LC3 II. In addition, feedback lysosomogenic signaling (TF-EB nuclear translocation) driven by lysosomal incompetence must be present because lysosomal/late endosomal proteins LAMP1 and -2 are upregulated in some cells that have accumulated the cationic drugs (Chen et al., 2011; Marceau et al., 2012). Quinacrine has specific pharmacologic actions of its own that are not related to ion trapping (e.g., direct binding to prion protein and DNA with possible antineoplasic applications; Ehsanian et al., 2011), but we aimed in the present text to promote its use as a superior probe for the assessment of cation trapping by cellular acidic compartments. It has been recently argued that the simple diffusion of drugs through the cell plasma membrane, a basic tenant of the model (Fig. 7.2, step 1), simply does not exist and that we just do not appreciate that multiple ATP-binding cassette, solute carrier, and other transporters are universally important in drug uptake (Kell, Dobson, Bilsland, & Oliver, 2013). Cation trapping may certainly operate in conjunction with other concentrating or nonconcentrating transport mechanisms that operate at the level of the plasma membrane, as discussed elsewhere (Marceau et al., 2012; Roy et al., 2013), and this may be the basis of the cell type/organ specificity in the distribution of amine drugs *in vivo*.

ACKNOWLEDGMENTS

Supported by an active operating grant from the Natural Sciences and Engineering Research Council of Canada and by a previous one from the Canadian Institutes of Health Research (MOP-74448). We thank our former laboratory colleagues Drs. Guillaume Morissette and Marie-Thérèse Bawolak for their important past contributions to this field and Dr. Alexandre Brunet for operating the cytofluorometry equipment.

REFERENCES

Ammoury, A., Michaud, S., Paul, C., Prost-Squarcioni, C., Alvarez, F., Lamant, L., et al. (2008). Photodistribution of blue-gray hyperpigmentation after amiodarone treatment: Molecular characterization of amiodarone in the skin. *Archives of Dermatology, 144*, 92–96.

Bawolak, M. T., Morissette, G., & Marceau, F. (2010). Vacuolar ATPase-mediated sequestration of local anesthetics in swollen macroautophagosomes. *Canadian Journal of Anesthesia, 57*, 230–239.

Chen, P. M., Gombart, Z. J., & Chen, J. W. (2011). Chloroquine treatment of ARPE-19 cells leads to lysosome dilation and intracellular lipid accumulation: Possible implications of lysosomal dysfunction in macular degeneration. *Cell & Bioscience, 1*, 10.

De Duve, C., de Barsy, T., Poole, B., Trouet, A., Tulkens, P., & Van Hoof, F. (1974). Lysosomotropic agents. *Biochemical Pharmacology, 23*, 2495–2531.

Doglia, S. M., Albinson, B., Hiort, C., Nordén, B., & Grässlund, A. (1993). Quinacrine: Spectroscopic properties and interactions with polynucleotides. *Biopolymers, 33*, 1431–1442.

Dohgu, S., Yamauchi, A., Takata, F., Sawada, Y., Higuchi, S., Naito, M., et al. (2004). Uptake and efflux of quinacrine, a candidate for the treatment of prion diseases, at the blood-brain barrier. *Cellular and Molecular Neurobiology, 24*, 205–217.

Ehsanian, R., Van Waes, C., & Feller, S. M. (2011). Beyond DNA binding—A review of the potential mechanisms mediating quinacrine's therapeutic activities in parasitic infections, inflammation, and cancers. *Cell Communication and Signaling, 9*, 13.

Gasparia, A. V., Burkart, C. A., Purmal, A. A., Brodsky, L., Pal, M., Saranadasa, M., et al. (2011). Curaxins: Anti-cancer compounds that simultaneously suppress NK-κB and activate p53 by targeting FACT. *Science Translational Medicine, 3*, 95ra74.

Huang, Y., Okochi, H., May, B. C., Legname, G., Prusiner, S. B., Benet, L. Z., et al. (2006). Quinacrine is mainly metabolized to mono-desethyl quinacrine by CYP3A4/5 and its brain accumulation is limited by P-glycoprotein. *Drug Metabolism and Disposition, 34*, 1136–1144.

Kaufmann, A. M., & Krise, J. P. (2007). Lysosomal sequestration of amine-containing drugs: Analysis and therapeutic implications. *Journal of Pharmaceutical Sciences, 96*, 729–746.

Kell, D. B., Dobson, P. D., Bilsland, E., & Oliver, S. G. (2013). The promiscuous binding of pharmaceutical drugs and their transporter-mediated uptake into cells: What we (need to) know and how we can do so. *Drug Discovery Today, 18*, 218–239.

Marceau, F., Bawolak, M.-T., Bouthillier, J., & Morissette, G. (2009). Vacuolar ATPase-mediated cellular concentration and retention of quinacrine: A model for the distribution of lipophilic cationic drugs to autophagic vacuoles. *Drug Metabolism and Disposition, 37*, 2271–2274.

Marceau, F., Bawolak, M.-T., Lodge, R., Bouthillier, J., Gagné-Henley, A., Gaudreault, R. C., et al. (2012). Cation trapping by cellular acidic compartments: Beyond the concept of lysosomotropic drugs. *Toxicology and Applied Pharmacology, 259*, 1–12.

Minta, J. O., & Pambrun, L. (1985). In vitro induction of cytologic and functional differentiation of the immature human monocyte-like cell line U-937 with phorbol myristate acetate. *American Journal of Pathology, 119,* 111–126.

Morissette, G., Lodge, R., & Marceau, F. (2008). Intense pseudotransport of a cationic drug mediated by vacuolar ATPase: Procainamide-induced autophagic cell vacuolization. *Toxicology and Applied Pharmacology, 228,* 364–377.

Reasor, M. J., Hastings, K. L., & Ulrich, R. G. (2006). Drug-induced phospholipidosis: Issues and future directions. *Expert Opinion on Drug Safety, 5,* 567–583.

Roy, C., Gagné, V., Fernandes, M. J. G., & Marceau, F. (2013). High affinity capture and concentration of quinacrine in polymormonuclear neutrophil via vacuolar ATPase-mediated ion trapping: Comparison with other peripheral blood leukocytes and implications for the distribution of cationic drugs. *Toxicology and Applied Pharmacology, 270,* 77–86.

Vallance, H., Chaba, T., Clarke, L., & Taylor, G. (2004). Pseudo-lysosomal storage disease caused by EMLA cream. *Journal of Inherited Metabolic Disease, 27,* 507–511.

CHAPTER EIGHT

Signaling Initiated by the Secretory Compartment

Carmen Ruggiero*, Jorge Cancino†, Monica Giannotta‡, Michele Sallese*,1

*Department of Cellular and Translational Pharmacology, Fondazione Mario Negri Sud, Unit of Genomic Approaches to Membrane Traffic, Santa Maria Imbaro (CH), Italy
†Department of Life Sciences, Institute of Protein Biochemistry, National Research Council and Telethon Institute of Genetics and Medicine, Naples, Italy
‡FIRC Institute of Molecular Oncology Foundation, Milan, Italy
1Corresponding author: e-mail address: sallese@negrisud.it

Contents

1. Introduction	134
2. ER-to-Golgi Traffic-Synchronization Protocols	136
2.1 PC-I traffic synchronization	136
2.2 PC-IV traffic synchronization	137
2.3 VSVG traffic synchronization	138
2.4 Traffic pulses generated by polymerization/depolymerization of cargo proteins	140
3. Read-Outs for Traffic-Generated Signaling	141
3.1 Detection and quantification of traffic-induced tyrosine phosphorylation by IF microscopy	141
3.2 Detection of traffic-induced tyrosine phosphorylation by Western blotting	143
3.3 Detection of traffic-induced signaling by FRET	144
4. KDELR Signaling	147
4.1 Activation of the KDELR signaling cascade by receptor overexpression	147
4.2 Activation of the KDELR signaling cascade using a genetically encoded artificial ligand	148
4.3 Activation of KDELR signaling by membrane-permeant peptides	148
5. The KDELR Transduction Machinery	149
5.1 Biochemical analysis of KDELR-induced [^{35}S]GTPγS loading on G$\alpha_{q/11}$	149
6. Conclusions	151
Acknowledgments	151
References	152

Abstract

Classical signal transduction is initiated at the plasma membrane by extracellular signals and propagates to the cytosolic face of the same membrane. Multiple studies have shown that endomembranes can act as signaling platforms for this plasma-membrane-originated signaling. Recent evidence has indicated that endomembranes can also trigger their own signaling cascades that involve some of the molecular players that are classically engaged in signal transduction at the plasma membrane. Endomembrane-initiated signaling is important for synchronization of the functioning of the secretory pathway and coordination of the activities of the secretory organelles with other cellular machineries. However, these endomembrane-initiated regulatory circuits are only partially understood to date. This novel field is slowed by a lack of specific tools and the objective difficulties in the study of signal transduction of endomembrane-localized receptors, as their accessibility is limited. For example, the ligand-binding site of the KDEL receptor (that transduces endomembrane signaling) is positioned in the lumen of the Golgi complex. Here we report some approaches that are suitable for the study of endomembrane-initiated signaling.

1. INTRODUCTION

The cellular endomembranes include the endosomes, the endoplasmic reticulum (ER), and the Golgi complex, and in recent years these have been emerging as novel signaling platforms. These membranes host different signaling molecules that can receive, integrate, and relay signaling that is initiated elsewhere. However, it is also clear that the ER and the Golgi complex trigger their own signals in the coordination of the secretory machinery (Sallese, Giannotta, & Luini, 2009). The flow of information and the identity of the molecular players in this endomembrane-initiated signaling are only partially understood, and their investigation requires specific approaches. For example, it is possible to study novel endomembrane regulatory circuits through the use of transport synchronization protocols.

Transport synchronization assays are based on both endogenous synchronizable secretory cargoes, for example, procollagen I (PC-I) and procollagen IV (PC-IV), and exogenous synchronizable secretory cargoes, for example, the temperature-sensitive mutant of the vesicular stomatitis virus G glycoprotein (VSVG; Bonfanti et al., 1998; Giannotta et al., 2012; Mironov et al., 2001; Pulvirenti et al., 2008; Trucco et al., 2004). When cells are at 40 °C, these proteins cannot fold completely, so their exit from the ER is blocked, while they are cleared from the rest of the secretory pathway. When the cells are then cooled to 32 °C, protein folding can occur in the ER, allowing these proteins to leave the ER, forming a "pulse" of secretory cargo that crosses the secretory pathway synchronously. Alternatively, membrane trafficking can be synchronized according to the polymerization/depolymerization of the FK506-binding protein (FKBP)-domain mutant (FM) (Rivera et al., 2000; Rollins et al., 2000; Volchuk et al.,

2000), which is modulated by rapamycin and/or rapamycin analogs (e.g., AP21998; Gordon, Bond, Sahlender, & Peden, 2010; Rivera et al., 2000; Volchuk et al., 2000).

The arrival of a traffic "pulse" at the Golgi complex elicits a signaling response on the Golgi itself (Pulvirenti et al., 2008). To detect the activation of such signaling, tyrosine phosphorylation provides a general readout, as it is involved in numerous signal transduction cascades. Phosphotyrosine (pTyr) can be revealed by semiquantitative immunofluorescence (IF) and immunoblotting. Alternatively, there are several biomolecular sensors to measure kinases, phosphatases, and GTPases by fluorescent resonance energy transfer (FRET).

Using these approaches, it has been shown that endomembrane signaling is based on specific receptor proteins (e.g., the KDEL receptor [KDELR]), the functioning of which is reminiscent of classical plasma membrane (PM) signaling (Giannotta et al., 2012). The KDELR cycles between the ER and Golgi compartments (Griffiths et al., 1994) for the retrieval from the Golgi complex to the ER of ER-luminal proteins that have a KDEL tetrapeptide in their carboxy-terminal sequence (Lewis & Pelham, 1992). Recent studies have demonstrated that the KDELR is a novel signaling receptor involved in the regulation of membrane trafficking (Capitani & Sallese, 2009; Giannotta et al., 2012; Pulvirenti et al., 2008).

Stimulation of the KDELR takes place from the luminal side of the Golgi complex due to the orientation of its ligand-binding pocket (Townsley, Wilson, & Pelham, 1993). This limitation can be largely overcome by overexpressing the KDELR, which in analogy with other receptors, autoactivates at high expression levels (Pulvirenti et al., 2008). In addition, the expression of a soluble secreted variant of horseradish peroxidase (ssHRP) bearing the KDEL motif at its carboxy-terminus (ssHRP^{KDEL}) (Connolly, Futter, Gibson, Hopkins, & Cutler, 1994) can work as a KDELR ligand (Pulvirenti et al., 2008). These provide chronic approaches to activate the KDELR (Pulvirenti et al., 2008). Conversely, acute delivery of KDEL peptides to the KDELR ligand-binding pocket requires that these peptides cross the PM and organelle membranes. BODIPY-conjugated KDEL peptides can achieve this (Pap, Dansen, van Summeren, & Wirtz, 2001), and their fluorescence can be visualized in living and fixed cells (Pap et al., 2001).

To investigate KDELR signaling mechanisms more deeply, our recent study (Giannotta et al., 2012) showed that the KDELR folding is reminiscent of the G-protein-coupled receptors (GPCRs) and activates the heterotrimeric G-protein $G\alpha_{q/11}$. This activation can be monitored using a modified version of the [^{35}S]GTPγS binding assay used for GPCRs (Giannotta et al., 2012; Harrison & Traynor, 2003).

Here, we report on a compendium of approaches that are currently available to investigate this novel endomembrane-initiated signaling.

2. ER-TO-GOLGI TRAFFIC-SYNCHRONIZATION PROTOCOLS

ER-to-Golgi traffic synchronization is often based on PC-I, PC-IV, and VSVG, as previously described (Bonfanti et al., 1998; Giannotta et al., 2012; Mironov et al., 2001; Pulvirenti et al., 2008; Trucco et al., 2004). Traffic pulses can also be generated by polymerization/depolymerization of FM, the FKBP-domain mutant (Gordon et al., 2010). The use of these methods is detailed in this section.

2.1. PC-I traffic synchronization

Human skin fibroblasts (HFs) produce and secrete large enough amounts of PC-I to be detected by IF analysis (Bonfanti et al., 1998; Mironov et al., 2001).

1. Plate the HFs between passages 3 and 8 on glass coverslips (10 mm^2; 50,000–70,000 cells) in 24-well plates, in Dulbecco's modified Eagle's medium (DMEM) supplemented with 10% calf serum, 4 mM L-glutamine, 100 U/ml penicillin and 100 μg/ml streptomycin, at 37 °C with a 5% CO_2 atmosphere. Grow the cells until they reach 90% confluence (generally in 24 h).

2. To eliminate the serum, wash the cells three times with 0.5 ml DMEM containing 20 mM HEPES, 4 mM L-glutamine, 100 U/ml penicillin and 100 μg/ml streptomycin, pH 7.4 (DMEM–HEPES), prewarmed to 40 °C.

3. To accumulate PC-I in the ER, incubate the cells with 0.5 ml prewarmed (40 °C) DMEM–HEPES containing 1% calf serum, at 40 °C for 3 h.

4. Change the cell medium for 0.5 ml prewarmed (32 °C) DMEM–HEPES containing 1% calf serum, 100 μg/ml ascorbic acid, and 100 μg/ml cycloheximide (to block protein synthesis). Ascorbic acid acts as a cofactor of the enzyme prolyl hydroxylase by maintaining its catalytic iron atom in the active $+2$ redox state (Peterkofsky & Udenfriend, 1965). Ascorbic acid supplementation promotes PC-I folding and exit from the ER (Bonfanti et al., 1998).

5. Incubate the cells at 32 °C for the desired times. PC-I usually reaches the Golgi complex in 10–30 min.
6. At the desired times during the cargo release at 32 °C, wash the cells twice with Dulbecco's phosphate-buffered saline, pH 7.2, with 0.1 mM Ca^{2+} and 1 mM Mg^{2+} (DPBS, Invitrogen).
7. To fix the cells, add 0.5 ml 4% paraformaldehyde in DPBS for 10 min at room temperature.
8. Incubate the cells in blocking/permeabilization solution (0.05% saponin, 0.5% bovine serum albumin, 50 mM NH$_4$Cl, in DPBS) for 45 min at room temperature.
9. Incubate the cells with an anti-PC-I antibody (e.g., the LF-68 monoclonal antibody; L.W. Fisher and J. Lippincott-Schwartz, National Institutes of Health, Bethesda, USA) diluted in blocking/permeabilization solution for 2 h at room temperature. The suggested antibody dilution is 1:200.
10. Wash the cells with DPBS three times, and then incubate them for 1 h at room temperature with an appropriate fluorophore-conjugate secondary antibody (e.g., Alexa Fluor 488-conjugated or Alexa Fluor 543-conjugated secondary antibodies, diluted 1:400 in blocking/permeabilization solution; Molecular Probes).
11. Wash the cells three times with DPBS.
12. To monitor PC-I arrival at the Golgi complex, costain the cells for a Golgi marker protein (e.g., the *cis*-Golgi marker GM130; BD Transduction Laboratories; catalog number, 610823) using the appropriate fluorophore-conjugated secondary antibody.
13. Mount the coverslips on microscope slides in mowiol or fluoromount-G (Electron Microscopy Sciences), and examine the cells using an inverted confocal microscope (Fig. 8.1).

2.2. PC-IV traffic synchronization

PC-IV is expressed by a number of epithelial and fibroblast cell lines, and cells expressing high levels of PC-IV provide optimal models to generate consistent pulses of synchronized cargo (Giannotta et al., 2012; Pulvirenti et al., 2008). As an example, human cervix squamous carcinoma HeLa cell variants (European Collection of Cell Cultures) express high levels of PC-IV. These have a "fibroblast-like" morphology, as compared to the "epithelial-like" morphology of other commercially available variants of HeLa cells, and they can be used to generate pulses of PC-IV.

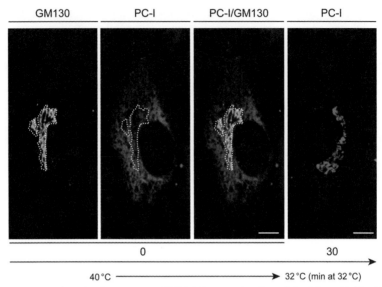

Figure 8.1 Synchronized transport of PC-I. HFs were incubated for 3 h at 40 °C, shifted to 32 °C for 30 min, and double stained for GM130 (green) and PC-I (red). Scale bars, 10 μm. *Modified from Pulvirenti et al. (2008).* (See color plate.)

1. Plate these HeLa cells between passages 3 and 15 on glass coverslips (10 mm^2; 70,000 cells/well) in 24-well plates, in minimum essential medium supplemented with 4 mM L-glutamine, 100 U/ml penicillin, 100 μg/ml streptomycin, 1% nonessential amino acids, and 10% fetal calf serum, at 37 °C with a 5% CO_2 atmosphere. Grow the cells to 80% confluence (generally in 24 h). For biochemical assays (see Section 3.2), 500,000 cells/well are plated in 6-well plates, without glass coverslips.
2. Proceed as detailed in Section 2.1 for the PC-I traffic pulse.
3. Monitor PC-IV transport using the M3F7 anti-PC-IV antibody (developed by H. Furthmayr, and obtained from Developmental Studies Hybridoma Bank, University of Iowa, USA). This antibody only recognizes the folded state of PC-IV, and thus does not label the unfolded PC-IV in the ER. An antibody dilution of 1:100 in blocking/permeabilization solution is suggested.

2.3. VSVG traffic synchronization

VSVG is a viral protein that can be expressed by cells upon infection of the vesicular stomatitis virus (VSV). A temperature-sensitive mutant VSV strain (ts045) is widely used to study membrane trafficking in mammalian cells (Beckers, Keller, & Balch, 1987; Bergmann, 1989; Doms, Keller, Helenius, & Balch, 1987; Hirschberg et al., 1998; Kreis & Lodish, 1986). At the nonpermissive temperature (40 °C), this mutant VSVG aggregates in the ER. At the permissive temperature (32 °C), the VSVG disaggregates and leaves the ER, moving to the Golgi complex and then the PM (Doms et al., 1987; Hirschberg et al., 1998).

1. Plate the HeLa cells as described in Section 2.2. The cell confluence is crucial for efficiency of infection: if they are too confluent, they will be poorly infected, if they are at a low confluence, they can detach during the infection procedure.
2. To eliminate the serum, wash the cells at least three times with 0.5 ml prewarmed (32 °C) DMEM–HEPES. This step is crucial, because serum can interfere with the viral infection.
3. Prepare the VSV dilution in prewarmed (32 °C) DMEM–HEPES, which must not contain serum. The appropriate amount of virus must be used for infecting the cells, which needs to be determined empirically. For high titer VSV ($>10^8$–10^9 PFU/ml), a 1:15 virus dilution is generally sufficient for 90% infection in HeLa cells.
4. Incubate the cells with 0.5 ml of this diluted virus for 45 min at 32 °C.
5. To completely remove excess virus, wash the cells at least three times with 0.5 ml prewarmed (32 °C) DMEM–HEPES.
6. To accumulate VSVG in the ER, incubate the cells in 0.5 ml prewarmed (40 °C) DMEM–HEPES containing 1% calf serum at 40 °C for 3 h.
7. Change the cell medium for 0.5 ml prewarmed (32 °C) DMEM–HEPES containing 1% calf serum and 100 µg/ml cycloheximide, and incubate the cells at 32 °C for the desired times. VSVG usually reaches the Golgi complex in 15–30 min.
8. Check the VSVG transport by IF, as detailed in Section 2.1. To stain VSVG, incubate the cells with the P5D4 anti-VSVG antibody (Sigma; catalog number, V4888) diluted 1:100 in blocking/permeabilization solution.

VSVG can also be expressed in cells by transfection. Generally, VSVG cDNA is expressed as the temperature-sensitive ts045 fusion protein tagged with GFP (Fig. 8.2; Presley et al., 1997). Addition of this GFP tag to the cytoplasmic tail of VSVG makes it also suitable for transport studies in living cells (Presley et al., 1997).

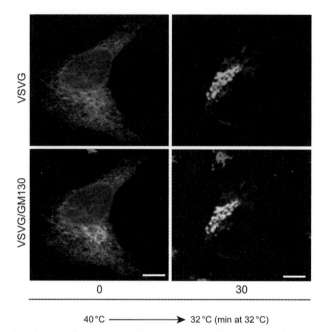

Figure 8.2 Synchronized transport of VSVG. HeLa cells were transfected with VSVG–GFP, incubated overnight at 40 °C, shifted to 32 °C for 30 min, and stained for GM130 (red). Scale bars, 10 μm. (See color plate.)

2.4. Traffic pulses generated by polymerization/depolymerization of cargo proteins

Polymerization/depolymerization of cargo proteins to generate traffic pulses can be used with any cargo that can be expressed as a fusion protein with FM, the FKBP-domain mutant. This method is based on a fusion protein between human growth hormone, green fluorescence protein, and FM (hGH–GFP–FM) (Gordon et al., 2010; Rivera et al., 2000; Rollins et al., 2000; Volchuk et al., 2000). Transfected hGH–GFP–FM polymerizes and cannot leave the ER until the cells are treated with the FM ligand AP21998 or D/D-Solubilizer, synthetic derivatives of rapamycin that acts as a solubilizer.

1. Plate the HeLa cells as described in Section 2.2. Cell confluence is important for the efficiency of transfection: if they are too confluent (90–100%), they will be poorly transfected.
2. Transfect the plated cells using an appropriate transfection method, and incubate them for 16–24 h at 37 °C. As an example, the JetPei (Polyplus)

transfection reagent (used according to the manufacturer instructions) gives high transfection efficiency.
3. Change the culture media by washing the cells three times with 0.5 ml DMEM–HEPES.
4. To release the cargo protein, change the culture media for prewarmed (37 °C) DMEM–HEPES with 50 μg/ml cycloheximide and 50 μM AP21998. hGH–GFP–FM reaches the Golgi in 5–10 min.
5. Monitor the arrival of hGH–GFP–FM at the Golgi complex as detailed in Section 2.1.

3. READ-OUTS FOR TRAFFIC-GENERATED SIGNALING

Signaling cascades generated by the arrival of a traffic pulse at the Golgi complex are usually detected using IF microscopy, FRET, and Western blotting. The advantage of the morphological approach is that events restricted to specific subcellular regions can also be monitored.

3.1. Detection and quantification of traffic-induced tyrosine phosphorylation by IF microscopy

The following procedures use a Zeiss LSM510 inverted confocal microscope system (Carl Zeiss, Gottingen, Germany), but they apply to any confocal microscope system.
1. At the desired times during the release of cargo from the 40 °C temperature block (generally 20–30 min), the cells can be processed for IF, as described in Section 2.1.
2. Incubate the cells with the PY99 anti-pTyr antibody (Santa Cruz Biotechnology; catalog number, sc-7020) or the 4G10 anti-pTyr antibody (Millipore Corporation, MA, USA; catalog number, 05-321) for 2 h at room temperature. The antibody dilution of 1:100 in blocking/permeabilization solution is suggested.
3. Costain the cells for the Golgi marker GM130, as described in Section 2.1. This staining is used as the reference Golgi area for quantification (Fig. 8.3).
4. Examine the cells under 63 × oil-immersion with the inverted confocal microscope system, with the pinhole at 1 airy unit to give the best signal-to-noise ratio, by filtering out most of the light scatter. This provides the most accurate acquisition of pTyr colocalization with the organelle of interest.

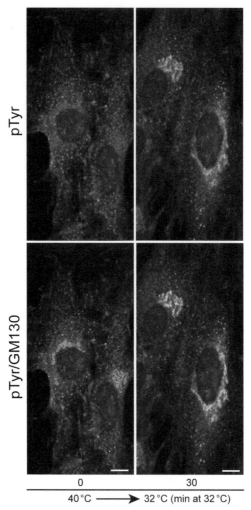

Figure 8.3 Transport-dependent tyrosine phosphorylation by immunofluorescence microscopy. HFs were incubated for 3 h at 40 °C, shifted to 32 °C for 30 min, and double stained for pTyr (green) and GM130 (red). Scale bars, 10 μm. (See color plate.)

5. Acquire unsaturated images (pixel fluorescence below 255 arbitrary units). Use the same acquisition settings for all samples, for direct comparison between them.
6. Acquire images from 50 to 100 cells per time point per experiment (from two wells of 24-well plates), with random cell sampling.

7. Use the acquired images to quantify the IF signal of interest (pTyr) in a specific area of the cell (e.g., the Golgi complex). Select the region of interest (ROI) according to the reference labeling (e.g., GM130). The software will show the size of the selected ROI.
8. Switch off the fluorescent channel of the reference signal (e.g., the GM130 channel). Use the histogram function to be certain there are no saturated pixels for the channel of interest (pTyr). If there are saturated pixels, the image is not suitable for quantification and will need to be acquired again using lower amplification gain. Visualize the frequency table of the pixel IF intensities within the ROI for the channel of interest (pTyr).
9. Copy the frequency table into an Excel worksheet and calculate the total IF intensity by integration of each IF pixel signal within the ROI. To obtain the fluorescence density, divide the total IF by the area of the ROI (from step 7). Repeat steps 7–9 for each cell. Average the fluorescence density across all cells analyzed.

3.2. Detection of traffic-induced tyrosine phosphorylation by Western blotting

Preparation of total lysates is a crucial step in phosphoprotein detection by Western blotting. Protease and phosphatase inhibitors must be used and added before the lysis procedure. All of the following procedures are carried out strictly at 4 °C, unless otherwise specified.

1. Thirty minutes after the traffic-block release (at 32 °C), harvest the cells in ice-cold lysis buffer (20 mM Tris–HCl, pH 7.6, 150 mM NaCl, 10 mM NaF, 5 mM Na$_3$VO$_4$, 30 mM β-glycerophosphate, and 5× protease cocktail inhibitor [Roche Applied Science]). Use 0.2 ml for each well of a 6-well plate. Any delays during cell harvesting or sample processing are deleterious to accurate and consistent detection of tyrosine phosphorylation.
2. Collect the cells using a rubber cell scraper, transfer the lysate into a 1.5-ml tube, and incubate this in ice for 5 min.
3. Centrifuge the cell lysate at 18,000 × g for 5 min and carefully transfer the supernatant into a new 1.5-ml tube.
4. Take an aliquot of the supernatant for determination of protein concentration, using, for example, the Bradford reagent (Bio-Rad), according to the manufacturer instructions.
5. Add 1/4 of the sample volume of 5× concentrated Laemmli buffer (10% sodium dodecylsulphate (SDS), 50% glycerol, 300 mM

Tris–HCl, pH 6.8, 0.05% (w/v) bromophenol blue) to each sample and incubate at 100 °C for 5 min.
6. Run the required SDS-PAGE gels and transfer the proteins to Protran nitrocellulose hybridization transfer membranes (0.2-μm pore size; Whatman).
7. To saturate the nitrocellulose membranes, incubate them with 5% (w/v) bovine serum albumin in 150 mM NaCl, 10 mM Tris–HCl, pH 7.4, 0.05% Tween (saturation solution) for 1 h at room temperature. Fat-free milk saturation needs to be avoided here as it contains phosphorylated casein, which will interfere with phosphoprotein detection.
8. Incubate the membranes with the primary anti-pTyr antibody (PY99, Santa Cruz Biotechnology; catalog number, sc-7020; or 4G10, Millipore Corporation; catalog number, 05-321), at 1:500 dilutions for 2 h at room temperature. Then incubate with the secondary HRP-conjugated anti-mouse antibody (1:10,000; Calbiochem; catalog number, 401215) for 1 h at room temperature (Fig. 8.4). Both antibodies are diluted in saturation solution.
9. For quantitative analysis, digital image acquisition is recommended. Use several acquisition times and select nonsaturated images for quantification.
10. The gel analysis tools provided by ImageJ (free software from NIH, USA) provide a useful and easy method for quantification.

3.3. Detection of traffic-induced signaling by FRET

The *in vivo* recording of traffic events became possible with the advent of fluorescent protein reporters (Lippincott-Schwartz, 2011). Several spectral variants are available, and some are suitable for specific fluorescent techniques. FRET analysis can be used to follow classical PM-initiated signaling dynamically (Aoki, Kiyokawa, Nakamura, & Matsuda, 2008; Miyawaki, 2003) and can be adapted to monitor traffic-induced signaling.
1. Plate HeLa cells (350,000 cells/well) in glass-bottomed 35-mm dishes (e.g., MatTek Corporation), as described in Section 2.2.
2. To follow cAMP signaling, transfect the cells (see Section 2.4) with a fluorescent traffic reporter (e.g., mCherry-VSVG) and the selected FRET-based sensor, YFP-regulatory/CFP-catalytic subunits of PKA (Zaccolo & Pozzan, 2002).

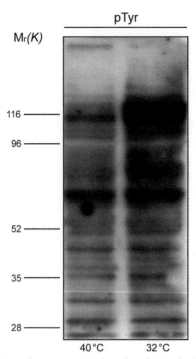

Figure 8.4 Transport-dependent tyrosine phosphorylation by Western blotting. HeLa cells were incubated for 3 h at 40 °C, shifted to 32 °C for 30 min, and then homogenized. The pTyr patterns were analyzed by immunoblotting. *Modified from Pulvirenti et al. (2008).*

3. Incubate the cells at 40 °C for 16 to 24 h. If a CO_2 supply is not available for maintenance of medium pH, change the medium to prewarmed (40 °C) DMEM–HEPES.
4. Preheat the temperature-controlled chamber on the microscope stage to 40 °C. Switch off the air conditioning in the microscope room. To ensure stable temperature over time, a thermal controller on the objective (as a collar) is highly recommended. A water-jacketed, temperature-controlled chamber allows rapid changes of temperature using separate water-bath temperature controllers.
5. Transfer the cells to the microscope stage, while avoiding any delay and consequent temperature drop. For temperature stabilization, leave the cells on the microscope stage for at least 5–10 min, while you prepare the microscope acquisition settings.
6. Ensure the appropriate laser selection. With CFP as a donor, 458-nm laser excitation is used, with acquisition at 460–480 nm. Emission of

YFP is collected at 510–530 nm. Acquisition of mCherry-VSVG uses 568 nm laser excitation, with the fluorescent signal collected at 600–620 nm. Images are acquired using 63× oil-immersion. The correct detector adjustment wavelength windows avoid excitation/emission bleed between the donor/acceptor and the cargo protein.

7. Select cells with a good level of, for example, mCherry-VSVG fluorescence in the ER.
8. Completely open the pinhole to guarantee good fluorescence capture and adjust the laser power to a minimum (5–10% of total) to minimize photobleaching of the FRET sensor and cargo. This will allow image acquisition with reasonably good fluorescence intensity, as normally 80–150 for 8-bit images (0-255 gray-scale level). Selection of fluorescence above 150 can hamper the recording of high increases in fluorescence due to FRET loss, especially for donor emission.
9. Adjust the digital gain as low as possible to reduce background fluorescence and allow capture of images using less averaging to improve acquisition speed. It is highly recommended to use line average instead of frame average. The capture of a 512 × 512 pixel image is a good compromise for spatial and time resolution.
10. Select the number of frames and the interval between them. For short recordings (e.g., 5–15 min), 5–10 s intervals is reasonable. For longer recordings (e.g., 30–60 min), 30–60 s intervals preserve the fluorescence until the end of the experiment and reduces loss by photobleaching.
11. Record the steady-state FRET signaling for at least 5 min during the 40 °C ER block of transport.
12. Release the cargo by changing the temperature to 32 °C. The experiment can be followed according to the appearance of mCherry-VSVG in the perinuclear region, while cAMP production is revealed by the CFP/YFP ratio of the selected ROI. The CFP/YFP average intensity ratio can be calculated for several ROI. The resulting images can be processed using any suitable image software (e.g., ImageJ or MetaMorph from Molecular Devices).
13. Plot the CFP/YFP ratio as a function of time. Use the normalized ratio by adjusting the first time point to define the ratio of 1.0, and divide the following time points by this first time point.

4. KDELR SIGNALING

The study of KDELR signal-transduction mechanisms requires specific approaches because the KDELR ligand-binding domain is in the lumen of the secretory organelles. Overexpression of the KDELR leads to its autoactivation and redistribution to the ER, with the consequent Src family kinase (SFK) activation in the Golgi area (Hsu, Shah, & Klausner, 1992; Pulvirenti et al., 2008). Alternatively, overexpression of a synthetic KDEL-containing polypeptide (e.g., ssHRP^{KDEL}) can stimulate the KDELR (Connolly et al., 1994). ssHRP lacking the KDEL motif provides the control.

To date, there are no chemical ligands available for KDELR stimulation. However, recent studies have demonstrated that KDEL peptides conjugated to BODIPY can cross biological membranes and overcome this limitation (Giannotta et al., 2012; Pap et al., 2001). Several fluorescent BODIPY derivatives are available, with selection of excitation/emission wavelengths depending on requirements. As an example, BODIPY630/650 derivatives allow FRET with the most commonly used FRET couple (CFP/YFP) and mCherry-VSVG.

4.1. Activation of the KDELR signaling cascade by receptor overexpression

1. Plate HeLa cells as described in Section 2.2.
2. Transfect the cells (see Section 2.4) with a GFP-tagged KDELR2 construct and incubate for a further 16–24 h.
3. To inhibit protein synthesis, change the growth medium for new prewarmed (37 °C) medium containing 100 µg/ml cycloheximide, and incubate at 37 °C for 1 h.
4. Wash the cells twice with DPBS, and then fix and stain them with a Golgi marker and an anti-phospho-SFK (pSFK) antibody, as described in Section 2.1. To follow SFK activation, use a polyclonal antibody that selectively recognizes pTyr at position 419 in human Src (or the equivalent tyrosines in other SFKs; Pulvirenti et al., 2008). Incubate the cells with the pSFK antibody (Invitrogen; catalog number, 44-660G) diluted 1:100 in blocking/permeabilization solution, for 16 h at 4 °C.
5. Analyze the pSFK levels in cells with high levels of the KDELR in the ER. The KDELR in the ER is recognized by a bright green fluorescent

ring around the nucleus, as the nuclear envelope. This pattern of KDELR distribution indicates that a portion of the total KDELR has been activated.

4.2. Activation of the KDELR signaling cascade using a genetically encoded artificial ligand

1. Plate the HeLa cells as described in Section 2.2.
2. Transfect the cells (see Section 2.4) with ssHRP (control) and ssHRP^{KDEL}, incubate them at 37 °C for 16–24 h, and then process them for IF, as detailed in Section 2.1.
3. Stain the HRP to identify transfected cells, using an anti-HRP monoclonal antibody (Abcam; catalog number, ab2110) for 2 h at room temperature. A good working dilution here is 1:100 in blocking/permeabilization solution.
4. Stain the cells for active SFKs, as reported in Section 4.1.
5. Stain a parallel set of cells for the KDELR using an anti-KDELR monoclonal antibody (clone KR-10; Stressgen Biotechnology; catalog number, ADI-VAA-PT048-D), diluted 1:100 in blocking/permeabilization solution, for 2 h at room temperature. This procedure is useful to monitor ssHRP^{KDEL}-dependent KDELR redistribution.
6. Analyze and quantify SFK activation in the Golgi area, as detailed in Section 3.1 for pTyr.

4.3. Activation of KDELR signaling by membrane-permeant peptides

1. Plate the HeLa cells as described in Section 2.2.
2. Prepare 3 μM solutions of BODIPY-KDEL or BODIPY-KDEA peptides in DMEM–HEPES, and warm the mixture to 37 °C. These BODIPY conjugated peptides are obtained according to Wirtz and coworkers (Pap et al., 2001).
3. Wash the cells three times with prewarmed (37 °C) DMEM–HEPES and replace this with KDEL-peptide- or KDEA-peptide-containing media.
4. Incubate the cells at 37 °C for the desired times. Generally, 10 min are sufficient for activation of the KDELR signaling cascade (Giannotta et al., 2012).
5. Fix and process the cells for IF, as described in Section 4.1.
6. Follow KDELR activation by monitoring its redistribution into the ER and SFKs activation at the Golgi complex (Fig. 8.5), as detailed in Sections 4.1 and 4.2.

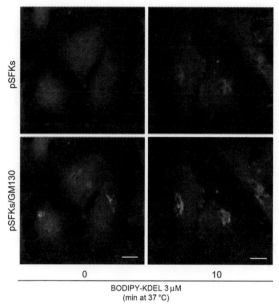

Figure 8.5 BODIPY-KDEL triggers Golgi-SFKs activation. HeLa cells were treated with the membrane-permeant KDELR agonist BODIPY-KDEL (3 μM) at 37 °C, for 10 min. The cells were then stained for active SFKs (pSFKs, red) and GM130 (green). Scale bars, 10 μm. *Modified from Giannotta et al. (2012).* (See color plate.)

5. THE KDELR TRANSDUCTION MACHINERY

KDELR-dependent loading of GTP on $G\alpha_{q/11}$ can be analyzed *in vitro* using Golgi-enriched membranes prepared from mice liver according to the procedure of Howell and coworkers (Taylor, Jones, Dahl, Nordeen, & Howell, 1997). This GTPγS loading method is modified from that described by Akam (Akam, Challiss, & Nahorski, 2001).

5.1. Biochemical analysis of KDELR-induced [^{35}S]GTPγS loading on $G\alpha_{q/11}$

1. To remove excess sucrose from the Golgi-enriched membranes dilute them 1:4 in assay buffer: 100 mM NaCl, 10 mM MgCl$_2$, 10 mM HEPES, pH 7.4, plus protease inhibitors (e.g., Roche) and centrifuge

at 18,000 × g for 10 min, discard the supernatant and repeat the washing. All of the following procedures are carried out in ice.

2. Resuspend the washed membranes in assay buffer (including complete protease inhibitors) with 2 μM GDP, to a final protein concentration of 1 μg/μl.
3. Prepare each sample in triplicate by adding 5 μg Golgi membranes to 50 μl assay buffer containing 2 nM [^{35}S]GTPγS (1000 Ci/mmol) (Perkin Elmer, USA; catalog number NEG030H001MC), 2 μM GDP, without or with BODIPY-KDEL or BODIPY-KDEA (3 μM).
4. To determine nonspecific binding, for each of the above conditions, prepare additional triplicates samples as in step 3, plus 50 μM non-radioactive GTPγS.
5. Incubate all the samples at 30 °C for 5 min; this temperature is critical for the outcome of the reaction.
6. Terminate the reactions by moving the samples to ice and immediately adding 1 ml ice-cold assay buffer.
7. Centrifuge the samples at 20,000 × g for 10 min, discard the supernatants, and solubilize the pellets using 100 μl ice-cold solubilization buffer: 100 mM Tris–HCl, 200 mM NaCl, 1 mM EDTA, 1.25% Igepal CA 630, 0.2% SDS, pH 7.4, plus complete protease inhibitors. Flick the tubes to help the solubilization, and avoid pipetting.
8. Once the pellets are completely solubilized, add 100 μl solubilization buffer without SDS to each tube.
9. Preclear the solubilized pellets with 1 μg/ml rabbit IgG and 70 μl protein A-Sepharose suspension (Sigma-Aldrich; catalog number, P3391; 30% [w/v] in 10 mM Tris–HCl, 10 mM EDTA, pH 8.0), for 1 h at 4 °C.
10. Centrifuge the samples at 20,000 × g for 5 min and transfer 100 μl supernatant to fresh tubes containing 8 μl anti-G$α_{q/11}$ antibody (200 μg/ml; Santa Cruz Biotechnology; catalog number, sc-392).
11. Incubate the samples on a rotating wheel for 2 h at 4 °C.
12. Add 140 μl protein A-Sepharose suspension, and rotate the samples for a further 1 h at 4 °C.
13. Pellet the protein A-Sepharose beads at 20,000 × g for 5 min, and wash them five times with 500 μl solubilization buffer without SDS.
14. Put the washed beads into vials, mix them with scintillation solution (Ultima Gold, Packard Bioscience), and determine the radioactivity in scintillation counter.

15. The mean $G\alpha_{q/11}$-associated radioactivity is calculated for each triplicate determination, and the mean of the corresponding triplicates (subjected to the same treatments) where nonradioactive GTPγS was added is subtracted.

6. CONCLUSIONS

The secretory pathway synthesizes, modifies, and transports about one third of all cellular proteins (Braakman & Bulleid, 2011; Lee, Miller, Goldberg, Orci, & Schekman, 2004; Nilsson, Au, & Bergeron, 2009; van Vliet, Thomas, Merino-Trigo, Teasdale, & Gleeson, 2003). To avoid dysfunctioning of this system, these activities must be coordinated with the rest of the cell and with environmental needs. Indeed, a large number of transduction molecules are localized on endomembranes, including G proteins (both small and heterotrimeric), phospholipases, kinases, and phosphatases (Sallese, Pulvirenti, & Luini, 2006). PM-initiated signaling cascades target this internal pool of signaling molecules to regulate both the secretory system (Buccione et al., 1996; De Matteis, Santini, Kahn, Di Tullio, & Luini, 1993; Fabbri, Bannykh, & Balch, 1994; Luton, Verges, Vaerman, Sudol, & Mostov, 1999) and nontrafficking functions, like cell growth, differentiation, and motility (Bivona & Philips, 2003; Ishikawa, Takeuchi, Haltiwanger, & Irvine, 2008; Preisinger et al., 2004; Sorkin, 2005; von Zastrow & Sorkin, 2007).

Endomembranes can also trigger their own signals for the organization of the secretory system as a whole, although we cannot formally exclude that internally initiated activities will influence cell interactions with the environment. There remain a number of endomembrane-localized signaling molecules that have poorly understood roles. These might be on endomembranes because they are simply being transported toward their final destinations, or they might belong to signaling cascades for which the flow of information is still not known. The experimental approaches described here will help in the study of these potential new endomembrane signaling cascades.

ACKNOWLEDGMENTS

We thank C. P. Berrie for editorial assistance and E. Fontana for artwork preparation. This study was supported by AIRC, Grant No. IG 11652, Telethon-Italy, Grant No. GGP12220 and Fondazione Cariplo, Grant No. 2010-0828.

REFERENCES

Akam, E. C., Challiss, R. A., & Nahorski, S. R. (2001). G(q/11) and G(i/o) activation profiles in CHO cells expressing human muscarinic acetylcholine receptors: Dependence on agonist as well as receptor-subtype. *British Journal of Pharmacology, 132,* 950–958.

Aoki, K., Kiyokawa, E., Nakamura, T., & Matsuda, M. (2008). Visualization of growth signal transduction cascades in living cells with genetically encoded probes based on Forster resonance energy transfer. *Philosophical Transactions of the Royal Society of London Series B, Biological Sciences, 363,* 2143–2151.

Beckers, C. J., Keller, D. S., & Balch, W. E. (1987). Semi-intact cells permeable to macromolecules: Use in reconstitution of protein transport from the endoplasmic reticulum to the Golgi complex. *Cell, 50,* 523–534.

Bergmann, J. E. (1989). Using temperature-sensitive mutants of VSV to study membrane protein biogenesis. *Methods in Cell Biology, 32,* 85–110.

Bivona, T. G., & Philips, M. R. (2003). Ras pathway signaling on endomembranes. *Current Opinion in Cell Biology, 15,* 136–142.

Bonfanti, L., Mironov, A. A., Jr., Martinez-Menarguez, J. A., Martella, O., Fusella, A., Baldassarre, M., et al. (1998). Procollagen traverses the Golgi stack without leaving the lumen of cisternae: Evidence for cisternal maturation. *Cell, 95,* 993–1003.

Braakman, I., & Bulleid, N. J. (2011). Protein folding and modification in the mammalian endoplasmic reticulum. *Annual Review of Biochemistry, 80,* 71–99.

Buccione, R., Bannykh, S., Santone, I., Baldassarre, M., Facchiano, F., Bozzi, Y., et al. (1996). Regulation of constitutive exocytic transport by membrane receptors. A biochemical and morphometric study. *Journal of Biological Chemistry, 271,* 3523–3533.

Capitani, M., & Sallese, M. (2009). The KDEL receptor: New functions for an old protein. *FEBS Letters, 583,* 3863–3871.

Connolly, C. N., Futter, C. E., Gibson, A., Hopkins, C. R., & Cutler, D. F. (1994). Transport into and out of the Golgi complex studied by transfecting cells with cDNAs encoding horseradish peroxidase. *Journal of Cell Biology, 127,* 641–652.

De Matteis, M. A., Santini, G., Kahn, R. A., Di Tullio, G., & Luini, A. (1993). Receptor and protein kinase C-mediated regulation of ARF binding to the Golgi complex. *Nature, 364,* 818–821.

Doms, R. W., Keller, D. S., Helenius, A., & Balch, W. E. (1987). Role for adenosine triphosphate in regulating the assembly and transport of vesicular stomatitis virus G protein trimers. *Journal of Cell Biology, 105,* 1957–1969.

Fabbri, M., Bannykh, S., & Balch, W. E. (1994). Export of protein from the endoplasmic reticulum is regulated by a diacylglycerol/phorbol ester binding protein. *Journal of Biological Chemistry, 269,* 26848–26857.

Giannotta, M., Ruggiero, C., Grossi, M., Cancino, J., Capitani, M., Pulvirenti, T., et al. (2012). The KDEL receptor couples to Galphaq/11 to activate Src kinases and regulate transport through the Golgi. *EMBO Journal, 31,* 2869–2881.

Gordon, D. E., Bond, L. M., Sahlender, D. A., & Peden, A. A. (2010). A targeted siRNA screen to identify SNAREs required for constitutive secretion in mammalian cells. *Traffic, 11,* 1191–1204.

Griffiths, G., Ericsson, M., Krijnse-Locker, J., Nilsson, T., Goud, B., Soling, H. D., et al. (1994). Localization of the Lys, Asp, Glu, Leu tetrapeptide receptor to the Golgi complex and the intermediate compartment in mammalian cells. *Journal of Cell Biology, 127,* 1557–1574.

Harrison, C., & Traynor, J. R. (2003). The [35S]GTPgammaS binding assay: Approaches and applications in pharmacology. *Life Sciences, 74,* 489–508.

Hirschberg, K., Miller, C. M., Ellenberg, J., Presley, J. F., Siggia, E. D., Phair, R. D., et al. (1998). Kinetic analysis of secretory protein traffic and characterization of golgi to plasma membrane transport intermediates in living cells. *Journal of Cell Biology, 143,* 1485–1503.

Hsu, V. W., Shah, N., & Klausner, R. D. (1992). A brefeldin A-like phenotype is induced by the overexpression of a human ERD-2-like protein, ELP-1. *Cell, 69*, 625–635.

Ishikawa, H. O., Takeuchi, H., Haltiwanger, R. S., & Irvine, K. D. (2008). Four-jointed is a Golgi kinase that phosphorylates a subset of cadherin domains. *Science, 321*, 401–404.

Kreis, T. E., & Lodish, H. F. (1986). Oligomerization is essential for transport of vesicular stomatitis viral glycoprotein to the cell surface. *Cell, 46*, 929–937.

Lee, M. C., Miller, E. A., Goldberg, J., Orci, L., & Schekman, R. (2004). Bi-directional protein transport between the ER and Golgi. *Annual Review of Cell and Developmental Biology, 20*, 87–123.

Lewis, M. J., & Pelham, H. R. (1992). Ligand-induced redistribution of a human KDEL receptor from the Golgi complex to the endoplasmic reticulum. *Cell, 68*, 353–364.

Lippincott-Schwartz, J. (2011). Emerging in vivo analyses of cell function using fluorescence imaging (*). *Annual Review of Biochemistry, 80*, 327–332.

Luton, F., Verges, M., Vaerman, J. P., Sudol, M., & Mostov, K. E. (1999). The SRC family protein tyrosine kinase p62yes controls polymeric IgA transcytosis in vivo. *Molecular Cell, 4*, 627–632.

Mironov, A. A., Beznoussenko, G. V., Nicoziani, P., Martella, O., Trucco, A., Kweon, H. S., et al. (2001). Small cargo proteins and large aggregates can traverse the Golgi by a common mechanism without leaving the lumen of cisternae. *Journal of Cell Biology, 155*, 1225–1238.

Miyawaki, A. (2003). Visualization of the spatial and temporal dynamics of intracellular signaling. *Developmental Cell, 4*, 295–305.

Nilsson, T., Au, C. E., & Bergeron, J. J. (2009). Sorting out glycosylation enzymes in the Golgi apparatus. *FEBS Letters, 583*, 3764–3769.

Pap, E. H., Dansen, T. B., van Summeren, R., & Wirtz, K. W. (2001). Peptide-based targeting of fluorophores to organelles in living cells. *Experimental Cell Research, 265*, 288–293.

Peterkofsky, B., & Udenfriend, S. (1965). Enzymatic hydroxylation of proline in microsomal polypeptide leading to formation of collagen. *Heilkunde - Heilwege, 53*, 335–342.

Preisinger, C., Short, B., De Corte, V., Bruyneel, E., Haas, A., Kopajtich, R., et al. (2004). YSK1 is activated by the Golgi matrix protein GM130 and plays a role in cell migration through its substrate 14-3-3zeta. *Journal of Cell Biology, 164*, 1009–1020.

Presley, J. F., Cole, N. B., Schroer, T. A., Hirschberg, K., Zaal, K. J., & Lippincott-Schwartz, J. (1997). ER-to-Golgi transport visualized in living cells. *Nature, 389*, 440–441.

Pulvirenti, T., Giannotta, M., Capestrano, M., Capitani, M., Pisanu, A., Polishchuk, R. S., et al. (2008). A traffic-activated Golgi-based signalling circuit coordinates the secretory pathway. *Nature Cell Biology, 10*, 912–922.

Rivera, V. M., Wang, X., Wardwell, S., Courage, N. L., Volchuk, A., Keenan, T., et al. (2000). Regulation of protein secretion through controlled aggregation in the endoplasmic reticulum. *Science, 287*, 826–830.

Rollins, C. T., Rivera, V. M., Woolfson, D. N., Keenan, T., Hatada, M., Adams, S. E., et al. (2000). A ligand-reversible dimerization system for controlling protein-protein interactions. *Heilkunde - Heilwege, 97*, 7096–7101.

Sallese, M., Giannotta, M., & Luini, A. (2009). Coordination of the secretory compartments via inter-organelle signalling. *Seminars in Cell and Developmental Biology, 20*, 801–809.

Sallese, M., Pulvirenti, T., & Luini, A. (2006). The physiology of membrane transport and endomembrane-based signalling. *EMBO Journal, 25*, 2663–2673.

Sorkin, A. (2005). TRKing signals through the Golgi. *Science's STKE, 267*, pe1.

Taylor, R. S., Jones, S. M., Dahl, R. H., Nordeen, M. H., & Howell, K. E. (1997). Characterization of the Golgi complex cleared of proteins in transit and examination of calcium uptake activities. *Molecular Biology of the Cell, 8*, 1911–1931.

Townsley, F. M., Wilson, D. W., & Pelham, H. R. (1993). Mutational analysis of the human KDEL receptor: Distinct structural requirements for Golgi retention, ligand binding and retrograde transport. *EMBO Journal, 12,* 2821–2829.

Trucco, A., Polishchuk, R. S., Martella, O., Di Pentima, A., Fusella, A., Di Giandomenico, D., et al. (2004). Secretory traffic triggers the formation of tubular continuities across Golgi sub-compartments. *Nature Cell Biology, 6,* 1071–1081.

van Vliet, C., Thomas, E. C., Merino-Trigo, A., Teasdale, R. D., & Gleeson, P. A. (2003). Intracellular sorting and transport of proteins. *Progress in Biophysics and Molecular Biology, 83,* 1–45.

Volchuk, A., Amherdt, M., Ravazzola, M., Brugger, B., Rivera, V. M., Clackson, T., et al. (2000). Megavesicles implicated in the rapid transport of intracisternal aggregates across the Golgi stack. *Cell, 102,* 335–348.

von Zastrow, M., & Sorkin, A. (2007). Signaling on the endocytic pathway. *Current Opinion in Cell Biology, 19,* 436–445.

Zaccolo, M., & Pozzan, T. (2002). Discrete microdomains with high concentration of cAMP in stimulated rat neonatal cardiac myocytes. *Science, 295,* 1711–1715.

CHAPTER NINE

Image-Based and Biochemical Assays to Investigate Endosomal Protein Sorting

Sophia Y. Breusegem[1], Matthew N.J. Seaman[1]
Department of Clinical Biochemistry, Cambridge Institute for Medical Research, University of Cambridge, Addenbrooke's Hospital, Cambridge, United Kingdom
[1]Corresponding authors: e-mail address: syab2@cam.ac.uk; mnjs100@cam.ac.uk

Contents

1. Introduction	156
2. Antibody-Uptake Assays	157
2.1 Antibody-uptake assay on a small scale	159
2.2 Antibody-uptake assay on a high-throughput scale	163
2.3 Antibody-uptake assay on a medium scale	169
3. Detailed Characterization of Endosomes	169
4. Endosome Recruitment and/or Association	171
4.1 Pellet versus supernatant	171
4.2 Native IP of endosomal proteins	172
4.3 Cost-efficient large-scale transient transfection for native immunoprecipitations	176
5. Summary	177
Acknowledgments	177
References	177

Abstract

The sorting of membrane proteins within the endosomal system occurs through a panoply of highly dynamic sequential molecular interactions that together govern many physiologically important processes. A key component of the endosomal protein sorting machinery is the retromer complex. Through two distinct subcomplexes, retromer operates to select cargo for endosome-to-Golgi retrieval and also drives membrane tubule formation. Many accessory proteins associate with retromer to facilitate protein sorting and/or tubule formation. The experience we have gained from studying retromer-mediated endosomal protein sorting and the assays developed and applied in the course of these studies can provide a template for researchers interested in related endosomal trafficking pathways.

Herein we describe image-based assays that can be applied to study endosomal protein sorting through the use of antibody-uptake assays in low-, medium-, and high-throughput formats. We additionally detail simple but effective native immunoprecipitation methods that can be employed to identify novel proteins that may interact transiently with a protein of interest within the endosomal pathway.

ABBREVIATIONS

CIMPR cation-independent mannose-6-phosphate receptor
GFP green fluorescent protein
IF immunofluorescence
IP immunoprecipitation
PBS phosphate-buffered saline
PEI polyethylenimine
PFA paraformaldehyde
ROI region of interest
SNX sorting nexin
TGN *trans*-Golgi network
VPS vacuole protein sorting

1. INTRODUCTION

The membrane-bound compartments that comprise the endosomal system function as an intersection where the biosynthetic cargo of the secretory pathway meets the endocytic cargo proteins that have been internalized from the plasma membrane. The correct functioning of the endosomal system is maintained through a myriad of highly dynamic molecular interactions that underpin and control diverse biological processes including nutrient uptake, downregulation of signaling receptors, synaptic transmission, response to infection, generation of immunity and autophagy, to name but a few.

Membrane proteins—often referred to as cargo—that traverse the endosomal system are acted on by sorting machinery that transiently recognize discrete motifs present in the cytoplasmic domain of the respective membrane protein. Following recognition by the various sorting machinery, cargo proteins are subsequently sorted into a specific trafficking pathway with essentially three possible destinations: the plasma membrane, the Golgi apparatus, or the lysosome (Seaman, 2008). The intersection of biosynthetic and recycling pathways at the endosome is exemplified by the rapid cycling of lysosomal hydrolase receptors between the Golgi and endosome as they mediate the delivery of newly synthesized hydrolases to the lysosome.

A critical component in the endosome-to-Golgi recycling pathway is the retromer complex, a conserved multimeric protein complex consisting of a cargo-recognition subcomplex (VPS26, VPS29, and VPS35) and a sorting nexin (SNX) dimer consisting of SNX1 or SNX2 paired with SNX5 or SNX6, to which a structural, membrane-deforming role is attributed (reviewed in Attar & Cullen, 2010). In the 15 years since its original description in yeast (Seaman, McCaffrey, & Emr, 1998), many different cargo molecules for retromer-mediated trafficking have been identified, as well as several accessory proteins that facilitate or contribute to retromer-mediated trafficking. Additionally, it has now become apparent that retromer is required not only for endosome-to-Golgi sorting but also for the efficient endosome-to-plasma membrane recycling of certain cargo molecules (see Seaman, 2012 for a review).

In this chapter, we present assays that can be employed to study retromer-mediated trafficking in mammalian cells. We describe in detail antibody-uptake assays to investigate the trafficking of model cargo molecules and we also discuss methods to characterize retromer-positive endosomes both by imaging methods and biochemical methods.

2. ANTIBODY-UPTAKE ASSAYS

In order to determine whether a protein of interest functions in the endosome-to-Golgi recycling pathway, a suitable assay is required. Described below are protocols for low-, medium-, and high-throughput antibody-uptake assays that can be employed to investigate the functioning of the endosome-to-Golgi pathway. Antibodies are well suited to this task as they can be modified by the covalent addition of fluorescent moieties and the high-affinity interaction of an antibody binding to its epitope will generally be resistant to the acidic environment present in the lumen of an endosomal compartment although other ligands including fluorescent Shiga toxin could also be employed (see Bujny, Popoff, Johannes, & Cullen, 2007).

The assays described here can be employed in either gain-of-function (e.g., overexpression) or loss-of-function (e.g., RNAi) experiments although studying the effect of overexpression of a gene (or genes) of interest on endosome-to-Golgi retrieval is more feasible using the low-throughput protocol. Specifically, the protocols detailed below monitor the trafficking of the cation-independent mannose-6-phosphate receptor (CIMPR) as prototypical cargo for the endosome-to-Golgi retrieval pathway in HeLa cells. If required, these protocols can be adapted to study the trafficking of other cargo proteins such as the cation-dependent mannose-6-phosphate receptor (CDMPR), sortilin or sortilin-related sorLA, or to study the pathway in different cell types.

Firstly, using standard fluorescent antibody staining and imaging, one can determine if the steady-state distribution and/or expression level of CIMPR is altered upon RNAi knockdown of a protein of interest. At steady state, most of the CIMPR is detected at the *trans*-Golgi network (TGN), but some is also present in endosomes and at the plasma membrane. RNAi-mediated knockdown of proteins important for its retrieval from endosomes will cause less CIMPR to be localized at the TGN and more CIMPR to be present in dispersed endosomal and lysosomal structures. Changes in CIMPR localization and intensity can be quantified using commercial software supplied with the fluorescence microscope or using freely available data analysis software such as ImageJ (http://rsbweb.nih.gov/ij/). Examples can be found in the literature where this method has been applied to study a small number of proteins of interest (e.g., Espinosa, Calero, Srivedi, & Pfeffer, 2009) but also for screening several RNAi-mediated knockdowns of related proteins in parallel, using automated imaging (Wassmer et al., 2007).

However, whilst a change in the steady-state distribution of CIMPR can be indicative of a defect in its endosome-to-Golgi trafficking, assays based on steady-state distribution can provide only indirect evidence of a defect in the endosome-to-Golgi pathway. We generally prefer to exploit the fact that a proportion of the CIMPR is transiently localized to the plasma membrane and its rapid internalization provides a means to directly assess the functioning of the plasma membrane to TGN pathway that also encompasses endosome-to-Golgi retrieval. Intervention by transient overexpression or RNAi of a gene or protein of interest then allows for the role of the gene or protein of interest in endosome-to-Golgi retrieval to be determined. The assay relies on the availability of an effective antibody that binds to an extracellular epitope of the receptor. Bound antibody–receptor complexes are then allowed to endocytose and traverse the endosomal system for a defined period of time, after which the cells are fixed, labeled and imaged, and the amount of antibody that has reached the TGN is quantified. Imaging will also reveal if the trafficking defect is in the initial endocytosis step (antibody remains on the plasma membrane) or in the endosome-to-Golgi trafficking step (more antibody is seen in endosomes and/or lysosomes). Furthermore, the antibody-uptake experiment can be performed for different times as well as at different temperatures to yield additional information on the kinetics of CIMPR retrieval under the given experimental conditions.

Anti-CIMPR antibodies that bind an extracellular epitope on the receptor are commercially available and can be used for an endosome-to-Golgi antibody-uptake assay (see, e.g., Robinson, Sahlender, & Foster, 2010). However, we routinely use HeLa cells stably expressing a CD8-CIMPR fusion protein consisting of the extracellular and transmembrane domain of the T-cell CD8α molecule and the cytoplasmic C-terminal tail sequence of bovine CIMPR. We have shown that the CD8-CIMPR fusion protein trafficking mimics the trafficking routes and kinetics of the endogenous protein (Seaman, 2004), yet it has the advantage that excellent antibodies against the CD8α extracellular domain can be purchased commercially or produced in large quantities at low cost using hybridoma cells. The use of CD8-reporter constructs is readily applicable for most type-I transmembrane proteins and has the potential advantage that the trafficking itinerary of several type-I membrane proteins can be investigated using the same assay(s). For example, CD8-reporters engineered to utilize the cytoplasmic domains of proteins such as sortilin, furin, LDLR, and sorLA have been employed in various image-based or biochemical assays (Fjorback et al., 2012; Motley, Bright, Seaman, & Robinson, 2003; Seaman, 2004). As the various CD8-reporters differ only in the cytoplasmic domain, the role and requirement for specific sorting motif can also be investigated and any variation in observed localization must be attributable to how the respective sorting motifs are recognized by the sorting machinery (e.g., retromer) responsible for endosomal protein sorting (see Fig. 9.1).

Below we describe the protocols for anti-CD8 antibody-uptake experiments in cells expressing CD8-CIMPR. In addition, the cells that we have used recently stably express a green fluorescent protein (GFP)-tagged TGN-localized protein, GOLPH3 (also known as Gmx33), the fluorescence of which we use to obtain a TGN mask in the image analysis. Alternatively, one can use an antibody against a Golgi-resident protein to define the organelle, taking care not to choose a protein that might itself depend on retrograde trafficking for its Golgi localization. A schematic of the antibody-uptake assay is depicted in Fig. 9.2.

2.1. Antibody-uptake assay on a small scale

This protocol uses standard laboratory equipment and reagents and is used to determine if a limited number of genes (typically 1–5) play a role in endosome-to-Golgi retrieval of the CIMPR.

Reagents required

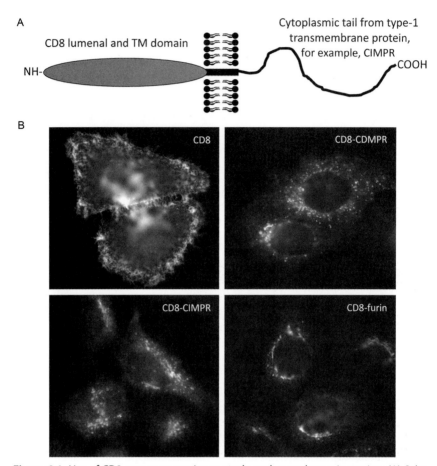

Figure 9.1 Use of CD8-reporter proteins to study endosomal protein sorting. (A) Schematic diagram depicting the topology of the CD8-CIMPR reporter. Similar reporters proteins can be constructed using the cytoplasmic tails of other type-I membrane proteins, for example, furin. (B) Immunofluorescence images of HeLa cells expressing either CD8 or CD8-CDMPR, CD8-CIMPR, and CD8-furin. Sorting motifs present in the respective cytoplasmic tail determine the steady-state localization of the CD8-reporter. (For color version of this figure, the reader is referred to the online version of this chapter.)

Ready-to-split cells stably expressing CD8-CIMPR and GFP-GOLPH3
siRNA and siRNA transfection reagents
Growth medium (DMEM with 10% fetal bovine serum and antibiotics)
Sterile phosphate-buffered saline (PBS)
Monoclonal anti-CD8 antibody, rabbit anti-GFP antibody
Paraformaldehyde (PFA) or methanol:acetone cell fixation solutions
Triton X-100

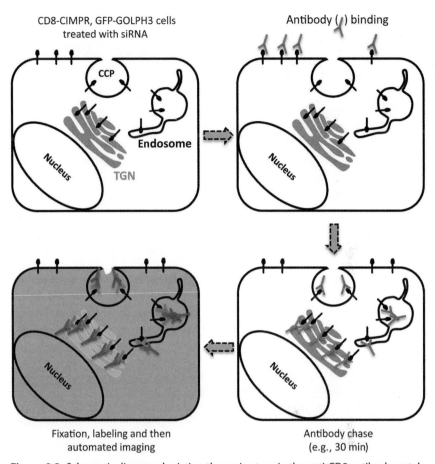

Figure 9.2 Schematic diagram depicting the main steps in the anti-CD8 antibody-uptake assay in cells stably expressing a CD8-CIMPR reporter protein and a GFP-tagged TGN-localized protein (GFP-GOLPH3, green). In the last step, internalized anti-CD8 antibody is labeled using a red fluorescent antibody, while the cell outline is defined by a blue whole cell stain. The GFP fluorescence is enhanced using a rabbit anti-GFP antibody and an Alexa Fluor 488 anti-rabbit secondary antibody. (For interpretation of the references to color in this figure legend, the reader is referred to the online version of this chapter.)

Immunofluorescence (IF) blocking buffer (3% bovine serum albumin, 0.02% sodium azide in PBS)

Fluorescently labeled secondary antibodies (we use Alexa Fluor–labeled antibodies from Life Technologies)

Mounting medium (e.g., Prolong Gold, Life Technologies)

Day 1: Seed cells in a suitable vessel, for example, in the wells of a 6-well plate.

Day 2: Knockdown protein(s) of interest using standard siRNA transfection protocol. We used Oligofectamine (Life Technologies) and On-Target-plus siRNA SMARTpools (Dharmacon) in Optimem (Life Technologies). For siRNA transfection of cells in a single well of a 6-well plate, we diluted 10 μL Oligofectamine in 20 μL Optimem and 10 μL of a 20 μ*M* siRNA solution in 160 μL Optimem. After 5 min we added the 30 μL Oligofectamine dilution to the 170 μL siRNA dilution to obtain 200 μL transfection mix. After 20 min at room temperature the transfection mix is diluted with 800 μL Optimem. The cells are washed once with PBS before adding the final 1 mL transfection solution. After 4 h incubation at 37 °C, we added 1 mL of growth medium containing twice the regular amount of growth serum (20% fetal bovine serum).

Day 3: Remove the transfection mix from the cells and replace with regular growth medium.

Day 4: Seed cells onto 22 mm square coverslips such that they will be 60–80% confluent on day 5.

Day 5: Antibody-uptake and staining:
  i. Wash cells once with ice-cold PBS.
  ii. Add 3 mL chilled medium to the cells and leave them at 4 °C (in a cold room or in a fridge) for 30 min to stop all trafficking.
  iii. Wash coverslip again with ice-cold PBS, blot-dry, and place on top of 100 μL medium containing 1 μg monoclonal anti-CD8 antibody for 30 min at 4 °C.
  iv. Wash coverslip with ice-cold PBS and transfer to a well of a 6-well plate containing 3 mL prewarmed medium. Incubate at 37 °C for a desired time, for example, 8, 16, and 24 min.
  v. Wash coverslips with room temperature PBS and fix using either 4% PFA or methanol/acetone for 10 min. If using PFA, permeabilize cells with 0.1% Triton X-100 in PBS for 10 min. Block nonspecific antibody binding by incubating for 30 min at room temperature with IF blocking buffer.
  vi. Label with anti-GFP rabbit antibody diluted 1:1000 in IF blocking buffer for 1 h at room temperature. Wash coverslips 3× with PBS.
  vii. Label with Alexa Fluor 488 anti-rabbit IgG and Alexa Fluor 555 anti-mouse IgG diluted 1:1000 for 1 h at room temperature. Wash coverslips 3× with PBS.
  viii. Mount coverslips.

Day 6: Imaging and analysis:

Take images of 25–75 cells per condition (wild-type cells vs. siRNA knockdown cells) using a standard epifluorescent microscope equipped with a two-channel filter setup for sequential collection of the fluorescence from Alexa Fluor 488 and Alexa Fluor 555. Using the software supplied with the image acquisition program, or freely available image analysis software such as ImageJ or Fiji (http://fiji.sc/), define a TGN mask using the Alexa Fluor 488 image and define the cell outline using the Alexa Fluor 555 image. It might be necessary to increase the contrast in the Alexa Fluor 555 image to clearly see the cell boundary. Calculate a background-corrected Alexa Fluor 555 anti-CD8 image and quantify the anti-CD8 retrieval to the TGN by the ratio of the total background-corrected Alexa Fluor 555 intensity in the TGN mask to the total background-corrected Alexa Fluor 555 intensity in the whole cell mask. Typically, in control cells, after a 30 min chase at 37 °C, up to 70% of the anti-CD8 antibody intensity is measured in the TGN mask.

Remark. As an alternative to using cells stably expressing both the reporter cargo and the TGN marker, one can label the TGN postfixation, for example, using antibodies against TGN46.

2.2. Antibody-uptake assay on a high-throughput scale

When assaying several tens of genes, for example, in a mini-library screen of genes of interest, or even thousands of genes in the case of a whole-genome screen, the above protocol becomes impractical, and automation is highly desirable at the sample preparation, assay and imaging steps, and for quantitative image analysis. We use a robotic liquid handling system (BiomekNX, Beckman Coulter) in a custom-made enclosure with airflow technology and a UV lamp (BigNeat Containment Technology, Hampshire) to aid with the siRNA transfections, a bench-top semi-automated liquid dispenser system (Matrix Wellmate, Thermo Scientific) to aid in the antibody-uptake steps as well as the fixing and staining steps, and an automated wide-field fluorescence microscope (Cellomics Arrayscan V^{TI}, Thermo Scientific) to aid in image acquisition and analysis.

Compared to the small scale protocol, a few differences are of particular note: Firstly, library or mini-library siRNAs are generally arrayed onto 96-well plates. Therefore, cells are reverse-transfected, that is, cells are seeded on top of siRNA transfection mixes into flat-bottomed 96-well cell culture plates suitable for imaging. The arrangement of the siRNAs on the plate will determine how and where control wells can be added. Ideally, each assay plate should have several positive and negative control wells. Figure 9.3A illustrates a typical layout with 80 sample siRNAs, 8 negative control wells, and 8 positive control wells. The positive and negative control wells are alternated in the first and second columns of the plate to cancel out possible plate edge effects.

To minimize the number of plate manipulations, the assay plates are left undisturbed for 72 h before the anti-CD8 antibody uptake is carried out. To produce the required large quantities of anti-CD8 antibody, we use mouse hybridoma cells (American Type Culture Collection). The hybridoma cells are grown in Excell medium (Sigma-Aldrich) supplemented with 10 mM L-glutamine, and the cell culture supernatant is collected and stored at $-20\,^\circ$C in the presence of sodium azide until needed. Prior to use in an antibody-uptake assay, the hybridoma culture supernatant is dialyzed against PBS to remove the sodium azide.

Further differences with the small scale protocol are a reduction in the number of washes in the antibody-uptake and staining steps and the addition of a whole cell stain to allow automatic marking of the cell outline by the image analysis software. Finally, it is worth noting that in the case of a whole-genome screen, it is important to have adequate stocks of frozen cells to avoid effects resulting from cells being cultured over prolonged periods of

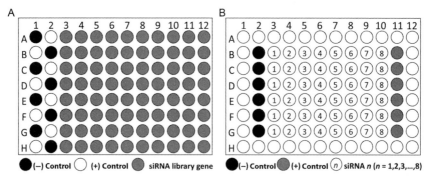

Figure 9.3 96-well plate layouts for high- or medium-throughput endosome-to-Golgi retrieval assays. (A) Typical plate layout for library siRNA screening. Black-filled wells have negative control (e.g., transfection reagent only and no siRNA). White-filled wells have positive control siRNA (e.g., SNX1 siRNA). Gray-filled wells contain library siRNA. (B) Typical plate layout for anti-CD8 uptake assay in five to eight protein knockdown conditions (medium scale). Only the central 60 wells of the plate are used, and six technical replicates are present for every knockdown. White-filled wells are not used.

time. Indeed, we observe that cells from older passages show more toxicity toward siRNA transfection compared to lower passage cells. To avoid these undesirable effects of cytotoxicity, we do not use cells beyond 10 passages.

Here we detail our procedure for assaying 4×80 siRNAs in duplicate.

Day 1: Cell seeding and siRNA transfection

Sterilize the liquid handling robot enclosure using UV light and ethanol for wiping of surfaces. Thaw four siRNA library plates containing 10 μL 1 μM siRNA in sample wells. Thaw control siRNAs, either prealiquoted in another 96-well plate or in separate vial(s). Spin siRNA plates for 5 min at $1000 \times g$. Bring the following reagents to room temperature: sterile PBS, trypsin solution, Optimem transfection reagent (Life Technologies), and DMEM growth medium (10% FBS and penicillin/streptomycin/L-glutamine).

In cell culture hood: Prepare a 1:10 Oligofectamine:Optimem dilution, diluting 550 μL Oligofectamine into 4.95 mL Optimem. Pipette up and down to mix. Leave for 8 min at room temperature, then aliquot 50 μL/well in all wells of a V-bottomed 96-well plate using a multichannel pipettor and a pipetting reservoir. Also transfer 10 μL 1 μM control siRNA from the control siRNA plate to the appropriate control wells on the four siRNA library plates. For the negative control wells, use 10 μL/well siRNA buffer, sterile water, or Optimem.

In robot enclosure: Prepare deck for preparation of transfection mixes. The transfection mix protocol will first dilute the siRNA on the library plates with 36 μL/well Optimem from a reservoir. Next it will add 9 μL diluted Oligofectamine to each siRNA and mix the two reagents together. The transfection mixes are then left at room temperature for 20 min to allow the siRNA:Oligofectamine complexes to form.

Meanwhile, in a cell culture hood, wash a confluent 75 cm^2 flask of cells with sterile PBS and trypsinize. Resuspend the cells in 10 mL complete medium. Use an automated cell counter to count the cell density. Dilute the cells as to obtain 100 mL of a cell suspension with a density of 80,000 cells/mL.

In robot enclosure: Prepare assay plates as follows: Remove the 96-well plate that contained the diluted Oligofectamine and place 8 flat-bottomed 96-well assay plates on the deck. These should be suitable for cell culture (sterile, surface-treated) and for automated imaging. Aliquot the siRNA:Oligofectamine complexes in two assay wells each (20 μL/well). Next, remove the Optimem reservoir and replace with a clean reservoir. Put the cell suspension into the reservoir. The last step in the robot protocol will add 100 μL of the cell suspension to each assay plate well.

Carefully place lids on all the assay plates and remove them from the robotic enclosure. Vortex very briefly (about 5 s) and place the plates in a humidified incubator at 37 °C.

Day 4: Antibody-uptake and staining

All procedures are done on a standard lab bench using a semi-automated liquid dispenser (Wellmate). The dispenser is primed and programmed to deliver volumes of 40, 50, 100, or 150 μL, depending on the step in the protocol, 40 μL being the minimum volume required to cover a 96-well plate well. To empty the plates of liquid, we firmly shake them out over a suitable container, one by one, and then turn them over on absorbent paper for a few seconds.

i. Prepare 1:5 dilution of dialyzed hybridoma cell culture supernatant in growth medium and bring to room temperature. Warm 100 mL growth medium to 37 °C.

ii. Wash the cells once with 100 μL/well PBS; then add 40 μL/well anti-CD8 antibody and leave to bind for 15 min at room temperature.

iii. Remove antibody solution and add 100 μL/well prewarmed growth medium. Immediately place the cells in a 37 °C incubator for 30 min.

iv. Wash the cells 1× with 50 μL/well PBS; next, fix the cells using 50 μL/well of a freshly prepared 4% PFA in PBS solution. This step and all subsequent steps are carried out at room temperature.

v. After 10 min remove PFA solution and add 50 μL/well 0.1% Triton X-100 in PBS to permeabilize the cells.

vi. After 10 min remove Triton X-100 solution and add 50 μL/well IF blocking buffer.

vii. After 30 min remove blocking buffer and add 40 μL/well primary antibodies diluted in IF blocking buffer. We use a rabbit anti-GFP antibody to enhance the fluorescent signal-to-noise in the compartment channel.

viii. After 1 h wash cells 1× with 100 μL/well PBS; then add 40 μL/well secondary antibodies diluted 1:1000 (for Alexa Fluor 555 anti-mouse IgG) or 1:2000 (Alexa Fluor 488 anti-rabbit IgG) in IF blocking buffer.

ix. After 1 h wash cells 1× with 100 μL/well PBS; then add 40 μL/well Whole Cell Stain Blue (Thermo Fisher) diluted 1:500 in PBS.

x. After 30 min wash cells 3× with 100 μL/well PBS. Add a final 150 μL/well PBS to each well.

xi. Image plates immediately or store them at 4 °C for later imaging.

2.2.1 Automated imaging

Several high-throughput fluorescence microscope systems based on either wide-field or confocal (spinning disk or laser-scanning) imaging are commercially available (Pepperkok & Ellenberg, 2006). We use a Cellomics Arrayscan V^{TI} high-content screening microscope both for imaging and analysis. Our Arrayscan is equipped with a CataLyst Express robotic plate loader that allows for automatic sequential imaging of multiple assay plates. To avoid evaporation, plates are kept lidded at all times, and care is taken to protect them from excessive light. Using the *Colocalization* bio-application, a plate protocol is setup for the acquisition of three images of every field using a 40 × 0.5 N.A. objective: the channel 1 image is of the whole cell stain, the channel 2 image is of GFP-GOLPH3, and the channel 3 image is of Alexa Fluor 555-labeled anti-CD8. Exposure times are adjusted on a regular basis to account for variation in the lamp intensity. Auto-focusing is performed every other field using the blue channel. For the channel 2 and channel 3 images, a *z*-offset is applied versus the channel 1 image to bring the TGN and endomembrane system into focus. The imaging protocol is time-efficient as the three images are acquired using the same dichroic mirror and multiband pass emission filter (XF93 filter set, Omega Optical) and exposure times are generally less than 0.5 s. Cells are identified as objects in the channel 1 image after smoothing of the pixel intensities, background correction, segmentation, and application of a fixed intensity threshold. Not all identified objects are selected for analysis, for example, cells undergoing mitosis are excluded based on their shape (very round) and intensity (very bright). Imaging in a well is continued until 250 cells are selected for analysis unless the well is sparse, for example, due to toxicity of the siRNA, and imaging is stopped when less than two objects are identified in each of six consecutive frames.

To determine the degree of overlap of the anti-CD8 antibody staining and the GFP-GOLPH3-defined TGN, we use the *Colocalization* bio-application to define two regions of interest (ROIs), ROI_A, defined by the GFP-GOLPH3 image (TGN mask), and the ROI_B, defined by the whole cell stain image (cell mask); and one target intensity channel, the channel 3 (anti-CD8) image. The GFP-GOLPH3 image is smoothened and a histogram-derived (isodata) threshold is applied to obtain ROI_A. The anti-CD8 image is background-corrected but not smoothened and all intensities above 0 are included in the analysis. We define an *anti-CD8 TGN retrieval ratio* as the ratio of the total intensity of the anti-CD8 image pixels in the TGN mask to the total intensity of the anti-CD8 image pixels in the whole cell mask. This anti-CD8 TGN retrieval ratio can be intuitively understood as the proportion of the total cell-associated anti-CD8 antibody that colocalizes with the TGN. The ratio is calculated by the software for every selected object as the ratio of the Target I total intensity in ROI_A to the Target I total intensity in ROI_B. Example images and the ROI overlays are shown in Fig. 9.4. The average anti-CD8 TGN retrieval ratios measured in the well with control cells was 0.74, while for the VPS26 KD cells and the SNX1 KD cells the average anti-CD8 TGN retrieval ratios were 0.47 and 0.48, respectively.

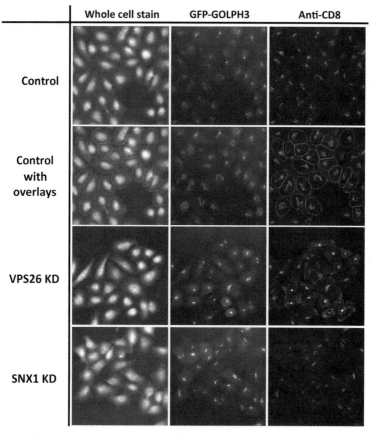

Figure 9.4 Representative images and software overlays of anti-CD8 antibody-uptake experiments performed in a 96-well plate and imaged using a Cellomics Arrayscan high-content screening microscope. The images in the second row are the same as the ones in the first row but they have been overlaid with the cell mask (blue line) and TGN mask (green). (For interpretation of the references to color in this figure legend, the reader is referred to the online version of this chapter.)

The images acquired can be analyzed and reanalyzed using a variety of bio-application assay protocols, and many additional parameters are measured by the software, for example, cell and TGN size and shape factors. If a second target (Target II) is defined, for example, the image of an additional marker in the fourth channel available with the XF93 filter set, the *Colocalization* bio-application also measures the Pearson's and Mander's correlation coefficients.

2.3. Antibody-uptake assay on a medium scale

To simultaneously assess and compare the effects of a number of different proteins on the endosomal trafficking of CD8-CIMPR, we find it useful to perform several knockdowns in parallel in a 96-well plate and to use automated imaging for quantification of the effects. The different siRNAs can be arrayed on the plate manually, while antibody-uptake and staining can be performed either using a manual multichannel pipettor or the semiautomatic Wellmate. Multiple technical replicates can be incorporated onto a single plate. An example layout is shown in Fig. 9.3B. The use of automated imaging has the advantage that the anti-CD8 antibody-uptake can be quantified in an unbiased way in a large number of cells. For example, in published work we have assessed the role of retromer and several retromer-associated proteins for their respective roles in endosome-to-Golgi retrieval (Harbour et al., 2010), analyzing both technical and biological replicates.

3. DETAILED CHARACTERIZATION OF ENDOSOMES

Endosomes are often classified by their distribution within the cell, that is, peripheral versus perinuclear. Automated microscopy is particularly suited to characterize endosomes although achieving a good signal-to-noise ratio requires an effective antibody that can label endosomes to high intensity. For example, we have successfully employed antibodies against endosomally localized proteins, either generated in-house (e.g., anti-VPS26) or commercially available (e.g., anti-EEA1, anti-SNX1) to investigate the endosomal localization of retromer proteins. With the Cellomics Arrayscan *Spotdetector* bio-application, one can count the number of endosomes per cell as well as quantify their (average) brightness (intensity) and size. This allows one to investigate morphological changes, for example, enlargement of endosomes, upon a particular cellular perturbation. Other bio-applications allow spot detection in addition to other measurements, for example, the *Colocalization* bio-application can determine the overlap between two endosomally localized labeled proteins, or the *Compartment Analysis* bio-application can, in the presence of a central intracellular marker stain, determine endosome dispersion.

We have used automated imaging and the *Spotdetector* bio-application to characterize endosomally localized VPS26 in wild-type cells and in cells in which retromer-associated proteins were knocked down using siRNA. Figure 9.5 shows some representative images of control cells as well as cells in which the retromer core component VPS26 or the sorting nexin SNX1 have been knocked down using siRNA. The whole cell stain image is used to determine the cell boundaries (overlaid in green). The VPS26-positive spots identified using the *Spotdetector* bio-application are indicated in red. The images illustrate the efficiency of the VPS26 KD and show that

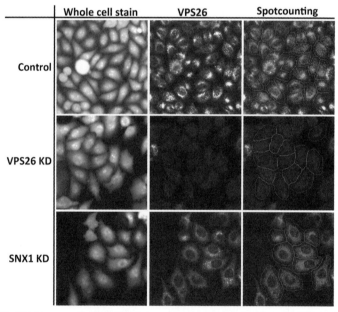

Figure 9.5 Wild-type and siRNA KD HeLa cells were stained with anti-VPS26 rabbit antibodies and Alexa 555 anti-rabbit antibodies as well as a blue whole cell stain. Two hundred and fifty cells were imaged on a Cellomics Arrayscan. Overlays were generated using the *Spotdetector* bio-application, where the whole cell stain defines the cell boundaries (green) and a fixed intensity threshold defines the VPS26-positive endosomes (red). (See color plate.)

SNX1 KD does not cause VPS26 to dissociate from endosomes. In fact, quantitation of the number of spots per cell area was reported in Vardarajan et al. (2012) and indicated no change in the number of VPS26-positive endosomes per cell area in SNX1 KD cells compared to wild-type cells.

4. ENDOSOME RECRUITMENT AND/OR ASSOCIATION

To determine changes in endosome recruitment or association of sorting machinery such as retromer, or to identify novel retromer-associated proteins, we use two complementary methods: (1) a simple pellet versus supernatant preparation followed by Western blotting and (2) native immunoprecipitation (IP) from HeLa cells expressing GFP-tagged endosomal proteins, in particular GFP-tagged subunits of the cargo-recognition retromer subcomplex.

4.1. Pellet versus supernatant

Proteins such as the retromer complex cycle rapidly between membrane-associated and soluble cytoplasmic pools. In the case of the cargo-selective retromer complex, this process is controlled through the action of the small GTPase Rab7a and the sorting nexin protein, Snx3 (Vardarajan et al., 2012). To augment the data obtained through image-based analyses, we have employed a simple biochemical assay to determine how much of a particular protein is membrane associated.

   i. Cells seeded onto 90 mm tissue culture dishes are grown to a confluency of $\sim$80–90%.
   ii. Remove the tissue culture media and wash the cells in 5 mL of chilled PBS.
   iii. Place the cells on ice at an angle to allow the PBS to drain. Remove the remaining PBS by pipetting or by tapping the inverted dish onto blotting paper or similar.
   iv. Using forceps to hold the dish by an edge, rapidly freeze the cells in the dish by "floating" the dish on liquid nitrogen—this will take only $\sim$10–15 s.
   v. Remove the dish from the liquid nitrogen and return to ice. Repeat freezing process for other dishes of cells but limit the number of dishes to no more than six to avoid prolonged processing times in subsequent steps.

vi. Remove dish from ice and rapidly thaw cells by holding bottom of dish against the palm of a hand.
vii. Add 0.4 mL of lysis buffer (0.1 M MES–NaOH, pH 6.5, 0.25 M sucrose, 1 mM Mg acetate, 200 μM Na vanadate).
viii. Scrape cells from dish using a cut rubber bung or similar cell scraper.
ix. Remove cells in lysis buffer using pipette and transfer to 1.5 mL tube on ice.
x. Repeat steps (vi)–(ix) for the other dishes. Rapidly processing the cells is important and hence trying to process more than six dishes at a time is not advised. We usually process four dishes at a time.
xi. Centrifuge the cells for 5 min at $10,000 \times g$ in a refrigerated centrifuge to separate the pelletable membrane-associated fraction from the soluble cytoplasmic fraction.
xii. Transfer supernatant (cytoplasmic fraction) to a fresh 1.5 mL tube on ice.
xiii. The pellet is then solubilized in lysis buffer to which Triton X100 and SDS have been added to concentrations of 1% and 0.1%, respectively. Resuspend pellet by pipetting up and down and allow pellet to solubilize on ice for 5 min.
xiv. Centrifuge the solubilized pellet fraction again at $10,000 \times g$ for 5 min. Transfer the supernatant (a small pellet of insoluble material will remain) to a fresh tube. This fraction contains transmembrane and membrane-associated proteins and is referred to as the pellet fraction.
xv. The pellet (membrane) and the supernatant (cytoplasmic) fractions generated in steps (xiv) and (xii), respectively, can now be prepared for analysis by SDS-PAGE and Western blotting.

4.2. Native IP of endosomal proteins

The identification of proteins that associate—often transiently—with endosomal sorting components such as the retromer complex can yield important insights into the mechanisms that govern endosomal protein sorting. In our studies of the retromer complex, we have successfully employed a simple native IP protocol to isolate retromer and identify associated proteins (see Harbour et al., 2010; Seaman, Harbour, Tattersall, Read, & Bright, 2009). We have mostly used GFP-tagged constructs stably expressed in HeLa cells and immunoprecipitated with anti-GFP for these experiments but transiently transfected cells can also be used and the tag employed need not be GFP. Important to success, however, is the use of a highly specific primary antibody that recognizes the respective tag. In this respect there are a great many commercially available antibodies against GFP or other tags. Set out below is a protocol that we have used in native IP of the cargo-selective retromer complex that comprises a trimer of VPS35, VPS29, and VPS26.

i. Stably transfected cells (or transiently transfected) are grown to near confluency in 140 mm tissue culture dishes.
ii. The media is removed and the cells are washed with 10 mL of chilled PBS. For poorly adherent cells, PBS containing calcium and magnesium should be used.
iii. Remove the PBS and leave the cells to drain on ice at an angle for 5 min.
iv. Remove any remaining PBS by pipette.
v. Add 1 mL of lysis buffer (20 mM HEPES–KOH, pH 7.2, 50 mM K acetate, 2 mM EDTA, 200 mM sorbitol, 0.1% Triton X-100 with protease inhibitors) and scrape up cells using a cut rubber bung.
vi. Transfer cells in lysis buffer (lysate) to 1.5 mL tube on ice.
vii. Centrifuge lysate for 5 min at 10,000 × g in a chilled centrifuge.
viii. Remove 950 μL of supernatant to avoid disturbing pellet.
ix. Transfer to fresh 1.5 mL tube containing 50 μL of protein-A sepharose (25% slurry).
x. Incubate lysate and protein-A sepharose on a rotating wheel for 30 min at 4 °C. This is a preclearing incubation to remove proteins that bind nonspecifically to protein-A sepharose.
xi. Repeat centrifugation described in step vii.
xii. Remove 900 μL of cleared lysate and transfer to fresh 1.5 mL tube on ice.
xiii. Add anti-GFP (or similar appropriate antibody); 1–5 μg of antibody should be sufficient.
xiv. Incubate on a rotating wheel at 4 °C for 90 min.
xv. Add 50 μL of protein-A sepharose (25% slurry).
xvi. Incubate on a rotating wheel at 4 °C for 60 min.
xvii. Collect the sepharose beads by centrifugation for 1 min at 10,000 × g in a chilled centrifuge. Remove supernatant by aspiration.
xviii. Add 1 mL of lysis buffer to wash the sepharose beads.
xix. Repeat steps xvii and xviii three more times.
xx. After removal of the supernatant, the sepharose beads can then be desiccated in a SpeedVac and prepared for analysis by SDS-PAGE and Western blotting.

The above protocol is a guide to successful native IPs but there are clearly many variations possible with the use of different buffers (e.g., MES, or PIPES) with increased or reduced salts and/or detergent to change the stringency of the conditions. If a monoclonal antibody is used then protein-G sepharose should be substituted for protein-A sepharose.

In order to identify retromer-associated proteins, we have usually combined native immunoprecipitations from 3 to 5 140-mm dishes of cells and used micro-spin columns and low pH elution to liberate the bound proteins from the protein-A sepharose beads.

Below are the steps that would be used if sufficient material were being prepared for analysis by SDS-PAGE and mass spectrometry.

The protocol is the same as above up to and including step xix:

- **xx.** Combine the beads from several IPs (e.g., 4) by adding 250 µL of lysis buffer to one tube and then transferring the buffer and beads to the next tube until all the beads are combined. Pipette beads and buffer into a micro-spin column placed in a 1.5 mL tube.
- **xxi.** Allow buffer to drain through column collecting the beads. If necessary centrifuge for 30 s at 5000 rpm in chilled centrifuge.
- **xxii.** Transfer spin column containing beads to fresh 1.5 mL tube.
- **xxiii.** Add 150 µL of low pH elution buffer (200 mM Glycine–HCl, pH 2.3). Incubate for 5 min.
- **xxiv.** Collect eluate as in step xxi.
- **xxv.** Repeat steps xxiii and xxiv.
- **xxvi.** There should now be 300 µL of eluate. Precipitate by adding 1.2 mL of ice-cold acetone.
- **xxvii.** Incubate for 60 min on ice.
- **xxviii.** Collect precipitated protein by centrifugation for 15 min at $16,000 \times g$ in chilled centrifuge.
- **xxix.** Carefully aspirate supernatant, wash pellet with 1 mL of fresh ice-cold acetone.
- **xxx.** Repeat so that pellet is washed twice with cold acetone.
- **xxxi.** Dry acetone pellet in SpeedVac.
- **xxxii.** The pellet can then be dissolved in SDS-PAGE sample buffer, analyzed by gel electrophoresis, and bands excised for protein identification by mass spectrometry. An example of a successful native IP of GFP-tagged retromer components is shown in Fig. 9.6A. Alternatively, the sample can be analyzed by LC–MS/MS where all the proteins are subjected to tryptic digest, liquid chromatography, and mass spectrometry. The sensitivity of this method can identify a

great many proteins that may be directly or indirectly associated with the tagged protein of interest. A suitable control for these experiments is the use of lysates from cells that are not expressing the tagged protein of interest.

Variations on the protocols above have been used to immunoprecipitate CD8-reporter proteins and demonstrate the association of the cargo-selective retromer complex with a CD8-CIMPR reporter protein (see Seaman, 2007). In those experiments, a lysis buffer comprising 0.1 M MES–NaOH, pH 6.5, 1 mM Mg acetate, 200 µM Na vanadate, and 1%

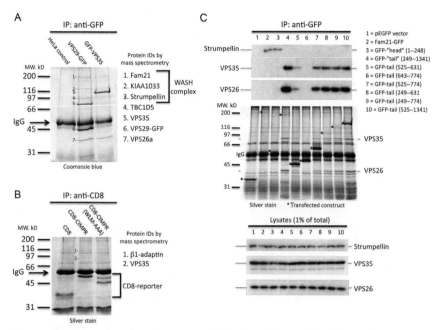

Figure 9.6 Use of native immunoprecipitation (IP) to identify novel retromer-interacting proteins. (A) Cells expressing GFP-tagged VPS29 or VPS35 were lysed and the GFP-tagged protein recovered by IP. Untransfected HeLa cells act as a control. The IPs were performed as in Section 4.2 and were subjected to SDS-PAGE. Bands 1–7 were excised and analyzed by mass spectrometry. Several retromer-associated proteins, including components of the WASH complex, are shown. (B) Native IP of CD8-reporter proteins. CD8-CIMPR immunoprecipitates with the clathrin adaptor AP-1 (β1-adaptin) and retromer (VPS35). CD8 and a mutated version of the CD8-CIMPR reporter serve as controls. Note that signal levels are low compared to the native IP shown in (A). (C) Native IP of GFP-tagged fragments of the WASH complex component, Fam21, transiently expressed in HeLa cells. Retromer proteins coIP with several of the Fam21 fragments. The expression and recovery of the Fam21 fragments was sufficient to allow analysis by both Western blotting and SDS-PAGE.

digitonin was used and a monoclonal anti-CD8 was employed to IP the CD8-reporter and associated proteins. In our experience, the association of sorting machinery such as retromer with "cargo" proteins, for example, the CD8-CIMPR reporter protein is much more transient and harder to detect than the interactions between retromer and other associated proteins such as TBC1D5 or the WASH complex and will usually require sensitive methods such as Western blotting to confirm the association. The results of a native IP of CD8-reporter proteins is shown in Fig. 9.6B. Note the signal levels are considerably lower than those observed for the native IP of the GFP-tagged retromer proteins.

4.3. Cost-efficient large-scale transient transfection for native immunoprecipitations

Whilst it is usually preferable to employ cells stably expressing the tagged protein of interest, this is not always possible and therefore an efficient method of transfecting cells for native IP experiments is required. The use of commercially available transfection reagents can be prohibitively expensive when contemplating conducting a number of native IP experiments to generate sufficient material for analysis by mass spectrometry. Set out below is a suitable protocol for transfecting HeLa cells in 140 mm dishes.

  i. Ahead of time prepare a 10 mM solution of PEI (polyethylenimine, Polysciences Inc.) in water. The PEI dissolves slowly in water at 45–50 °C. Add HCl gradually as the PEI dissolves so that pH is 7.0 when PEI is completely dissolved. Filter sterilize and aliquot into 1.5 mL tubes. Store at −80 °C for use in transient transfections.
 ii. Seed HeLa cells into 140 mm tissue culture dishes at a confluency ~30–40%.
iii. The next day transfect using the following method:
 iv. For each dish prepare 1.5 mL of Optimem with 60 μL of the 10 mM PEI solution. Incubate for 5 min. At the same time prepare another tube containing 1.5 mL of Optimem and add 24 μg of the DNA you wish to transfect into HeLa cells.
  v. Add the Optimem containing PEI to the Optimem containing the DNA and incubate for 20 min at room temperature.
 vi. Remove dish of cells from incubator. Replace media with 14 mL of prewarmed media.
vii. Add the 3 mL of Optimem containing the PEI and DNA dropwise to the cells. Swirl to mix and return dish of cells to incubator.

viii. Allow the cells to grow and divide for 48 h, then perform the native IP as described above.

This method has worked well and proven to be very cost-effective in native IPs using GFP-tagged fragments of the Fam21 C-terminal region to determine which part binds to the cargo-selective retromer complex. It may not be necessary to use 24 μg of DNA for each transfection and some preliminary experiments to titrate the amount of DNA to use for each transfection is advisable. Shown in Fig. 9.6C is the results of a native IP of GFP-tagged constructs comprising fragments of the Fam21 protein. The Fam21 protein is part of the WASH complex and associates directly with the retromer complex via VPS35. The GFP-tagged constructs were expressed transiently in HeLa cells using the protocol above. Similar data has recently been reported using RFP-tagged Fam21 fragments in Helfer et al. (2013).

5. SUMMARY

The endosomal system is very complex and highly dynamic. Therefore, to characterize the itinerary and understand the fate of a protein of interest traveling through it, a multitude of methods are often necessary. Here we have shown how important insights into the mechanisms of endosomal protein sorting can be obtained through a combination of cell biological imaging-based methods and biochemical methods. These are by no means the only methods that can be used but they provide a good starting point and should be accessible to most researchers.

ACKNOWLEDGMENTS

This work was supported by a Medical Research Council Senior Research Fellowship awarded to M. N. J. S. (G0701444). We thank Dr. Lena Wartosch for the PEI-mediated transient transfection protocol.

REFERENCES

Attar, N., & Cullen, P. J. (2010). The retromer complex. *Advances in Enzyme Regulation, 50*, 216–236.

Bujny, M. V., Popoff, V., Johannes, L., & Cullen, P. J. (2007). The retromer component sorting nexin-1 is required for efficient retrograde transport of Shiga toxin from early endosome to the trans Golgi network. *Journal of Cell Science, 120*, 2010–2021.

Espinosa, E. J., Calero, M., Srivedi, K., & Pfeffer, S. R. (2009). RhoBTB3: A Rho GTPase-family ATPase required for endosome to Golgi transport. *Cell, 137*, 938–948.

Fjorback, A. W., Seaman, M., Gustafsen, C., Mehmedbasic, A., Gokool, S., Wu, C., et al. (2012). Retromer binds the FANSHY sorting motif in SorLA to regulate amyloid precursor protein sorting and processing. *Journal of Neuroscience, 32*, 1467–1480.

Harbour, M. E., Breusegem, S. Y. A., Antrobus, R., Freeman, C., Reid, E., & Seaman, M. N. J. (2010). The cargo-selective retromer complex is a recruiting hub for protein complexes that regulate endosomal tubule dynamics. *Journal of Cell Science*, *123*, 3703–3717.

Helfer, E., Harbour, M. E., Henriot, V., Lakisic, G., Sousa-Blin, C., Volceanov, L., et al. (2013). Endosomal recruitment of the WASH complex: Active sequences and mutations impairing interaction with the retromer. *Biology of the Cell*, *105*, 191–207.

Motley, A., Bright, N. A., Seaman, M. N., & Robinson, M. S. (2003). Clathrin-mediated endocytosis in AP-2-depleted cells. *The Journal of Cell Biology*, *162*, 909–918.

Pepperkok, R., & Ellenberg, J. (2006). High-throughput fluorescence microscopy for systems biology. *Nature Reviews Molecular Cell Biology*, *7*, 690–696.

Robinson, M. S., Sahlender, D. A., & Foster, S. A. (2010). Rapid inactivation of proteins by rapamycin-induced rerouting to mitochondria. *Developmental Cell*, *18*, 324–331.

Seaman, M. N. J., McCaffrey, J. M., & Emr, S. D. (1998). A Membrane Coat Complex Essential for Endosome to Golgi Retrograde Transport in Yeast. *The Journal of Cell Biology*, *142*(3), 665–681.

Seaman, M. N. J. (2004). Cargo-selective endosomal sorting for retrieval to the Golgi requires retromer. *The Journal of Cell Biology*, *165*, 111–122.

Seaman, M. N. (2007). Identification of a novel conserved sorting motif required for retromer-mediated endosome-to-TGN retrieval. *Journal of Cell Science*, *120*, 2378–2389.

Seaman, M. N. J. (2008). Endosome protein sorting: Motifs and machinery. *Cellular and Molecular Life Sciences*, *65*, 2842–2858.

Seaman, M. N. J. (2012). The retromer complex—Endosomal protein recycling and beyond. *Journal of Cell Science*, *125*, 4693–4702.

Seaman, M. N., Harbour, M. E., Tattersall, D., Read, E., & Bright, N. (2009). Membrane recruitment of the cargo-selective retromer subcomplex is catalysed by the small GTPase Rab7 and inhibited by the Rab-GAP TBC1D5. *Journal of Cell Science*, *122*, 2371–2382.

Vardarajan, B. N., Breusegem, S. Y., Harbour, M. E., St. George-Hyslop, P., Seaman, M. N. J., & Farrer, L. A. (2012). Identification of Alzheimer disease-associated variants in genes that regulate retromer function. *Neurobiology of Aging*, *33*, 2231.e15–2231.e30.

Wassmer, T., Attar, N., Bujny, M. V., Oakley, J., Traer, C. J., & Cullen, P. J. (2007). A loss-of-function screen reveals SNX5 and SNX6 as potential components of the mammalian retromer. *Journal of Cell Science*, *120*, 45–54.

SECTION II

Transport and Transfer

CHAPTER TEN

Cytokines, Polarity Proteins, and Endosomal Protein Trafficking and Signaling—The Sertoli Cell Blood–Testis Barrier System *In Vitro* as a Study Model

Xiang Xiao[*,†], Elissa W.P. Wong[*], Pearl P.Y. Lie[*], Dolores D. Mruk[*], Chris K.C. Wong[‡], C. Yan Cheng[*,1]

[*]The Mary M. Wohlford Laboratory for Male Contraceptive Research, Center for Biomedical Research, Population Council, New York, USA
[†]Department of Reproductive Physiology, Zhejiang Academy of Medical Sciences, Hangzhou, Zhejiang, China
[‡]Department of Biology, Hong Kong Baptist University, Hong Kong, China
[1]Corresponding author: e-mail address: Y-Cheng@popcbr.rockefeller.edu

Contents

1. Introduction	182
2. Endocytosis Assay	184
3. Materials	187
4. Buffers	187
5. Methods	188
6. Cell Staining to Assess Endocytosis	190
7. Results	190
8. Summary	191
Acknowledgments	192
References	192

Abstract

Endosomal signaling is emerging as one of the most important cellular events that regulate signaling function in mammalian cells or an epithelium in response to changes in environment such as the presence of stimuli mediated by cytokines, toxicants, heat, ions during growth and development, and other cellular processes such as cytokinesis and spermatogenesis. Recent studies have shown that protein endocytosis—the initial step of endosomal signaling—involves the participation of polarity proteins, such as partitioning defective protein 6 (Par6), Cdc42 and 14-3-3 (also known as Par5), which in turn is regulated by cytokines (e.g., TGF-β2, TGF-β3) and testosterone at the Sertoli cell blood–testis barrier (BTB) in the mammalian testis. In this short method paper, we provide a detailed protocol of assessing protein endocytosis, the initial and also the most critical step of endosomal signaling at the Sertoli cell BTB. This biochemical endocytosis assay summarizes our experience for the last decade, which should likely be performed in conjunction with the dual-labeled immunofluorescence analysis to assess protein endocytosis. While we are using a Sertoli cell *in vitro* system that mimics the BTB *in vivo*, this approach should be applicable to virtually all mammalian cells.

1. INTRODUCTION

Endosomal signaling is a rapidly evolving field. It refers to the transmission of incoming signals outside a mammalian cell via endosome-mediated trafficking, so that outside signals can be appropriately responded physiologically which are mediated by changes in specific microdomain(s) (e.g., basolateral or apical region of an epithelial cell) through re-arrangement of proteins through transcytosis, recycling or degradation within the cell, such that cells in an epithelium can make appropriate responses to incoming signals (Gonnord, Blouin, & Lamaze, 2012; Le Roy & Wrana, 2005a; Miaczynska & Bar-Sagi, 2010; Palfy, Remenyi, & Korcsmaros, 2012) during growth, development, and/or in response to stimuli from the environment, such as the presence of growth factors and toxicants. Endosomal signaling is known to affect multiple cellular events including cell migration, metabolism, survival, cell division, and proliferation (Le Roy & Wrana, 2005b; Leto & Saltiel, 2012; Neto, Collins, & Gould, 2011; Palfy et al., 2012; Polo & Di Fiorce, 2006; Schiefermeier, Teis, & Huber, 2011), which may also involve protein ubiquitination (Haglund & Dikic, 2012; Marchese & Trejo, 2013). Once inside the cell, endosomes can also serve as signaling platforms to mediate crosstalk between signaling pathways, so that appropriate responses can be made in response to changes in environment, during growth and development, or pathogenesis (Palfy et al., 2012). Furthermore, endosomes are also being used to deliver other intracellular proteins, such as Src family of non-receptor protein tyrosine kinases (Sandilands & Frame, 2008), Rho GTPase (Falkenberg & Loew, 2013) to specific microdomain of the plasma membrane to regulate cellular functions in response to changes in environment (Sandilands & Frame, 2008). The initial response of endosomal signaling is mediated by endosome-mediated internalization of cell surface proteins via endocytosis. While these changes can be correctly captured via dual-labeled immunofluorescence using confocal or regular fluorescence microscopy, such analysis can only be considered an initial step of assessment of endosomal signaling since these data are qualitative in nature. In order to provide a better semiquantitative data analysis, biochemical-based endocytosis assay provides better analysis in particular when the kinetics of endocytosis is being assessed. Biochemical analysis of cellular events pertinent to endocytosis, which include pinocytosis and phagocytosis, was first reported in the 1970s and 1980s (Bode, Baumann, & Kinne, 1975; Bode, Pockrandt-Hemstedt, Baumann, & Kinne, 1974; Daukas & Zigmond, 1985; Henning, Kaulen, & Stoffel, 1970; Loose, Megirian, & Turinsky, 1984; Quie, 1977). Since then, biochemical assay to assess protein endocytosis has been better developed with the initial use of radiolabeled proteins involving tedious biochemical steps such as ultracentrifugation (Wiley et al., 1991) to the use of protein biotinylation involving simple steps of lysate preparation and protein extraction with avidin-conjugated beads, to be following by immunoblot analysis using different specific antibodies (Le, Yap, & Stow, 1999; Morimoto et al., 2005) to track the events or kinetics of endocytosis.

In this short review, we provide a detailed step-by-step protocol based on our earlier experience using Sertoli cells in the rat testis as a study model to study protein endocytosis (Lie, Cheng, & Mruk, 2011; Wong, Mruk, Lee, & Cheng, 2010; Wong, Sun, Li, Lee, & Cheng, 2009; Yan, Mruk, Lee, & Cheng, 2008). While our experience is limited to the Sertoli cells in the testis, this approach, however, is applicable to other mammalian cells. Nonetheless, using this assay, we have shown that cellular events that occur in the seminiferous epithelium during spermatogenesis are coordinated by endocytic vesicle-mediated protein trafficking mediated by polarity proteins and under the influence of both cytokines and/or testosterone. For instance, testosterone and cytokines (e.g., TGF-β2) that have antagonistic effects on the blood–testis barrier (BTB) permeability function were found to have differential effects on endosome-based intracellular signaling in which testosterone facilitates endosome-mediated protein transcytosis/recycling, whereas TGF-β2 promotes endosome-mediated protein degradation based on biochemical assays (Yan et al., 2008). These findings have been subsequently confirmed using a fluorescence-based approach by staining Sertoli cells with specific markers of endocytosis, transcytosis, and recycling (Su, Mruk, Lee, & Cheng, 2010). Protein endocytosis was also shown to be crucial to TGF-β3-mediated disruptive effects on BTB function via the use of a Cdc42-dominant negative mutant for studies (Wong et al., 2010) in which Cdc42 is a critical component of the Par-based polarity protein complex (Wong & Cheng, 2009). For instance, it was shown that TGF-β3-mediated acceleration of protein endocytosis at the BTB is mediated by active Cdc42 since the deletion of Cdc42 functionality in the Sertoli cell epithelium via an overexpression of a dominant negative mutant of Cdc42 would insensitize these cells to TGF-β3 treatment (Wong et al., 2010), illustrating the Par6-Cdc42 complex is crucial in regulating protein endocytosis. This conclusion is also supported by findings in which a knockdown of Par3 or Par6 by RNAi in Sertoli cell epithelium was shown to induce mislocalization of integral membrane proteins at the Sertoli cell BTB (e.g., N-cadherin, JAM-A, nectin-2), with these proteins being redistributed, moved from the cell–cell interface and into the cell cytosol (Wong, Mruk, Lee, & Cheng, 2008), possibly via an increase in protein endocytosis. The notion that polarity proteins are crucial regulators of protein endocytosis is further supported by a study in which a knockdown of Par5 (14-3-3) was found to accelerate the kinetics of endocytosis of BTB integral membrane proteins JAM-A (junctional adhesion molecule-A) and N-cadherin (Wong et al., 2009). Taken collectively, these findings suggest that the initial event of endosomal signaling, namely, protein endocytosis, that occurs at the microenvironment of the BTB in the seminiferous epithelium of mammalian testis is primarily regulated by cytokines and testosterone, and with the involvement of polarity proteins such as Par and Cdc42, so that internalized proteins can be targeted to specific cellular domain(s) via transcytosis and recycling to mediate signaling function in response to changes in the environment, such as during the epithelial cycle of spermatogenesis or following exposure to toxicants and/or drugs. Results of these findings helped us to provide a biochemical-based model to study the processes of transport of preleptotene spermatocytes across the BTB during spermatogenesis (Cheng & Mruk, 2010; Cheng et al., 2011; Su, Mruk, & Cheng, 2013).

The assay protocol detailed below summarized our findings in assessing the effects of IL-1α, a cytokine in the testis known to accelerate protein endocytosis at the Sertoli cell BTB (Lie et al., 2011), on the kinetics of protein endocytosis using the *in vitro* Sertoli cell culture system which closely mimics the BTB *in vivo*. This assay, however, can be easily modified to assess changes in endocytosis following overexpression of any target proteins (e.g., Cdc42, a Par-based polarity protein complex component) using a mammalian cell expression vector (e.g., pCI-neo) (Wong et al., 2010), treatment of cells with toxicants (e.g., bisphenol A, cadmium) (Li, Mruk, Lee, & Cheng, 2009; Siu et al., 2009) as described in recently published reports. Thus, besides cytokines, steroids, and polarity proteins, other proteins that may regulate cellular functions via endosomal signaling can be rapidly assessed using this protocol.

2. ENDOCYTOSIS ASSAY

This assay is based on the use of biotin (also known as vitamin H) via biotinylation of cell surface proteins in which sulfo-NHS-biotin can be covalently and spontaneously conjugated to primary amines (such as lysine side chain ε-amines and N-terminal α-amines) in the amino acid residues of BTB integral membrane proteins such as CAR (coxsackievirus and adenovirus receptor) and JAM-A. Since sulfo-NHS-SS-biotin is water soluble and carries a charge, rendering it impermeable to plasma membranes, it only biotinylates the extracellular domains of BTB integral membrane proteins. Furthermore, a disulfide bond (S—S) is present in Sulfo-NHS-SS-biotin, which can be cleaved under reducing conditions in the presence of 2-mercaptoethanol to release the target protein in SDS-sample buffer using SDS-PAGE to be followed by immunoblotting using a specific antibody (see Fig. 10.1). Following biotinylation at 4 °C during which endocytosis fails to occur, nonbound biotins will be removed by quenching, and endocytosis will be allowed to take place by placing cells in

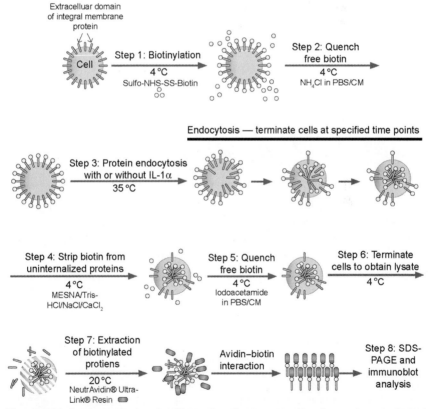

Figure 10.1 A schematic drawing illustrating the concept of endocytosis assay based on the use of biotinylation of cell surface proteins. The basic concept and detailed information of the various buffers and reagents can be found in the text. (For color version of this figure, the reader is referred to the online version of this chapter.)

dishes in a CO_2 incubator (5% CO_2/95% air, v/v) at 35 °C (which is the optimal temperature for testicular cells such as Sertoli cells—for other mammalian cells, 37 °C should be used instead) and cells will be terminated at specified time points as shown in Fig. 10.1 (see also Fig. 10.2). At termination, biotins on the uninternalized cell surface proteins will be stripped with 50 mM sodium 2-mercaptoethanesulfonate (MESNA) in 100 mM Tris–HCl, 100 mM NaCl, and 2.5 mM $CaCl_2$, pH 8.6 at 4 °C and quenched with 5 mg/ml iodoacetamide in PBS/CM buffer at 4 °C for 15 min, so that only endocytosed biotinylated proteins will be subsequently analyzed (Fig. 10.1). Cell lysates will be obtained

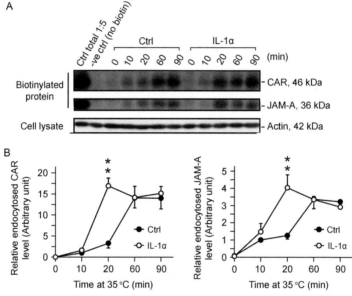

Figure 10.2 Effects of IL-1α on the kinetics of endocytosis of CAR and JAM-A in Sertoli cells cultured *in vitro* with a functional tight junction-permeability barrier. Sertoli cells were cultured at 0.5×10^6 cells/cm^2 for 4.5-day on Matrigel-coated dishes to allow the establishment of a functional TJ-permeability barrier. Thereafter, cells were subjected to biotinylation at 4 °C as described in text, and protein endocytosis was monitored at specified time points at 10, 20, 60, and 90 min versus time 0 in the absence (control, Ctrl) or presence of IL-1α (100 pg/ml) at 35 °C. (A) Endocytosed proteins at specified time points were monitored by extracting biotinylated proteins in cell lysates by using avidin-based resin for SDS-PAGE and immunoblot analysis after stripping of biotins from uninternalized biotinylated proteins on cell surface and quenching of the stripped free biotins in media. Following treatment of Sertoli cells with IL-1α which is known to perturb the Sertoli cell TJ-barrier function (Lie et al., 2011), an enhancement in endocytosis of CAR and JAM-A was noted. (B) Data were plotted against time to illustrate an increase in the kinetics of protein endocytosis following IL-1α treatment. Each data point is a mean ± SD of 3 replicates of a typical experiment, and this experiment was repeated three times using different batches of Sertoli cells. **$P < 0.01$.

and biotinylated proteins will be pulled down with UltraLink® Immobilized NeutrAvidin® beads since avidin has an extraordinary affinity for biotin ($K_D = \sim 10^{-14}\ M$) and the interaction between biotin and avidin is one of the strongest noncovalent molecular interactions in nature. Thereafter, the target proteins (e.g., CAR, JAM-A) are harvested with RIPA buffer and can be analyzed by SDS-PAGE following cleavage of the biotin in SDS sample buffer [0.125 M Tris, pH 6.8 at 22 °C containing 1% SDS (w/v), 1.6% 2-mercaptoethanol (v/v), 10% glycerol (v/v)], to be followed by immunoblot analysis.

3. MATERIALS

1. Ten 20-day-old male Sprague–Dawley rats (Charles River Laboratories)
2. 6-Well Culture Plate (Corning, 3516) coated with BD Matrigel™ Basement Membrane Matrix (BD Biosciences, 354234) diluted at 1:7 with DME/F-12
3. 10 ml Stripette Serological Pipets (Corning, 4488)
4. EZ-Link™ Sulfo-NHS-SS-Biotin (Thermo Scientific, 21331)
5. NeutrAvidin® UltraLink® Resin (Thermo Scientific, 53151)
6. Recombinant Rat IL-1α (R&D Systems, 500-RL-005), stored in 5 μg/ml aliquots in sterile PBS containing 0.1% BSA at −20 °C
7. Dulbecco's Modified Eagle's Medium/Nutrient Mixture F-12 Ham (DME/F-12) (Sigma-Aldrich, D2906) supplemented with 10 μg/ml insulin, 5 μg/ml human transferrin, 2.5 ng/ml EGF and 5 μg/ml bacitracin
8. Antibodies for immunoblotting analysis: rabbit anti-CAR (Santa Cruz Biotechnology, sc-15405, 1:200 dilution); rabbit anti-JAM-A (Life Technologies Corporation, 36-1700, 1:250 dilution)

4. BUFFERS

All chemicals listed below were obtained from Sigma-Aldrich unless otherwise noted.

Buffers should be made fresh each time and stored at 4 °C prior to use.
1. PBS: 10 mM NaH$_2$PO$_4$, 0.15 M NaCl, pH 7.4 at 22 °C
2. PBS/CM: 10 mM NaH$_2$PO$_4$, 0.15 M NaCl, 0.9 mM CaCl$_2$, 0.33 mM MgCl$_2$, pH 7.4 at 22 °C
3. Labeling buffer: 0.5 mg/ml EZ-Link™ Sulfo-NHS-SS-Biotin in PBS/CM

4. Quenching buffer 1: 50 mM NH$_4$Cl in PBS/CM
5. Stripping buffer: 50 mM MESNA, 100 mM Tris–HCl, 100 mM NaCl, 2.5 mM CaCl$_2$, pH 8.6 at 22 °C
6. Quenching buffer 2: 5 mg/ml iodoacetamide in PBS/CM
7. RIPA buffer: 50 mM Tris–HCl, 150 mM NaCl, 5 mM EGTA, 0.2% SDS (i.e., 0.2 g/100 ml), 1% Triton X-100 (v/v), 1% Na deoxycholate (i.e., 1 g/100 ml), 2 mM N-ethylmaleimide, pH 8.0 at 22 °C, cleared by filtration through a 0.2-μm filtering unit. Freshly add protease inhibitors (2 mM PMSF, 1 μg/ml aprotinin and leupeptin) as well as Phosphatase Inhibitor Cocktail 2 (P5726) and 3 (P0044) at 1:100 dilution prior to use.

5. METHODS

1. Isolate Sertoli cells from ten 20-day-old male Sprague–Dawley rats and plate them in six 6-well culture plates at a high cell density (0.5×10^6 cells/cm^2) as earlier described (Mruk & Cheng, 2011). Considering the 9.4-cm^2 growth area in each well of the plate, one can anticipate isolating $\sim 144 \times 10^6$ cells (which is the routine yield of Sertoli cells from 10 male pups) to seed in at least 24 wells, with four in each of the six plates. Culture the cells in DME/F-12 for 4.5 days to allow the establishment of a functional permeability barrier as detailed earlier (Mruk & Cheng, 2011).

    Note: Each time point of the endocytosis assay requires a separate 6-well culture plate so that cells to be terminated later will not be disturbed.

2. Before experiment starts, warm 50 ml DME/F-12 containing IL-1α at 100 pg/ml (*Note*: IL-1α is dissolved in 0.1% BSA-PBS at 5 μg/ml as a stock) and another 50 ml DME/F-12 without cytokine to serve as control to 37 °C, respectively.

3. Immediately before use, weigh 6 mg Sulfo-NHS-SS-Biotin and dissolve it in 1 ml Milli-Q water (Model Advantage A10, Millipore). Use 1:12 dilution in PBS/CM to get 0.5 mg/ml Sulfo-NHS-SS-Biotin as the labeling buffer.

4. Take cells out of the incubator and put on ice, wash the cells with ice-cold PBS/CM twice.

    Note: From now on, all steps should be done at 4 °C or on ice (so that endocytosis ceases to take place) unless otherwise noted.

5. Add labeling buffer from step 3 to each well (1 ml/well) except the negative-control well (−ve Ctrl) in which cells are incubated with plain PBS/CM. Incubate cells at 4 °C for 30 min with gentle agitation.
6. After the incubation, wash cells with PBS/CM once, and remove the free/excess Sulfo-NHS-SS-Biotin with quenching buffer 1 at 4 °C for 15 min with gentle agitation.
7. Wash with PBS/CM once. Lyse one well of cells as "total" biotinylated surface proteins with 1 ml RIPA buffer, and also harvest the −ve Ctrl cells.
8. Put cells in 35 °C incubator (to initiate endocytosis) with 5 ml/well DME/F-12 [with IL-1α and without (control)]. Take cells out (one plate at a time) at specified time points (with 0, 10, 20, 60, 90 min respective duration in 35 °C incubator, and each time point should have at least duplicate culture well), and put back on ice to proceed with step 9.
9. Wash with PBS/CM twice. Incubate in stripping buffer at 4 °C for 30 min with gentle agitation to remove biotin from uninternalized biotinylated cell surface proteins.
10. Wash with PBS/CM once. Incubate in quenching buffer 2 at 4 °C for 15 min with gentle agitation to quench free biotins.
11. Wash with PBS/CM once. Lyse cells with RIPA buffer.
12. Sonicate cells, obtain cell lysates in clear supernate by centrifugation at 14,000 × g for 10 min at 4 °C, and determine the protein concentration (routinely around 1 mg/ml).

    Note: Cell lysates can be stored at −20 °C with or without sonication/centrifugation until use.
13. Add ~20 μl NeutrAvidin® UltraLink® Resin to ~300 μg protein of cell lysate (~300 μl) for each reaction in a 0.6-ml microcentrifuge tube. Mix on a rocking platform for 4–6 h at room temperature.

    Note: For optimization, wash the resin with RIPA buffer three times (washing is done by resuspending resin in RIPA buffer, gently mix the sample in a microcentrifuge tube and collect resin by centrifugation at 3000 g, 1 min each - do not vortex resin to avoid damaging its physicochemical properties).
14. Centrifuge the tube for 1 min at 3000–5000 × g to obtain the resin-bound complex (i.e., avidin-biotinylated protein complexes) and discard supernate.
15. Wash the resin-bound complexes with RIPA buffer four times. In between spin down for 1 min at 3000–5000 × g and discard supernate.
16. Heat the resin-bound complex in SDS sample buffer [0.125 M Tris, pH 6.8 at 22 °C containing 1% SDS (w/v), 1.6% 2-mercaptoethanol

(v/v), 20% glycerol (v/v)] so that biotins can be cleaved from the proteins. Following SDS-PAGE, proteins will be analyzed by immunoblotting using specific antibodies.

6. CELL STAINING TO ASSESS ENDOCYTOSIS

1. Isolate Sertoli cells from 20-day-old male Sprague–Dawley rats and plate them onto glass coverslips (Thomas Scientific, 6662F43) inserted in 12-Well Culture Plate (Corning, 3513) coated with BD Matrigel™ Basement Membrane Matrix at a low cell density (0.05×10^6 cells/cm^2). Culture the cells in DME/F-12 for 4.5 days to allow the establishment of a functional permeability barrier.
2. Fix cells with methanol at $-20\ °C$ for 5 min.
3. Wash with PBS three times. Incubate with blocking solution (1% BSA in PBS) for 30 min at room temperature.
4. Remove blocking solution by vacuum aspiration. Add ~50 µl of rabbit anti-CAR or anti-JAM-A using 1:100 dilution in PBS. Incubate for 4–6 h at room temperature.
5. Wash with PBS three times. Add ~50 µl of Alexa Fluor® 555 goat anti-rabbit IgG (H+L) (Life Technologies Corporation, A-21429) using 1:200 dilution in PBS. Incubate for 1 h at room temperature.
6. Wash with PBS three times. Dry coverslips in dark. Mount onto microscope slides (Thomas Scientific, 6686S50) with ProLong® Gold antifade reagent with DAPI (Life Technologies Corporation, P36935).

7. RESULTS

Figure 10.2A illustrates the results of a typical endocytosis assay in which exposure of Sertoli cells to IL-1α was found to enhance the kinetics of endocytosis of CAR (Fig. 10.2B, left panel) and JAM-A (Fig. 10.2B, right panel), which are integral membrane proteins at the Sertoli cell BTB, consistent with a recent report from our laboratory (Lie et al., 2011). These findings were confirmed by immunofluorescence microscopy as shown in Fig. 10.3, in which CAR and JAM-A were also found to become mislocalized following exposure of these cells to IL-1α wherein these proteins redistributed from the cell surface and moved into the cell cytosol, destabilizing the Sertoli cell tight junction barrier. Thus, these findings shown in Figs. 10.2 and 10.3 have unequivocally demonstrated that

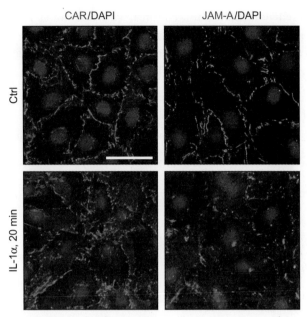

Figure 10.3 A study by immunofluorescence microscopy to assess changes in protein distribution at the Sertoli cell cell–cell interface following treatment with IL-1α. Sertoli cells were cultured at 0.05×10^6 cells/cm^2 for 4.5-day on Matrigel-coated coverslips which were placed in 12-well dishes with 2-ml DME/F-12 medium per well (supplemented with growth factors) to allow the establishment of a functional TJ-permeability that mimicked the Sertoli cell BTB *in vivo* as described (Xiao, Cheng, & Mruk, 2013; Xiao, Mruk, Lee, & Cheng, 2011). Thereafter, cells were treated without (control, Ctrl) or with IL-1α (100 pg/ml) for 20 min at 35 °C in a CO$_2$ incubator. Sertoli cells were then harvested by fixing cells in methanol at −20 °C for 5 min and stained for either CAR or JAM-A (red fluorescence) as described in the text. Sertoli cell nuclei were visualized by DAPI (4′,6-diamidino-2-phenylindole) staining. It is noted that treatment of Sertoli cells accelerated the internalization of CAR and JAM-A, with these proteins redistributed from the cell surface into the cell cytosol, confirming data shown in Fig. 10.2. Scale bar, 60 μm, which applies to all other micrographs. (See color plate.)

IL-1α perturbs Sertoli cell TJ-permeability barrier function via an increase in protein endocytosis, thereby disrupting the BTB function. Thus, a combination of the endocytosis assay and the fluorescence microscopy is a powerful technique to study endosomal signaling regulation in mammalian cells.

8. SUMMARY

This brief chapter has provided a detailed step-by-step protocol of studying protein endocytosis—an initial step in endosomal signaling function—in mammalian cells using Sertoli cells culture in vitro as a model. These studies are efficient, cost effective, and highly reproducible, and they can be performed in virtually any modern biochemistry and cell biology laboratory without expensive equipment and/or setup, applicable to all mammalian cells.

ACKNOWLEDGMENTS
This work was supported by grants from the National Institutes of Health (NICHD R01 HD056034 to C.Y.C.; U54 HD029990, Project 5 to C.Y.C.), National Science Foundation of China (NSFC 31371176 to X.X.), and The Hong Kong General Research Fund (GRF HKBU261812 to C.K.C.W.)
Disclosure: The authors have nothing to declare

REFERENCES
Bode, F., Baumann, K., & Kinne, R. (1975). Biochemical aspects of pinocytosis in kidney. *Contributions to Nephrology*, *1*, 21–27.
Bode, F., Pockrandt-Hemstedt, H., Baumann, K., & Kinne, R. (1974). Analysis of the pinocytic process in rat kidney. I. Isolation of pinocytic vesicles from rat kidney cortex. *Journal of Cell Biology*, *63*, 998–1008.
Cheng, C. Y., & Mruk, D. D. (2010). A local autocrine axis in the testes that regulates spermatogenesis. *Nature Reviews Endocrinology*, *6*, 380–395.
Cheng, C. Y., Wong, E. W. P., Lie, P. P. Y., Li, M. W. M., Mruk, D. D., Yan, H. H. N., et al. (2011). Regulation of blood-testis barrier dynamics by desmosome, gap junction, hemidesmosome and polarity proteins: An unexpected turn of events. *Spermatogenesis*, *1*, 105–115.
Daukas, G., & Zigmond, S. H. (1985). Inhibtion of receptor-mediated but not fluid-phase endocytosis in polymorphonuclear keukocytes. *Journal of Cell Biology*, *101*, 1673–1679.
Falkenberg, C. V., & Loew, L. M. (2013). Computational analysis of Rho GTPase cycling. *PLoS Computational Biology*, *9*, e1002831.
Gonnord, P., Blouin, C. M., & Lamaze, C. (2012). Membrane trafficking and signaling: Two sies of the same coin. *Seminars in Cell & Developmental Biology*, *23*, 154–164.
Haglund, K., & Dikic, I. (2012). The role of ubiquitylation in receptor endocytosis and endosomal sorting. *Journal of Cell Science*, *125*, 265–275.
Henning, R., Kaulen, H. D., & Stoffel, W. (1970). Biochemical analysis of the pinocytotic process. I. Isolation and chemical composition of the lysosomal and the plasma membrane of the rat liver cell. *Hoppe-Seyler's Zeitschrift für Physiologische Chemie*, *351*, 1191–1199.
Le Roy, C., & Wrana, J. L. (2005a). Clathrin- and non-clathrin-mediated endocytic regulation of cell signalling. *Nature Reviews Molecular Cell Biology*, *6*, 112–126.
Le Roy, C., & Wrana, J. L. (2005b). Signaling and endocytosis: A team effort for cell migration. *Developmental Cell*, *9*, 167–168.
Le, T. L., Yap, A. S., & Stow, J. L. (1999). Recycling of E-cadherin: A potential mechanism for regulating cadherin dynamics. *Journal of Cell Biology*, *146*, 219–232.
Leto, D., & Saltiel, A. R. (2012). Regulation of glucose transport by insulin: Traffic control of GLUT4. *Nature Reviews Molecular Cell Biology*, *13*, 383–396.
Li, M. W. M., Mruk, D. D., Lee, W. M., & Cheng, C. Y. (2009). Disruption of the blood-testis barrier integrity by bisphenol A *in vitro*: Is this a suitable model for studying blood-testis barrier dynamics? *International Journal of Biochemistry & Cell Biology*, *41*, 2302–2314.

Lie, P. P. Y., Cheng, C. Y., & Mruk, D. D. (2011). Interleukin-1α is a regulator of the blood-testis barrier. *FASEB Journal*, *25*, 1244–1253.
Loose, L. D., Megirian, R., & Turinsky, J. (1984). Biochemical and functional alterations in macrophages after thermal injury. *Infection and Immunity*, *44*, 554–558.
Marchese, A., & Trejo, J. (2013). Ubiquitin-dependent regulation of G protein-coupled receptor trafficking and signaling. *Cellular Signalling*, *25*, 707–716.
Miaczynska, M., & Bar-Sagi, D. (2010). Signaling endosomes: Seeing is believing. *Current Opinion in Cell Biology*, *22*, 535–540.
Morimoto, S., Nishimura, N., Terai, T., Manabe, S., Yamamoto, Y., Shinahara, W., et al. (2005). Rab13 mediates the continuous endocytic recycling of occludin to the cell surface. *Journal of Biological Chemistry*, *280*, 2220–2228.
Mruk, D. D., & Cheng, C. Y. (2011). An *in vitro* system to study Sertoli cell blood-testis barrier dynamics. *Methods in Molecular Biology*, *763*, 237–252.
Neto, H., Collins, L. L., & Gould, G. W. (2011). Vesicle trafficking and membrane remodelling in cytokinesis. *Biochemical Journal*, *437*, 13–24.
Palfy, M., Remenyi, A., & Korcsmaros, T. (2012). Endosomal crosstalks: Meeting points for signaling pathways. *Trends in Cell Biology*, *22*, 447–456.
Polo, S., & Di Fiorce, P. P. (2006). Endocytosis conducts the cell signaling orchestra. *Cell*, *124*, 897–900.
Quie, P. G. (1977). Disorders of phagocyte function: Biochemical aspects. *Progress in Clinical and Biological Research*, *13*, 157–169.
Sandilands, E., & Frame, M. C. (2008). Endosomal trafficking of Src tyrosine kinase. *Trends in Cell Biology*, *18*, 322–329.
Schiefermeier, N., Teis, D., & Huber, L. A. (2011). Endosomal signaling and cell migration. *Current Opinion in Cell Biology*, *23*, 615–620.
Siu, E. R., Wong, E. W. P., Mruk, D. D., Sze, K. L., Porto, C. S., & Cheng, C. Y. (2009). An occludin-focal adhesion kinase protein complex at the blood-testis barrier: A study using the cadmium model. *Endocrinology*, *150*, 3336–3344.
Su, W. H., Mruk, D. D., & Cheng, C. Y. (2013). Regulation of actin dynamics and protein trafficking during spermatogenesis—insights into a complex process. *Critical Reviews in Biochemistry and Molecular Biology*, *48*, 153–172.
Su, L., Mruk, D. D., Lee, W. M., & Cheng, C. Y. (2010). Differential effects of testosterone and TGF-β3 on endocytic vesicle-mediated protein trafficking events at the blood-testis barrier. *Experimental Cell Research*, *316*, 2945–2960.
Wiley, H. S., Herbst, J. J., Walsh, B. J., Lauffenburger, D. A., Rosenfeld, M. G., & Gill, G. N. (1991). The role of tyrosin kinase activity in endocytosis, compartmentation, and down-regulation of the epidermal growth factor receptor. *Journal of Biological Chemistry*, *266*, 11083–11094.
Wong, E. W. P., & Cheng, C. Y. (2009). Polarity proteins and cell-cell interactions in the testis. *International Review of Cell and Molecular Biology*, *278*, 309–353.
Wong, E. W. P., Mruk, D. D., Lee, W. M., & Cheng, C. Y. (2008). Par3/Par6 polarity complex coordinates apical ectoplasmic specialization and blood-testis barrier restructuring during spermatogenesis. *Proceedings of the National Academy of Sciences of the United States of America*, *105*, 9657–9662.
Wong, E. W. P., Mruk, D. D., Lee, W. M., & Cheng, C. Y. (2010). Regulation of blood-testis barrier dynamics by TGF-β3 is a Cdc42-dependent protein trafficking event. *Proceedings of the National Academy of Sciences of the United States of America*, *107*, 11399–11404.
Wong, E. W. P., Sun, S., Li, M. W. M., Lee, W. M., & Cheng, C. Y. (2009). 14-3-3 protein regulates cell adhesion in the seminiferous epithelium of rat testes. *Endocrinology*, *150*, 4713–4723.
Xiao, X., Cheng, C. Y., & Mruk, D. D. (2013). Intercellular adhesion molecule (ICAM)-1 is a regulator of blood-testis barrier function. *Journal of Cell Science*, *125*, 5677–5689.

Xiao, X., Mruk, D. D., Lee, W. M., & Cheng, C. Y. (2011). c-Yes regulates cell adhesion at the blood-testis barrier and the apical ectoplasmic specialization in the seminiferous epithelium of rat testes. *International Journal of Biochemistry & Cell Biology, 43*, 651–665.

Yan, H. H. N., Mruk, D. D., Lee, W. M., & Cheng, C. Y. (2008). Blood-testis barrier dynamics are regulated by testosterone and cytokines via their differential effects on the kinetics of protein endocytosis and recycling in Sertoli cells. *FASEB Journal, 22*, 1945–1959.

CHAPTER ELEVEN

Methods of Analysis of the Membrane Trafficking Pathway from Recycling Endosomes to Lysosomes

Takahide Matsui, Mitsunori Fukuda[1]
Laboratory of Membrane Trafficking Mechanisms, Department of Developmental Biology and Neurosciences, Graduate School of Life Sciences, Tohoku University, Sendai, Miyagi, Japan
[1]Corresponding author: e-mail address: nori@m.tohoku.ac.jp

Contents

1. Introduction	196
2. Degradation of TfR in Lysosomes	197
2.1 Materials	197
2.2 Analysis of constitutive degradation of TfR protein by Western blotting	197
2.3 Analysis of lysosomal degradation of TfR protein by Western blotting	199
2.4 Analysis of TfR protein trafficking to lysosomes by an immunofluorescence method	199
3. Screening Methods for Rab Proteins Involved in Lysosomal Degradation of TfR	200
3.1 Determination of the colocalization rates between EGFP-Rabs and TfR	200
3.2 Determination of the colocalization rates between the candidate EGFP-Rabs and Lamp-1	202
3.3 Effect of knockdown of each candidate Rab on the TfR expression level	202
4. Effect of Rab12 Knockdown on an EGFR Endocytic Pathway and a Tf Recycling Pathway	203
4.1 Materials required to perform the EGFR degradation assay	204
4.2 EGFR degradation assay in Rab12-knockdown cells	204
4.3 Materials for the Tf recycling assay	204
4.4 Tf recycling assay in Rab12-knockdown cells	205
5. Concluding Remarks	205
Acknowledgments	205
References	205

Abstract

The transferrin receptor (TfR) is responsible for iron uptake through its trafficking between the plasma membrane and recycling endosomes, and as a result it has become a well-known marker for recycling endosomes. Although the molecular basis of the TfR recycling pathway has been thoroughly investigated, the TfR degradation mechanism has been poorly understood. Exposure of cultured cells to two drugs, the protein synthesis inhibitor cycloheximide and the V-ATPase inhibitor bafilomycin A1, recently showed that TfR is not only recycled back to the plasma membrane after endocytosis but is constitutively transported to lysosomes for degradation. The results of genome-wide screening of mouse Rab small GTPases (common regulators of membrane trafficking in all eukaryotes) have indicated that Rab12 regulates TfR trafficking to lysosomes independently of the known membrane trafficking pathways, for example, the conventional endocytic pathway and recycling pathway. This chapter summarizes the methods that the authors used to analyze the membrane trafficking pathway from recycling endosomes to lysosomes that is specifically regulated by Rab12.

1. INTRODUCTION

Because plasma membrane receptor proteins play a key role in signal transduction and/or nutrient uptake, the quality control of receptor proteins is one of the most important issues in cellular homeostasis. It is well known that the iron-binding protein transferrin (Tf) and its receptor, the transferrin receptor (TfR), form a complex that promotes iron uptake by endocytosis (Dautry-Varsat, Ciechanover, & Lodish, 1983; Klausner et al., 1983). After endocytosis, the complex is recycled back to the plasma membrane where it is used in the next round of iron uptake. Many proteins that are involved in TfR recycling, including small GTPases Rab4, Rab11, and Rab22A, have already been identified (Daro, van der Sluijs, Galli, & Mellman, 1996; Ren et al., 1998; Schlierf, Fey, Hauber, Hocke, & Rosorius, 2000; Ullrich, Reinsch, Urbé, Zerial, & Parton, 1996; Weigert, Yeung, Li, & Donaldson, 2004), but the mechanisms of the quality control of TfR, especially the mechanism by which it is degraded, have been poorly understood. We have recently shown that the TfR protein incorporated into cells by endocytosis is not only recycled back to the plasma membrane but is transported to lysosomes for degradation (Matsui, Itoh, & Fukuda, 2011). We have also performed genome-wide screening for Rabs that are involved in TfR degradation in lysosomes and identified Rab12 as a novel regulator of TfR trafficking. Intriguingly, Rab12 regulates TfR degradation through a novel membrane trafficking pathway "from recycling endosomes to lysosomes," but it is not involved in the conventional endocytic pathway (e.g., epidermal growth factor receptor (EGFR) degradation) or recycling pathway (e.g., Tf recycling) (Matsui & Fukuda, 2011). In this chapter, we summarize the methods that we have used to analyze the Rab12-mediated TfR degradation pathway.

2. DEGRADATION OF TfR IN LYSOSOMES

Constitutive degradation of TfR protein in lysosomes has been demonstrated by exposing cells to cycloheximide (CHX), a protein synthesis inhibitor, and/or bafilomycin A1, V-ATPase inhibitor that blocks lysosomal degradation of target proteins.

2.1. Materials

1. A 10 mg/mL CHX (Sigma-Aldrich, C1988) solution in dimethyl sulfoxide (DMSO). A 20-μM bafilomycin A1 (Calbiochem, 196000) solution in DMSO. The stock solutions can be stored frozen at $-30\,^{\circ}$C until used.
2. Phosphate-buffered saline (PBS): 137 mM NaCl, 2.7 mM KCl, 4.3 mM Na$_2$HPO$_4$, 1.4 mM KH$_2$PO$_4$, pH 7.3.
3. Lysis buffer: 150 mM NaCl, 50 mM HEPES–KOH, pH 7.2, 1% Triton X-100 (Nacalai Tesque, 35501-15), complete protease inhibitor cocktail (Roche, 05 056 489 001).
4. Antibodies: anti-TfR mouse monoclonal antibody (Zymed, 13-6800), anti-β-actin mouse monoclonal antibody (Applied Biological Materials, G043), and anti-Lamp-1 rat monoclonal antibody (BD Biosciences, 553792).

2.2. Analysis of constitutive degradation of TfR protein by Western blotting

1. Seed mouse embryonic fibroblasts (MEFs) in 35-mm dishes and culture under 5% CO$_2$ at 37 $^{\circ}$C in Dulbecco's modified Eagle's medium (DMEM) (Wako, 044-29765) containing 10% fetal bovine serum (FBS; Gibco) and antibiotics. The cells can be used in the experiment when cells have reached 80% confluency.
2. Discard the medium and add fresh culture medium containing 50 μg/mL CHX or DMSO alone (control) to the cell culture. Then incubate the cells for 2, 4, and 6 h.
3. Discard the medium and rinse the cells with ice-cold PBS once. Place the dishes on ice and add 500 μL of ice-cold PBS to each cell culture. Scrape the cells off with a cell scraper and transfer each cell suspension into a microcentrifuge tube on ice. Collect the cells by centrifuging at $700 \times g$ for 5 min at 4 $^{\circ}$C. Discard the supernatant.

4. Suspend the cells in 50 μL of ice-cold lysis buffer with a microtube mixer (TOMY). Place the tube on ice for 30 min. Centrifuge each tube at $17,400 \times g$ for 10 min at 4 °C and recover the supernatant as the sample in a new microcentrifuge tube.
5. Measure the protein concentration of each sample by the Bradford method using the Bio-Rad protein assay (Bio-Rad, 500-0006). Add 1/4 volume of 5× sample buffer (10% SDS, 0.3125 M Tris–HCl pH 6.8, 0.1% bromophenol blue, 50% glycerol, 10% 2-mercaptoethanol) to each sample, and boil for 5 min.
6. Analyze each sample by 10% sodium dodecyl sulfate polyacrylamide gel electrophoresis (SDS-PAGE).
7. Transfer the proteins from the gels to a PVDF membrane (Millipore, IPVH304F0) at a constant voltage of 10 V for 1 h by electroblotting with a semidry transfer system (Bio-Rad, Trans Blot SD cell) in transfer buffer (25 mM Tris, 192 mM glycine, 20% methanol).
8. Incubate the membrane for 30 min at room temperature with 1% (w/v) skim milk in PBS-T (0.1% Tween 20 in PBS) with constant agitation to block the membrane.
9. Incubate the membrane for 1 h at room temperature with anti-TfR antibody (1/5000 dilution in 1%, w/v, skim milk in PBS-T) or anti-β-actin antibody (1/10,000 dilution in 1%, w/v, skim milk in PBS-T), with constant agitation. Wash the membrane three times with PBS-T for 5 min at room temperature, with constant agitation.
10. Incubate the membrane for 45 min at room temperature with horseradish peroxidase (HRP)-conjugated anti-mouse IgG (Sigma-Aldrich) (1/10,000 dilution in 1%, w/v, skim milk in PBS-T), with constant agitation. Wash the membrane three times with PBS-T for 8 min at room temperature, with constant agitation.
11. Immunoreactive bands are detected by enhanced chemiluminescence. Note that CHX reduces the amount of TfR protein in a time-dependent manner. The half-life of TfR protein (i.e., the constitutive degradation rate of TfR protein) appears to differ considerably according to the cell line (estimated half-life of TfR protein in MEF and NIH3T3 cells: 4 h as opposed to over 12 h in Neuro2a, B16F1, and HeLa cells) (Matsui et al., 2011; Rutledge, Mikoryak, & Draper, 1991; Ward, Kushner, & Kaplan, 1982). Also note that the constitutive degradation rate of TfR protein is often affected by the passage number of the cells. Under our experimental conditions, the constitutive degradation rate of TfR protein in MEF cells is slower when the passage number is more than 30. Thus, it is strongly recommended

that the conditions, including the CHX exposure time and the passage number of the cells, be considered before performing the experiment.

2.3. Analysis of lysosomal degradation of TfR protein by Western blotting

1. Culture MEF cells as described in Step 1, Section 2.2.
2. Preincubate the cells for 9 h in culture medium containing 100 nM bafilomycin A1 or DMSO alone (control). Add CHX to the medium to a final concentration of 50 μg/mL and then incubate the cells for 4 h.
3. Follow Steps 3–11 in Section 2.2.

Note that the CHX-induced reduction in the amount of TfR protein is prevented by simultaneous treatment with bafilomycin A1, indicating that TfR protein is constitutively degraded in lysosomes.

2.4. Analysis of TfR protein trafficking to lysosomes by an immunofluorescence method

1. Seed MEF cells on 12 mm-round coverslips (Matsunami Glass, No. 1 thickness) in 35-mm dishes and culture as described in Step 1, Section 2.2.
2. Discard the medium and add fresh culture medium containing 100 nM bafilomycin A1 or DMSO alone (control). Culture the cells for 9 h.
3. Wash the cells once with PBS at room temperature, and after fixing for 20 min at room temperature with 4% paraformaldehyde (Wako Pure Chemical Industries, 168-20955) in 0.1 M sodium phosphate buffer, wash three times with PBS.
4. Permeabilize the cells with 0.3% Triton X-100 in PBS for 2.5 min and immediately wash them three times with the blocking solution (1% bovine serum albumin fraction V (Sigma, A7030-100G) and 0.1% Triton X-100 in PBS). Then incubate the cells in the blocking solution for 30 min at room temperature.
5. Incubate the cells on coverslips for 1 h at room temperature with 50 μl of anti-TfR antibody and anti-Lamp-1 antibody (1/300 and 1/500 dilution, respectively) in the blocking solution. Wash the samples three times with the blocking buffer.
6. Incubate the cells on coverslips for 1 h at room temperature with 50 μl of Alexa 488-conjugated anti-mouse IgG antibody and Alexa 594-conjugated anti-rat IgG antibody (1/5000 dilution) in the blocking buffer. Wash the samples four times with the blocking buffer and once with PBS.

7. Mount on glass slides with 6 μl of Gold Antifade Reagent (Molecular Probes, P36934). Examine for the localization of the TfR and Lamp-1 signals with a confocal laser-scanning fluorescence microscope equipped with a 100 × objective lens.

Note that inhibition of lysosomal function by bafilomycin A1 induces colocalization of TfR with Lamp-1 (Matsui et al., 2011), indicating that TfR is constitutively degraded in lysosomes.

3. SCREENING METHODS FOR Rab PROTEINS INVOLVED IN LYSOSOMAL DEGRADATION OF TfR

A three-step screening method has been described to identify Rabs involved in the lysosomal degradation of TfR (Fukuda, 2010): Step 1) screening for EGFP-tagged Rabs that colocalize with TfR, Step 2) screening for EGFP-Rabs that colocalize with Lamp-1, and Step 3) screening for Rabs that affect the TfR expression level by knocking down candidate Rabs.

3.1. Determination of the colocalization rates between EGFP-Rabs and TfR

1. Seed an appropriate cell number of MEF cells on 12 mm-round coverslips in 35-mm dishes and incubate until they reach 50% confluence the following day.
2. Transfect the cells with pEGFP-C1-mouse Rab1A–43 (Tsuboi & Fukuda, 2006) by LipofectAMINE 2000 (Invitrogen, 11668-019) according to the manufacturer's protocol, and then culture for 24 h.
3. Follow Steps 3 and 4 in Section 2.4.
4. Incubate the cells on coverslips for 1 h at room temperature with 50 μl of anti-TfR antibody (1/300 dilution) in the blocking solution. Wash the samples three times with the blocking buffer.
5. Incubate the cells on coverslips for 1 h at room temperature with 50 μl of Alexa 594-conjugated anti-mouse IgG antibody (1/5000 dilution) in the blocking buffer. Wash the samples four times with the blocking buffer and once with PBS.
6. Follow Step 7 in Section 2.4.
7. Quantify the colocalization rate between EGFP-Rabs and TfR (>50 cells) with the RG2B colocalization plug-in of Image J software (NIH, version 1.42q) as follows. The colocalization measurements are made by using a minimum threshold pixel intensity adjusted to ~150 and set equivalently for both channels to correspond to the image

intensity. The RGB output of the plug-in is split into green images (i.e., EGFP-Rabs images), red images (i.e., TfR images), and blue images (merged images), and each image is quantified by measuring the total pixel area. Colocalization rates are calculated by dividing the area of the colocalization pixels by the total pixel area and multiplying by 100. Rab proteins whose colocalization rate with TfR is greater than 20% are selected as the initial candidates (Fig. 11.1A).

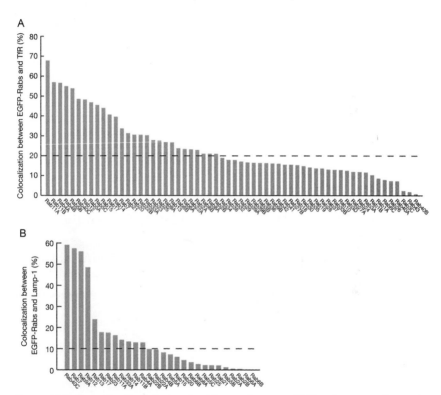

Figure 11.1 Screening for Rabs involved in membrane trafficking form recycling endosomes to lysosomes. (A) MEF cells transiently expressing EGFP-tagged mouse Rabs (Rab1A–43) were immunostained with anti-TfR antibody. The colocalization rates between the EGFP-Rabs and TfR were measured with the RG2B colocalization plug-in of Image J software as described in Step 7, Section 3.1. In the first screening, we selected the 28 Rabs that were found to have a greater than 20% colocalization rate (broken line) as initial candidates. (B) MEF cells transiently expressing the initial candidate Rabs with EGFP-tag were immunostained with anti-Lamp-1 antibody. The colocalization rates were measured as described in Step 7, Section 3.2. In the second screening, we selected the 12 Rabs (Rab4A, 7, 9A, 11A, 11B, 12, 13, 14 17, 23, 33A, and 40C) that were found to have a greater than 10% colocalization rate (broken line) as second-tier candidates.

3.2. Determination of the colocalization rates between the candidate EGFP-Rabs and Lamp-1

1. Prepare MEF cells as described in Step 1, Section 3.1.
2. Transfect the cells with pEGFP-C1 harboring the initial candidate mouse Rabs as described in Step 2, Section 3.1.
3. Follow Steps 3 and 4 in Section 2.4.
4. Incubate the cells on coverslips for 1 h at room temperature with 50 μl of anti-Lamp-1 antibody (1/300 dilution) in the blocking solution. Wash the samples three times with the blocking buffer.
5. Incubate the cells on coverslips for 1 h at room temperature with 50 μl of Alexa 594-conjugated anti-rat IgG antibody (1/5000 dilution) in the blocking buffer. Wash the samples four times with the blocking buffer and once with PBS.
6. Follow Step 7 in Section 2.4.
7. Quantify the colocalization rate between EGFP-Rabs and Lamp-1 as described in Step 7, Section 3.1. Rab proteins whose colocalization rate with Lamp-1 is greater than 10% are selected as second-tier candidates (Fig. 11.1B).

3.3. Effect of knockdown of each candidate Rab on the TfR expression level

1. Generate at least three independent short hairpin RNAs (shRNAs) targeting each candidate Rab as described previously (Kuroda & Fukuda, 2004). To evaluate efficiency of the shRNAs, express each EGFP-Rab and its corresponding shRNA (or control shRNA) in COS7 cells and evaluate its effect on the expression of EGFP-Rab by Western blotting with anti-GFP rabbit polyclonal antibody (MBL, 598) (see Fig. 11.2A). Select the most effective shRNA and synthesize small interfering RNA (siRNA) based on the shRNA sequence.
2. The day before transfection, seed an appropriate number of MEF cells in 35-mm dishes.
3. An hour before transfection, discard the culture medium and add DMEM containing 10% FBS (without antibiotics) to the cell culture. Transfect siRNA with RNAiMAX (Invitrogen, 13778-150) according to the manufacturer's protocol. At 24 h after transfection, replace the medium with fresh normal culture medium and culture for an additional 24 h.
4. Follow Steps 3–11 in Section 2.2.

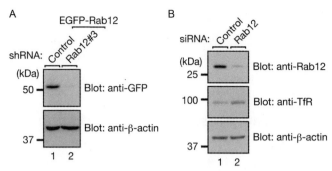

Figure 11.2 Effect of Rab12 knockdown on the amount of TfR protein. (A) Cell lysates from COS7 cells that had been cotransfected with pEGFP-C1-Rab12 and pSilencer-Neo empty vector or pSilencer-Neo-Rab12#3 were analyzed by Western blotting with anti-GFP antibody and anti-β-actin antibody. (B) Cell lysates from MEF cells that had been transfected with control siRNA and *Rab12* siRNA were analyzed by Western blotting with anti-Rab12 antibody, anti-TfR antibody, and anti-β-actin antibody. Note that Rab12 knockdown increased the amount of TfR protein. The positions of the molecular mass markers (in kilodaltons) are shown on the left.

5. The intensity of immunoreactive bands of TfR is measured with Image J software and normalized to the amount of β-actin.

As a result of using these three screening methods in our study, Rab12 was selected as the prime candidate for a Rab that is specifically involved in lysosomal degradation of TfR (Matsui et al., 2011). Since Rab12 is localized both at TfR-positive recycling endosomes and at lysosomes and Rab12 knockdown increased the amount of TfR protein (Fig. 11.2B), we have proposed that Rab12 regulates lysosomal degradation of TfR through a novel membrane trafficking pathway from recycling endosomes to lysosomes.

4. EFFECT OF Rab12 KNOCKDOWN ON AN EGFR ENDOCYTIC PATHWAY AND A Tf RECYCLING PATHWAY

Before proposing the existence of a novel membrane trafficking pathway from recycling endosomes to lysosomes that is regulated by Rab12, it was important to show that Rab12 is not involved in the conventional endocytic pathway or TfR recycling pathway. To do so, we performed an EGFR degradation assay and fluorescent transferrin (Tf) recycling assay in Rab12-knockdown MEF cells.

4.1. Materials required to perform the EGFR degradation assay

1. A 50 μg/mL EGF (Invitrogen, E3476) solution in PBS. The stock solution can be stored frozen at −30 °C until used.
2. Anti-EGFR sheep polyclonal antibody (Fitzgerald, 20-ES04).

4.2. EGFR degradation assay in Rab12-knockdown cells

1. MEF cell culture, medium change, and siRNA transfection are performed as described in Steps 2 and 3, Section 3.3.
2. At 36 h after transfection, the cells are washed twice with PBS and the culture for an additional 12 h in DMEM alone (serum- and antibiotic-free). Discard the medium and add DMEM (serum- and antibiotic-free) containing 50 ng/mL EGF to the cell culture. Then incubate the cells for 15, 45, and 90 min.
3. Discard the medium and wash twice with ice-cold PBS. Immediately after washing, add 100 μL of ice-cold lysis buffer directly into each dish. Transfer each cell suspension into a microcentrifuge tube on ice, and after allowing the tubes to stand on ice for 30 min, centrifuge the tubes at $17,400 \times g$ for 10 min at 4 °C, and recover the supernatants in a new microcentrifuge tube.
4. Follow Step 5 in Section 2.2.
5. Analyze the samples by 8% SDS-PAGE.
6. Follow Steps 7 and 8 in Section 2.2.
7. Incubate the membrane for 1 h at room temperature with anti-EGFR antibody (1/5000 dilution in 1%, w/v, skim milk in PBS-T) or anti-β-actin antibody (1/10,000 dilution in 1%, w/v, skim milk in PBS-T), with constant agitation. Wash the membrane three times with PBS-T for 5 min at room temperature, with constant agitation.
8. Incubate the membrane for 45 min at room temperature with HRP-conjugated anti-sheep IgG antibody (1/10,000 dilution in 1%, w/v, skim milk in PBS-T) or HRP-conjugated anti-mouse IgG antibody (1/10,000 dilution in 1%, w/v, skim milk in PBS-T), with constant agitation. Wash the membrane three times with PBS-T for 8 min at room temperature, with constant agitation.
9. Follow Step 11 in Section 2.2.

4.3. Materials for the Tf recycling assay

1. Alexa Fluor 594-conjugated Tf (Molecular Probes, T13343).

4.4. Tf recycling assay in Rab12-knockdown cells

1. Prepare MEF cells as described in Step 1, Section 3.1.
2. Transfect siRNA into the cells as described in Step 3, Section 3.3.
3. Prior to the assay, wash the cells twice with PBS and add DMEM (serum- and antibiotic-free) to the cell culture. After incubation for 4 h, change the medium to DMEM containing 20 μg/mL of Alexa Fluor 594-conjugated Tf and incubate the cells at 37 °C for 5 min.
4. After washing the cells twice with cold culture medium, immediately fix the cells (sample time: 0 min) or incubate the cells in the culture medium (without Alexa Fluor 594-conjugated Tf) at 37 °C for 30 min. After the incubation, fix the cells (sample time: 30 min).
5. Follow Step 7 in Section 2.4.
6. Measure the intracellular intensity of Alexa 594-Tf immunofluorescence signals (>40 cells) with ImageJ software. To evaluate the efficiency of Tf recycling, normalize the intensity of data obtained when time = 30 to the data obtained when time = 0.

5. CONCLUDING REMARKS

TfR is a well-known marker for recycling endosomes, and it has already been used to study the recycling pathway (Dautry-Varsat et al., 1983; Maxfield & McGraw, 2004; Ren et al., 1998). However, our approach has yielded a new insight into TfR trafficking, namely, that TfR can be used as a marker not only of the recycling pathway but of a novel membrane trafficking pathway from recycling endosomes to lysosomes. In this chapter we have described the detailed protocols that have been used to analyze this novel membrane trafficking pathway. These methods will be useful in investigating the fate, that is, recycling or degradation, of plasma membrane receptors and channels.

ACKNOWLEDGMENTS

This work was supported in part by Grants-in-Aid for Scientific Research from the Ministry of Education, Culture, Sports, and Technology (MEXT) of Japan (to M. F.). T. M. was supported by the Japan Society for the Promotion of Science (JSPS).

REFERENCES

Daro, E., van der Sluijs, P., Galli, T., & Mellman, I. (1996). Rab4 and cellubrevin define different early endosome populations on the pathway of transferrin receptor recycling. *Proceedings of the National Academy of Sciences of the United States of America*, *93*, 9559–9564.

Dautry-Varsat, A., Ciechanover, A., & Lodish, H. F. (1983). pH and the recycling of transferrin during receptor-mediated endocytosis. *Proceedings of the National Academy of Sciences of the United States of America, 80,* 2258–2262.

Fukuda, M. (2010). How can mammalian Rab small GTPases be comprehensively analyzed?: Development of new tools to comprehensively analyze mammalian Rabs in membrane traffic. *Histology and Histopathology, 25,* 1473–1480.

Klausner, R. D., van Renswoude, J., Ashwell, G., Kempf, C., Schechter, A. N., Dean, A., et al. (1983). Receptor-mediated endocytosis of transferrin in K562 cells. *Journal of Biological Chemistry, 258,* 4715–4724.

Kuroda, T. S., & Fukuda, M. (2004). Rab27A-binding protein Slp2-a is required for peripheral melanosome distribution and elongated cell shape in melanocytes. *Nature Cell Biology, 6,* 1195–1203.

Matsui, T., & Fukuda, M. (2011). Small GTPase Rab12 regulates transferrin receptor degradation: Implications for a novel membrane trafficking pathway from recycling endosomes to lysosomes. *Cellular Logistics, 1,* 155–158.

Matsui, T., Itoh, T., & Fukuda, M. (2011). Small GTPase Rab12 regulates constitutive degradation of transferrin receptor. *Traffic, 12,* 1432–1443.

Maxfield, F. R., & McGraw, T. E. (2004). Endocytic recycling. *Nature Reviews Molecular Cell Biology, 5,* 121–132.

Ren, M., Xu, G., Zeng, J., De Lemos-Chiarandini, C., Adesnik, M., & Sabatini, D. D. (1998). Hydrolysis of GTP on rab11 is required for the direct delivery of transferrin from the pericentriolar recycling compartment to the cell surface but not from sorting endosomes. *Proceedings of the National Academy of Sciences of the United States of America, 95,* 6187–6192.

Rutledge, E. A., Mikoryak, C. A., & Draper, R. K. (1991). Turnover of the transferrin receptor is not influenced by removing most of the extracellular domain. *Journal of Biological Chemistry, 266,* 21125–21130.

Schlierf, B., Fey, G. H., Hauber, J., Hocke, G. M., & Rosorius, O. (2000). Rab11b is essential for recycling of transferrin to the plasma membrane. *Experimental Cell Research, 259,* 257–265.

Tsuboi, T., & Fukuda, M. (2006). Rab3A and Rab27A cooperatively regulate the docking step of dense-core vesicle exocytosis in PC12 cells. *Journal of Cell Science, 119,* 2196–2203.

Ullrich, O., Reinsch, S., Urbé, S., Zerial, M., & Parton, R. G. (1996). Rab11 regulates recycling through the pericentriolar recycling endosome. *Journal of Cell Biology, 135,* 913–924.

Ward, J. H., Kushner, J. P., & Kaplan, J. (1982). Regulation of HeLa cell transferrin receptors. *Journal of Biological Chemistry, 257,* 10317–10323.

Weigert, R., Yeung, A. C., Li, J., & Donaldson, J. G. (2004). Rab22a regulates the recycling of membrane proteins internalized independently of clathrin. *Molecular Biology of the Cell, 15,* 3758–3770.

CHAPTER TWELVE

Measurement of Intercellular Transfer to Signaling Endosomes

Kristopher D. Marjon, Jennifer M. Gillette[1]

Department of Pathology, University of New Mexico Health Sciences Center, Albuquerque, New Mexico, USA
[1]Corresponding author: e-mail address: JGillette@salud.unm.edu

Contents

1. Introduction	208
2. Labeling of Transferred Signals	209
2.1 Cellular proteins and lipids	209
2.2 Exosomes/microvesicles	211
2.3 Nucleic acids	212
3. Measuring ICT Using Fluorescence Techniques	214
3.1 Flow cytometry	214
3.2 Confocal microscopy	214
4. Signaling Endosomes	216
4.1 Examples of signaling endosomes: Labeling and signaling pathways	216
5. Experimental Example: ICT to SARA-Positive Signaling Endosomes	217
6. Summary	218
References	219

Abstract

Cell–cell communication is essential for an abundance of physiological processes. As such, various mechanisms have evolved to regulate and ensure proper cell-to-cell signaling. Recently, a novel mechanism of cell communication has emerged which involves the physical transfer of proteins, lipids, and nucleic acids between cells. Following this process termed intercellular transfer (ICT), the transferred molecules can signal within recipient cells by entering the endosomal system and trafficking to signaling endosomes. Signaling endosomes can modulate signal localization within the cell as well as the specificity of, and cross talk between, signaling pathways. As such, ICT into signaling endosomes has the potential to modify the signaling profile of the recipient cell. In this chapter, we describe the different methods of ICT as well as how transfer to signaling endosomes can be visualized and quantified.

1. INTRODUCTION

The ability of cells to receive signals from neighboring cells and correctly respond to their surrounding microenvironment is critical for early development, immunity, and normal tissue homeostasis and repair. Furthermore, disruption in the exchange or processing of intercellular signals is responsible for a number of disease states such as cancer, diabetes, and autoimmunity. Classical methods of intercellular communication mediated by cell–cell or cell–matrix interactions involve membrane protein binding, secreted signaling molecules, and gap junctions. More recently, a novel mechanism of cell communication has emerged whereby proteins, lipids, and even nucleic acids are physically exchanged between cells in a process of intercellular transfer (ICT).

Cell communication through ICT is mediated by a diverse set of mechanisms. Currently, two main subtypes of ICT have been described based upon whether a direct physical connection between cells is required. Contact-dependent transfer mechanisms include the use of membrane bridges, cellular phagocytosis or trogocytosis, and a combination of endocytosis and protein proteolysis (Ahmed, Munegowda, Xie, & Xiang, 2008; Davis & Sowinski, 2008; Joly & Hudrisier, 2003; Kramer, 2000; Nichols et al., 2007; Sherer & Mothes, 2008). At present, only one mechanism of ICT exists in which the transfer of cargo can occur between cells not in direct contact. This type of transfer requires the release of material in a membrane-bound vesicle structure, like microvesicles or exosomes (Al-Nedawi, Meehan, & Rak, 2009; Cocucci, Racchetti, & Meldolesi, 2009; Lykke-Andersen, Brodersen, & Jensen, 2009; Pap, Pallinger, Pasztoi, & Falus, 2009; Thery, Ostrowski, & Segura, 2009).

In addition to the diverse methods used by cells to transfer cellular material, a variety of "transferable" cargos have been identified. Studies have established the transfer of a number of proteins such as major histocompatibility complex-II, P-glycoprotein, and a truncated form of the epidermal growth factor receptor (EGFR) (Al-Nedawi et al., 2008; Ambudkar, Sauna, Gottesman, & Szakacs, 2005; Levchenko et al., 2005; Patel, Arnold, White, Nardella, & Mannie, 1999). The transfer of nucleic acids including small RNAs and microRNAs (miRs) has also been identified indicating transcriptional communication between cells (Dinger, Mercer, & Mattick, 2008; Mittelbrunn et al., 2011; Simpson, Lim, Moritz, & Mathivanan, 2009; Valadi et al., 2007). Additionally, the transfer of lipids and even organelles has been observed between cells (Islam et al., 2012; Spees, Olson, Whitney, & Prockop, 2006), thus emphasizing the broad potential of signal activation downstream of ICT.

Once transferred to a recipient cell, the molecules can be processed in a variety of ways to stimulate a signaling response. Regulation of signaling by endocytosis has functional importance in many biological and physiological processes and increasing evidence suggests that endosomes can regulate specific signaling pathways within cells (Miaczynska, Pelkmans, & Zerial, 2004; Murphy, Padilla, Hasdemir, Cottrell, & Bunnett, 2009; Palfy, Remenyi, & Korcsmaros, 2012; Sadowski, Pilecka, & Miaczynska, 2009). Therefore, in this chapter, we concentrate on endosomal-mediated signaling with a particular focus on signaling endosomes. Signaling endosomes serve as a physical platform for the assembly of signaling complexes where endosomal scaffold proteins can facilitate the signaling reactions between recruited components. A recent review highlights many of the identified endosomal scaffolds, which include SMAD Anchor for Receptor Activation (SARA) and APPL (Adaptor protein containing PH domain, PTB domain, and Leucine zipper motif) (Miaczynska, Christoforidis, et al., 2004; Palfy et al., 2012). For example, SARA-positive endosomes are responsible for propagating the extracellular TGFβ signal through to the nucleus by the recruitment of SMAD proteins, whereas APPL-positive endosomes can regulate Wnt signaling through interactions with GSK3β (Panopoulou et al., 2002; Schenck et al., 2008). Furthermore, transferred molecules have been shown to associate with endosomal scaffold proteins suggesting a signaling link through the endocytic pathway (Gillette, Larochelle, Dunbar, & Lippincott-Schwartz, 2009). Here, we describe methods to visualize ICT and quantify the association of transferred cargo with signaling endosomes.

2. LABELING OF TRANSFERRED SIGNALS

Proteins, lipids, nucleic acids, organelles, and membrane-bound vesicles such as exosomes and microvesicles have all been described to transfer between cells. This diversity in the types of transferrable cargos identified to this point illustrates that the appropriate labeling procedure will depend strongly upon the type of signal that is being transferred.

2.1. Cellular proteins and lipids

In order to visualize the transfer of proteins, such as EGFR (Al-Nedawi et al., 2008) or P-glycoprotein (Levchenko et al., 2005) between cells, one can most efficiently quantify this process by using fluorescent protein (FP) tags. The noninvasive nature of FPs makes them ideal for visualizing dynamic, live cell processes, such as ICT. Furthermore, simple transfection techniques enable the cellular expression of a protein of interest labeled with any number of FPs (Day & Davidson, 2009), which can be tracked between cells by fluorescence methods such as microscopy and flow cytometry (described below). The primary drawback to using FPs to follow ICT is the fact that it requires the proteins to be overexpressed. As such, the transfer of endogenous proteins cannot be visualized with this labeling method. An additional consideration for this type of labeling is the topology of the protein of interest and the cellular location of the FP. For example, the transfer of the extracellular fragment of Notch occurs following an enzymatic cleavage step that occurs at an extracellular site within Notch (Nichols et al., 2007). As such, if Notch were tagged at its cytoplasmic tail with an FP, ICT would be undetected. To visualize the transfer of endogenous proteins, other tags such as quantum dots or fluorescently conjugated antibodies can be used to track the transfer of surface proteins. However, because these labels cannot penetrate the plasma membrane, this approach is only suitable for imaging live transfer of surface proteins. Finally, plasma membrane glycoproteins can be labeled using fluorescently conjugated lectins such as wheat germ agglutinin (WGA) (Invitrogen) (Fig. 12.1).

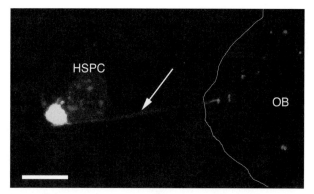

Figure 12.1 Intercellular transfer via membrane nanotubes. The hematopoietic stem/progenitor cell line KG1a (HSPC) was transfected with YFP-CD82 (Addgene) and then labeled on ice with Alexa-555-WGA (Invitrogen). The labeled KG1a cells were cocultured with osteoblastic cells (SaOS$_2$) for 1 h and imaged by confocal microscopy. The arrow designates the membrane nanotube formed from the KG1a cell to the OB. Both Alexa-555-WGA and YFP-CD82 are observed within the osteoblastic cell (outlined with a white line—OB) indicating intracellular transfer of protein from the KG1a cell to the osteoblastic cell. Scale bar—10 μm. (For color version of this figure, the reader is referred to the online version of this chapter.)

In addition to protein, the ICT of lipids has been reported to occur between various cells including many cells of the immune system (Ahmed et al., 2008; Gillette et al., 2009; Gutierrez-Vazquez, Villarroya-Beltri, Mittelbrunn, & Sanchez-Madrid, 2013). Membrane dyes such as PKH-26 (Sigma) and the lipophilic tracers (DiI, DiO, DiD, DiA, and DiR—Invitrogen) can be easily used to label cellular membranes and then visualize the cellular exchange of lipids. Furthermore, these methods of fluorescently labeling membrane lipids are critical for observing the ICT that occurs through membrane bridges and/or membrane nanotubes (Davis & Sowinski, 2008).

2.2. Exosomes/microvesicles

When the limiting membrane of multivesicular bodies (MVBs) fuses with the plasma membrane, the intraluminal vesicles localized within the MVBs are released into the extracellular space. These 10–100 nm membrane bound structures have been termed exosomes and can serve to signal between cells at a distance. In contrast, microvesicles are membrane-bound vesicles that bud from the plasma membrane into the extracellular space in a manner similar to viral budding. Both of these membrane-bound vesicles can be exchanged between cells transferring proteins, lipids, and nucleic acids (Gutierrez-Vazquez et al., 2013; Lee, El Andaloussi, & Wood, 2012).

2.2.1 Exosome labeling de novo
1. Wash cells of interest into HBSS.
2. With a Hamilton syringe inject labeled lipids N-(lissamine rhodamine B sulfonyl)-phosphatidylethanolamine (N-Rh-PE) from Avanti Polar Lipids into Hanks buffer (pH 7.4) while vortexing.
3. Incubate cells with the labeled lipids for 60 min at 4 °C.
4. Wash cells extensively with cold HBSS (Booth et al., 2006; Vidal, Mangeat, & Hoekstra, 1997).
5. Culture cells for 1–2 days for exosome enrichment in the conditioned medium, which can be isolated as described below.

2.2.2 Exosome/microvesicle isolation
1. Grow cells of interest in serum-free medium or with exosome-free/exosome-depleted medium. It is important to note that standard serum contains exosomes that can contaminate the isolation prep. Therefore,

media that has been precleared of serum-derived exosomes by ultracentrifugation should be used.
2. After the desired culture time, transfer the conditioned medium to conical tubes.
3. If working with suspension cells, centrifuge at $300 \times g$ for 10 min at 4 °C to the pellet cells.
4. Remove supernatant and transfer to ultracentrifuge tubes, balancing the tubes with PBS if necessary.
5. Centrifuge the tubes at $16,500 \times g$ for 20 min at 4 °C to remove cellular debris.
6. Filter supernatant through a 0.2-μm filter to enrich for particles smaller than 200 nm.
7. Transfer filtered supernatant to a new ultracentrifuge tube and centrifuge at $120,000 \times g$ for 70 min at 4 °C
8. Remove supernatant keeping the exosome/microvesicle pellet undisturbed.
9. Resuspend the pellet (50–150 μl) in the buffer intended for downstream applications.
10. To further purify exosomes from the microvesicle fraction, the pellet can be floated on a sucrose gradient and the exosomes will be found primarily in the fraction representing a density of 1.13–1.19 g/ml (Chaput & Thery, 2011). Electron microscopy can be an additional technique used to characterize the size and heterogeneity of the isolated exosome/microvesicle prep (Fig. 12.2).
11. As a control, exosome release can be inhibited by treating cells with 10 μM GW4869 (neutral sphingomylinase inhibitor, Calbiochem) for 24 h (Kosaka et al., 2010).

2.2.3 Exosome/microvesicle labeling after purification

1. Purified exosomes or microvesicles can be labeled with 4.8 μM BODIPY-ceramide, BODIPY-PC or 2-6-(7-nitrobenz-2-oxa-1,3-diazol-4-yl) amino (NBD)-PC in PBS (Molecular Probes).
2. Incubate for 30 min at 37 °C protected from light.
3. Remove unbound lipids with two ultracentrifugation spins at $110,000 \times g$ for 70 min at 4 °C.

2.3. Nucleic acids

In addition to the transfer of proteins, genetic materials such as miRs have been shown to transfer between cells through both exosomes and

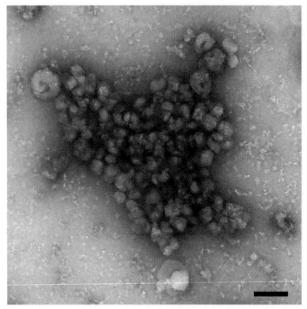

Figure 12.2 Electron microscopy characterization of isolated exosomes and microvesicles. Following the ultracentrifugation protocol in Section 2.2.2, the isolated membrane vesicle fraction was visualized by electron microscopy. Vesicles ranging from approximately 20 to 100 nm are detected. Scale bar—100 nm.

microvesicles (Ismail et al., 2013; Umezu, Ohyashiki, Kuroda, & Ohyashiki, 2013). To quantify miR transfer, fluorescent tags can be added to the miRs of interest, which can then be transfected into cells.

2.3.1 Labeling of miRs
1. Identify miRNA of interest and obtain pre-miR miRNA precursor (PRE-miR miRNA precursor, Ambion).
2. Label pre-miR miRNA precursor with Label IT siRNA tracker CY3 Kit according to manufacturer's instructions.
3. Transfect 1×10^5 cells with 10 nM of Cy3- pre-miR miRNA precursor.
4. Twenty-four hours after transfection, wash cells with PBS and incubate with exosome-free, serum containing medium or serum-free medium for 24 h.
5. Exosomes with labeled miRs can be isolated from the conditioned medium as described earlier.

After labeling and isolation, these vesicles can be used to treat recipient cells such as endothelial cells and osteoblasts (Gillette et al., 2009; Sahoo et al., 2011) and transfer can be quantified by the techniques described below.

3. MEASURING ICT USING FLUORESCENCE TECHNIQUES

Once the transferable signals are labeled within the cells, the coculture system can be set up to allow for the exchange of molecules to recipient cells. The time frame for coculture can range from 1 h to overnight depending upon the transferred molecules and the efficiency of transfer. ICT can be most readily quantified by fluorescence techniques such as flow cytometry and confocal microscopy.

3.1. Flow cytometry

Following coculture, recipient cells should be washed well to remove interacting cells and transferred signals that have not been internalized. If recipient cells are adherent, they will need to be trypsinized, and then cells can be resuspended in cold PAB Buffer (1× PBS, 0.1% Sodium Azide, and 3% BSA) at a cell concentration of 1×10^6–10×10^6 cells/ml for flow cytometry analysis. Cells containing FP-tagged signals transferred during the coculture will be detected (Mittelbrunn et al., 2011). As such, it is possible to calculate the percent of recipient cells that are positive for a transfer event. A critical control includes measuring recipient cells without coculture treatment, which allows a clear negative population of cells to be identified by the cytometer and compared to treatment conditions. While flow cytometry provides critical quantification, it does not allow for the assessment of ICT specifically to signaling endosomes. For this type of analysis, confocal microscopy is ideal.

3.2. Confocal microscopy

Confocal microscopy is implemented so that signaling endosomes can be distinguished within the volume of the cell and it allows for the localization of ICT signals to specific types of endocytic populations. By transfecting FP-tagged endosomal scaffold proteins into recipient cells, it is possible to follow, in real time, the trafficking of transferred signals to specific signaling endosomes (Fig. 12.3). However, if protein overexpression is not experimentally appropriate, then the recipient cells can be fixed and stained with signaling endosome-specific antibodies following ICT (see below). In addition to allowing for the detection of transferred molecules to signaling endosomes, confocal microscopy enables the quantification of colocalization

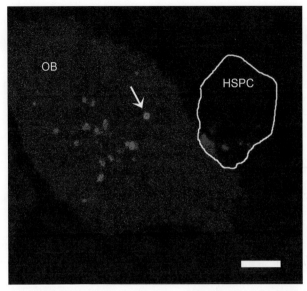

Figure 12.3 Intercellular transfer to endosomes. Osteoblastic cells (SaOS$_2$) were transfected with the 2×FYVE-GFP construct and then cocultured with KG1a cells labeled with N-Rh-PE (HSPC) as indicated in Section 2.2.1. The KG1a cell is outline with a white line. Following 1 h of coculture, N-Rh-PE is detected within a 2×FYVE positive endosome (arrow) in the osteoblastic cell indicating intracellular transfer to the endosomal system. Scale bar—5 μm. (For color version of this figure, the reader is referred to the online version of this chapter.)

between specific endosomes and transferred signals (Gillette et al., 2009). Calculation of fluorescence overlaps between signaling endosomes and transferred signals can be completed by software on most microscopes. The most common correlation in statistics is the Pearson correlation, which in the case of fluorescence imaging shows the correlation between intensity distributions. The Pearson coefficient can be easily affected by the addition of noncolocalizing signals. An alternative analysis uses the Manders coefficient, which indicates the proportion of overlap between different fluorescence channels. While Manders coefficient is sensitive to background signals within the image, it is not sensitive to the intensity of overlapping pixels. For more information about colocalization analysis and its application in confocal imaging, please review the following references (Bolte & Cordelieres, 2006; Zinchuk, Zinchuk, & Okada, 2007) http://www.olympusconfocal.com/java/colocalization/.

4. SIGNALING ENDOSOMES

While endocytosis was originally viewed as a mechanism to degrade or recycle receptors, it is now clear that diverse families of molecules can signal from the endosomal network to control essential cellular responses and behaviors. As such, endosomes can serve as a highly flexible scaffold for mediating the precise spatiotemporal control and transport of diverse biological signals.

4.1. Examples of signaling endosomes: Labeling and signaling pathways

The endosomal membrane that faces the cytoplasm can serve as a recruitment platform for signaling complexes that are regulated by scaffold proteins. Similar to scaffolds in the plasma membrane, endocytic scaffold proteins can facilitate downstream signal activation and serve to regulate interactions between different signaling pathways. Next, we describe two examples of signaling endosomes, their signaling pathways, and how they can be visualized.

4.1.1 APPL endosomes

Recent studies have shown that when the scaffold protein APPL is localized to endosomes, it can regulate the specificity of Akt signaling. More specifically, APPL on endosomes interacts with Akt and GSK3β thereby serving as a hub for the cross talk between insulin signaling (Akt) and Wnt signaling (GSK3β). APPL-mediated cross talk between these signaling pathways can regulate cellular functions such as survival, proliferation, and metabolism (Schenck et al., 2008; Wen et al., 2010). Trafficking of transferred cargo to APPL-positive endosomes can be visualized by immunofluorescence imaging. Following transfer, recipient cells can be fixed, permeabilized, and stained with anti-APPL antibodies (Abcam, ab59592). For live cell imaging of APPL endosomes, the recipient cells can be transfected with APPL-GFP prior to coculture (Zoncu et al., 2009).

4.1.2 SARA endosomes

SARA is an endosome-specific scaffold protein that is most widely known to regulate the TGF-β signaling pathway by serving to recruit and interact with SMAD2/3. However, in addition to TGF-β signaling, a recent study identified the involvement of SARA endosomes in Wnt signaling. In a yeast-two hybrid screen, SARA was found to interact with β-catenin (Colland et al., 2004). As such, endosomally localized SARA may serve to connect the TGF-β and Wnt signaling pathways through the combined recruitment of SMADs and β-catenin (Zhang et al., 2010). For live-cell trafficking of transfer to SARA endosomes, the SARA-GFP construct can be transfected into recipient cells (Coumailleau, Furthauer, Knoblich, & Gonzalez-Gaitan, 2009). In addition, anti-SARA antibodies can be used for immmunofluorescence labeling and confocal imaging of SARA endosomes (Santa Cruz Biotechnology, sc-9135).

5. EXPERIMENTAL EXAMPLE: ICT TO SARA-POSITIVE SIGNALING ENDOSOMES

In the example presented here, we describe the experimental processes of measuring ICT to SARA-positive signaling endosomes. We begin by plating recipient cells in a Labtek Chambered-Coverglass from Nunc, which has a coverslip bottom that is compatible with confocal imaging.

1. Plate recipient cells in the chambered-coverglass and grow for 18 h.
2. Transfect transfer cells with an FP-tagged molecule of interest 18 h before coculturing.
3. Before imaging, rinse the recipient cells with fresh medium and cover the cells with CO_2 independent medium (unless a CO_2 chamber is used for the microscope).
4. Add the transfected transfer cells to the well. Usually at a concentration of around 1:5 compared to the recipient cells, so 10,000 transfer cells for every 50,000 recipient cells.
5. ICT can be imaged live on the microscope for the time frame of the experiment, usually 1–5 h.
6. Following the allowed transfer time, the transfer cells can be washed away and the recipient cells can be fixed and processed for immunofluorescence labeling.
7. Using antibodies specific to a signaling endosome scaffold protein, such as SARA, the signaling endosome population of interest can be labeled.
8. Once labeled, the chambered-coverglass can be taken back to the confocal microscope to image the colocalization between the FP-tagged transferred protein and the antibody labeled SARA signaling endosomes.
9. Quantification of colocalization between two different fluorophores can be completed by calculating the Pearson correlation coefficient or the Manders coefficient. These calculations can be completed using most microscope softwares or the colocalization plugin within ImageJ (Fig. 12.4).

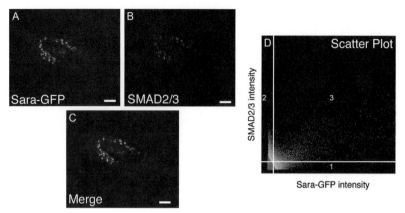

Figure 12.4 Measurement of colocalization within signaling endosomes. Osteoblastic cells (SaOS$_2$) were transfected with Sara-GFP (A) and then fixed and stained with SMAD2/3 (B). The merge image (C) indicates the localization of endogenous SMAD2/3 to the Sara-GFP endosomes. Scale bar—10 μm. The scatter plot (D) indicates the Sara-GFP colocalization with SMAD2/3 (Section 3). Calculations using the Zeiss LSM software indicate an overlap coefficient of 0.9, whereas the correlation $R = 0.61$. These data suggest colocalization between the two fluorescent signals. (See color plate.)

6. SUMMARY

Here, we provide an overview of ICT as a new method of intercellular communication. We describe techniques that can be used to label and visualize various transfer processes. In addition, we describe how these transferred cargos can be tracked to, and measured within, signaling endosomes. As this field of study is in its infancy, much remains to be determined regarding the activation and propagation of signaling mechanisms downstream of ICT to signaling endosomes. Substantial evidence exists confirming ICT as a method of communication and the role of the endosomal system in cellular signaling. However, future studies are required to identify the mechanisms that result in the specific localization of transferred signals to signaling endosomes. Furthermore, we still know very little about how molecules stimulate cellular signaling from within endosomes. Future challenges include a better understanding of the physiological relevance of endosomal signaling and whether endosomal initiated signals transmit a unique and functionally distinct signaling response.

REFERENCES

Ahmed, K. A., Munegowda, M. A., Xie, Y., & Xiang, J. (2008). Intercellular trogocytosis plays an important role in modulation of immune responses. *Cellular & Molecular Immunology, 5*, 261–269.

Al-Nedawi, K., Meehan, B., Micallef, J., Lhotak, V., May, L., Guha, A., et al. (2008). Intercellular transfer of the oncogenic receptor EGFRvIII by microvesicles derived from tumour cells. *Nature Cell Biology, 10*, 619–624.

Al-Nedawi, K., Meehan, B., & Rak, J. (2009). Microvesicles: Messengers and mediators of tumor progression. *Cell Cycle, 8*, 2014–2018.

Ambudkar, S. V., Sauna, Z. E., Gottesman, M. M., & Szakacs, G. (2005). A novel way to spread drug resistance in tumor cells: Functional intercellular transfer of P-glycoprotein (ABCB1). *Trends in Pharmacological Sciences, 26*, 385–387.

Bolte, S., & Cordelieres, F. P. (2006). A guided tour into subcellular colocalization analysis in light microscopy. *Journal of Microscopy, 224*, 213–232.

Booth, A. M., Fang, Y., Fallon, J. K., Yang, J. M., Hildreth, J. E., & Gould, S. J. (2006). Exosomes and HIV Gag bud from endosome-like domains of the T cell plasma membrane. *Journal of Cell Biology, 172*, 923–935.

Chaput, N., & Thery, C. (2011). Exosomes: Immune properties and potential clinical implementations. *Seminars in Immunopathology, 33*, 419–440.

Cocucci, E., Racchetti, G., & Meldolesi, J. (2009). Shedding microvesicles: Artefacts no more. *Trends in Cell Biology, 19*, 43–51.

Colland, F., Jacq, X., Trouplin, V., Mougin, C., Groizeleau, C., Hamburger, A., et al. (2004). Functional proteomics mapping of a human signaling pathway. *Genome Research, 14*, 1324–1332.

Coumailleau, F., Furthauer, M., Knoblich, J. A., & Gonzalez-Gaitan, M. (2009). Directional Delta and Notch trafficking in Sara endosomes during asymmetric cell division. *Nature, 458*, 1051–1055.

Davis, D. M., & Sowinski, S. (2008). Membrane nanotubes: Dynamic long-distance connections between animal cells. *Nature Reviews Molecular Cell Biology, 9*, 431–436.

Day, R. N., & Davidson, M. W. (2009). The fluorescent protein palette: Tools for cellular imaging. *Chemical Society Reviews, 38*, 2887–2921.

Dinger, M. E., Mercer, T. R., & Mattick, J. S. (2008). RNAs as extracellular signaling molecules. *Journal of Molecular Endocrinology, 40*, 151–159.

Gillette, J. M., Larochelle, A., Dunbar, C. E., & Lippincott-Schwartz, J. (2009). Intercellular transfer to signalling endosomes regulates an ex vivo bone marrow niche. *Nature Cell Biology, 11*, 303–311.

Gutierrez-Vazquez, C., Villarroya-Beltri, C., Mittelbrunn, M., & Sanchez-Madrid, F. (2013). Transfer of extracellular vesicles during immune cell-cell interactions. *Immunological Reviews, 251*, 125–142.

Islam, M. N., Das, S. R., Emin, M. T., Wei, M., Sun, L., Westphalen, K., et al. (2012). Mitochondrial transfer from bone-marrow-derived stromal cells to pulmonary alveoli protects against acute lung injury. *Nature Medicine, 18*, 759–765.

Ismail, N., Wang, Y., Dakhlallah, D., Moldovan, L., Agarwal, K., Batte, K., et al. (2013). Macrophage microvesicles induce macrophage differentiation and miR-223 transfer. *Blood, 121*, 984–995.

Joly, E., & Hudrisier, D. (2003). What is trogocytosis and what is its purpose? *Nature Immunology, 4*, 815.

Kosaka, N., Iguchi, H., Yoshioka, Y., Takeshita, F., Matsuki, Y., & Ochiya, T. (2010). Secretory mechanisms and intercellular transfer of microRNAs in living cells. *Journal of Biological Chemistry, 285*, 17442–17452.

Kramer, H. (2000). RIPping notch apart: A new role for endocytosis in signal transduction? *Science's STKE, 2000*, PE1.
Lee, Y., El Andaloussi, S., & Wood, M. J. (2012). Exosomes and microvesicles: Extracellular vesicles for genetic information transfer and gene therapy. *Human Molecular Genetics, 21*, R125–R134.
Levchenko, A., Mehta, B. M., Niu, X., Kang, G., Villafania, L., Way, D., et al. (2005). Intercellular transfer of P-glycoprotein mediates acquired multidrug resistance in tumor cells. *Proceedings of the National Academy of Sciences of the United States of America, 102*, 1933–1938.
Lykke-Andersen, S., Brodersen, D. E., & Jensen, T. H. (2009). Origins and activities of the eukaryotic exosome. *Journal of Cell Science, 122*, 1487–1494.
Miaczynska, M., Christoforidis, S., Giner, A., Shevchenko, A., Uttenweiler-Joseph, S., Habermann, B., et al. (2004). APPL proteins link Rab5 to nuclear signal transduction via an endosomal compartment. *Cell, 116*, 445–456.
Miaczynska, M., Pelkmans, L., & Zerial, M. (2004). Not just a sink: Endosomes in control of signal transduction. *Current Opinion in Cell Biology, 16*, 400–406.
Mittelbrunn, M., Gutierrez-Vazquez, C., Villarroya-Beltri, C., Gonzalez, S., Sanchez-Cabo, F., Gonzalez, M. A., et al. (2011). Unidirectional transfer of microRNA-loaded exosomes from T cells to antigen-presenting cells. *Nature Communications, 2*, 282.
Murphy, J. E., Padilla, B. E., Hasdemir, B., Cottrell, G. S., & Bunnett, N. W. (2009). Endosomes: A legitimate platform for the signaling train. *Proceedings of the National Academy of Sciences of the United States of America, 106*, 17615–17622.
Nichols, J. T., Miyamoto, A., Olsen, S. L., D'Souza, B., Yao, C., & Weinmaster, G. (2007). DSL ligand endocytosis physically dissociates Notch1 heterodimers before activating proteolysis can occur. *Journal of Cell Biology, 176*, 445–458.
Palfy, M., Remenyi, A., & Korcsmaros, T. (2012). Endosomal crosstalk: Meeting points for signaling pathways. *Trends in Cell Biology, 22*, 447–456.
Panopoulou, E., Gillooly, D. J., Wrana, J. L., Zerial, M., Stenmark, H., Murphy, C., et al. (2002). Early endosomal regulation of Smad-dependent signaling in endothelial cells. *Journal of Biological Chemistry, 277*, 18046–18052.
Pap, E., Pallinger, E., Pasztoi, M., & Falus, A. (2009). Highlights of a new type of intercellular communication: Microvesicle-based information transfer. *Inflammation Research, 58*, 1–8.
Patel, D. M., Arnold, P. Y., White, G. A., Nardella, J. P., & Mannie, M. D. (1999). Class II MHC/peptide complexes are released from APC and are acquired by T cell responders during specific antigen recognition. *Journal of Immunology, 163*, 5201–5210.
Sadowski, L., Pilecka, I., & Miaczynska, M. (2009). Signaling from endosomes: Location makes a difference. *Experimental Cell Research, 315*, 1601–1609.
Sahoo, S., Klychko, E., Thorne, T., Misener, S., Schultz, K. M., Millay, M., et al. (2011). Exosomes from human CD34(+) stem cells mediate their proangiogenic paracrine activity. *Circulation Research, 109*, 724–728.
Schenck, A., Goto-Silva, L., Collinet, C., Rhinn, M., Giner, A., Habermann, B., et al. (2008). The endosomal protein Appl1 mediates Akt substrate specificity and cell survival in vertebrate development. *Cell, 133*, 486–497.
Sherer, N. M., & Mothes, W. (2008). Cytonemes and tunneling nanotubules in cell-cell communication and viral pathogenesis. *Trends in Cell Biology, 18*, 414–420.
Simpson, R. J., Lim, J. W., Moritz, R. L., & Mathivanan, S. (2009). Exosomes: Proteomic insights and diagnostic potential. *Expert Review of Proteomics, 6*, 267–283.
Spees, J. L., Olson, S. D., Whitney, M. J., & Prockop, D. J. (2006). Mitochondrial transfer between cells can rescue aerobic respiration. *Proceedings of the National Academy of Sciences of the United States of America, 103*, 1283–1288.
Thery, C., Ostrowski, M., & Segura, E. (2009). Membrane vesicles as conveyors of immune responses. *Nature Reviews Immunology, 9*, 581–593.

Umezu, T., Ohyashiki, K., Kuroda, M., & Ohyashiki, J. H. (2013). Leukemia cell to endothelial cell communication via exosomal miRNAs. *Oncogene, 32*, 2747–2755.

Valadi, H., Ekstrom, K., Bossios, A., Sjostrand, M., Lee, J. J., & Lotvall, J. O. (2007). Exosome-mediated transfer of mRNAs and microRNAs is a novel mechanism of genetic exchange between cells. *Nature Cell Biology, 9*, 654–659.

Vidal, M., Mangeat, P., & Hoekstra, D. (1997). Aggregation reroutes molecules from a recycling to a vesicle-mediated secretion pathway during reticulocyte maturation. *Journal of Cell Science, 110*(Pt. 16), 1867–1877.

Wen, L., Yang, Y., Wang, Y., Xu, A., Wu, D., & Chen, Y. (2010). Appl1 is essential for the survival of Xenopus pancreas, duodenum, and stomach progenitor cells. *Developmental Dynamics, 239*, 2198–2207.

Zhang, M., Wang, M., Tan, X., Li, T. F., Zhang, Y. E., & Chen, D. (2010). Smad3 prevents beta-catenin degradation and facilitates beta-catenin nuclear translocation in chondrocytes. *Journal of Biological Chemistry, 285*, 8703–8710.

Zinchuk, V., Zinchuk, O., & Okada, T. (2007). Quantitative colocalization analysis of multicolor confocal immunofluorescence microscopy images: Pushing pixels to explore biological phenomena. *Acta Histochemica et Cytochemica, 40*, 101–111.

Zoncu, R., Perera, R. M., Balkin, D. M., Pirruccello, M., Toomre, D., & De Camilli, P. (2009). A phosphoinositide switch controls the maturation and signaling properties of APPL endosomes. *Cell, 136*, 1110–1121.

CHAPTER THIRTEEN

Liposome-Based Assays to Study Membrane-Associated Protein Networks

Christian Niehage, Christoph Stange, Mihaela Anitei, Bernard Hoflack[1]

Biotechnology Center, Dresden University of Technology, Dresden, Germany
[1]Corresponding author: e-mail address: bernard.hoflack@biotec.tu-dresden.de

Contents

1. Introduction — 224
2. Isolation of Core Machineries Required for Carrier Biogenesis on Synthetic Membranes — 226
   2.1 Preparation of mouse brain cytosol — 227
   2.2 Expression and purification of 6xHis-MBP-Tev-Cys-tagged proteins — 227
   2.3 Preparation of liposomes — 229
   2.4 Coupling of peptides to liposomes via cysteine — 230
   2.5 Coupling of peptides to liposomes via an alpha-hydrazino acetyl group — 231
   2.6 Binding of cytosolic proteins to functionalized liposomes — 233
   2.7 Purification of coated liposomes — 233
3. Identification of Core Machineries by Mass Spectrometry-Based, Label-Free Quantitative Proteomics — 234
   3.1 SDS-PAGE and in-gel digestion — 234
   3.2 LC–MS acquisition — 235
   3.3 Postacquisition workflow (MaxQuant, Perseus, STRING) — 236
4. Visualization of Protein Dynamics on Giant Unilamellar Vesicles by Fluorescence Microscopy — 238
   4.1 Preparation of cytosol from cells stably expressing fluorescently tagged proteins — 239
   4.2 Preparation of GUVs from liposomes — 239
   4.3 Imaging of protein dynamics — 241
5. Summary — 241
Acknowledgments — 242
References — 242

Abstract

Transport carriers regulate the bidirectional flow of membrane between the compartments of the secretory and endocytic pathways. Their biogenesis relies on the recruitment of a number of cytosolic proteins and protein complexes on specific membrane microdomains with defined protein and lipid compositions. The timely assembly of these cellular machines onto membranes involves multiple protein–protein and protein–lipid interactions and is necessary to select membrane proteins and lipids into nascent carriers, to bend the flat membrane of the donor compartment, to change the shape of this nascent carrier into a tubular–vesicular structure, and to operate its scission from the donor compartment. A challenge in this field of membrane cell biology has been to identify these machineries and to understand their precise function, in particular by studying their spatial and temporal dynamics during carrier biogenesis. During the past years, liposome-based synthetic biology fully recapitulating the fidelity of carrier biogenesis as seen *in vivo* has proved to be instrumental to identify these key cytosolic components using mass spectrometry and their dynamics using fluorescence microscopy. We describe here the methods to isolate on synthetic membranes the protein networks needed for carrier biogenesis, to identify them using label-free quantitative proteomics, and to visualize their dynamics on giant unilamellar vesicles.

1. INTRODUCTION

It is now well established that transport carriers connect the compartments of the secretory and the endocytic pathways. They control the flow of membrane components within these pathways thereby maintaining the identity of the different organelles (Bonifacino & Glick, 2004; Rothman & Wieland, 1996). Their formation, which occurs within few minutes relies on the recruitment of a number of cytosolic proteins and protein complexes, which assemble onto specific membrane microdomains where they segregate selected transmembrane proteins and lipids into nascent carriers, deform flat membranes into curved structures. The curved structures elongate into tubular–vesicular elements, which then detach from the donor compartment. These transport carriers of $\approx$ 100–200 nm in diameter are then free to move along cytoskeleton tracks and fuse with a specific acceptor compartment.

During the last decades, yeast genetics, subcellular proteomics, and cell biological approaches have been instrumental to identify the core machineries involved in the different elementary steps of carrier biogenesis, their interactors, and the key molecules regulating their interaction with membranes. Among those coat protein complexes are found COP I, COP II functioning in bidirectional transport between the endoplasmic reticulum and the Golgi (Duden, 2003; Spang, 2002; Stagg, LaPointe, & Balch, 2007), the clathrin adaptor protein complexes (AP-1 to AP-5) mediating post-Golgi transport, endocytosis or yet unknown steps of membrane traffic (Hirst et al., 2011; Robinson & Bonifacino, 2001), the retromer complex mediating membrane recycling from endosomes (Bonifacino & Hurley, 2008), and the ESCRT complexes functioning in multivesicular body biogenesis (Henne, Buchkovich, & Emr, 2011). These coats recognize sorting motifs in the cytosolic domain of cargo proteins (Bonifacino & Traub, 2003). Other classes of proteins are those, which, together with coats, bend membranes or sense membrane curvature (Itoh & De Camilli, 2006; McMahon & Gallop, 2005) and core machineries regulating actin polymerization, which sustain carrier formation (Anitei & Hoflack, 2012). Finally, there are the GTPases coordinating the assembly of the supramolecular structures needed for vesicle formation and scission (Campelo & Malhotra, 2012).

Transport carrier biogenesis is the result of a timely controlled series of low-affinity protein–protein and protein–lipid interactions occurring on specific membrane microdomains with defined protein and lipid compositions. Among these lipids are the phosphoinositides. Their synthesis, which is catalyzed by specific kinases and phosphatases (Krauss & Haucke, 2007a), is highly compartmentalized and therefore phosphoinositides represent one of the hallmarks to specify the identity of the different compartments of the secretory and endocytic pathways (De Matteis & Godi, 2004; Di Paolo & De Camilli, 2006). Phosphoinositides are used as binding platforms by several components involved in membrane traffic such as some coats or a number of other cytosolic proteins with PH, PX, or FYVE domains. The binding kinetics of proteins binding to phosphoinositides can be greatly enhanced by a coincidence detection, meaning a simultaneous binding to other proteins present (Krauss & Haucke, 2007b).

If the different methods used thus far to explore membrane traffic have been instrumental to identify a plethora of proteins, they also have some limitations. For example, genetic screens in yeast may not identify components, which modulate membrane traffic. Subcellular proteomics of isolated transport carriers may not identify key molecules involved in their biogenesis. Furthermore, *in vivo* assays or *in vitro* systems based on the use of isolated organelles limit the manipulation of the chemical and the physical properties of membranes. Such limitations can be overcome by the use of liposome-based synthetic biology fully recapitulating precise steps of membrane traffic, in which the biochemical and physical properties of membranes can be controlled. These protein complexes functioning under defined conditions could then be investigated using targeted Western blot analysis or quantitative mass spectrometry, which offers an untargeted way to identify unknown proteins (Hubner et al., 2010). The precise spatial and temporal distribution of fluorescently tagged proteins or protein complexes can also be investigated under well-defined conditions using fluorescence microscopy. The biological importance of interesting candidates can be validated in multiple ways like colocalization studies using confocal microscopy and gene silencing using appropriate *in vivo* or *in vitro* assays.

In this chapter, we will summarize the different technological aspects of liposome-based assays used to reconstitute key steps of transport carrier biogenesis, to identify key components using mass spectrometry-based label-free quantitative proteomics, and to visualize their spatial and temporal behavior. The clathrin/AP-1-dependent post-Golgi transport to endosomes and the retromer-dependent transport from endosomes are used as examples.

2. ISOLATION OF CORE MACHINERIES REQUIRED FOR CARRIER BIOGENESIS ON SYNTHETIC MEMBRANES

Because sorting motifs present in cytoplasmic domains of transmembrane proteins interact with coat components and therefore are part of the membrane components stabilizing their interactions with membranes, we have developed chemical methods to couple these cytoplasmic domains onto liposomes (Baust, Czupalla, Krause, Bourel-Bonnet, & Hoflack, 2006; Bourel-Bonnet et al., 2005; Pocha, Wassmer, Niehage, Hoflack, & Knust, 2011). In addition, the use of the same chemistry allows the covalent coupling of wild-type or mutated cytoplasmic domains with the same efficiency. Depending on the length of the cytoplasmic domains to be anchored onto liposomes, we use either a chemically synthesized or in *Escherichia coli* expressed cytosolic domain of a membrane protein. As functional reactive groups, either an amino-terminal hydrazino acetyl group is added to the N-terminus of chemically synthesized peptides or a cysteine is introduced at the N-terminus of cytoplasmic domain of type I transmembrane proteins. A cysteine can also be added at the C-terminus of the cytoplasmic domain of type II transmembrane proteins. The coupling to synthetic liposomes requires an aldehyde-lipid anchor when hydrazino-peptides are used or a maleimido-functionalized lipid anchor for cysteine-targeted polypeptides. These functionalized liposomes are then incubated with cytosol in the presence of the slowly hydrolyzable GTP analog GTP-γ-S to stabilize small GTPases in an active GTP-bound form. Liposomes and their bound proteins are then purified by flotation on density gradients. Sedimentation is an alternative, but should only be used for Western blot analyses as the separation of liposomes from cytosolic proteins is not sufficient for mass spectrometric analysis. A potential obstacle in using flotation density centrifugation is that the density of liposomes may become too high, especially when actin polymerizes on liposomes. Therefore, latrunculin B can be added during the incubation of liposomes with cytosol to prevent extensive actin polymerization.

2.1. Preparation of mouse brain cytosol

Mass spectrometric approaches require the use of well-annotated proteomes. In the nonredundant UniprotKB, the best annotated organism is mouse beside humans. Mouse brain cytosol is easy to prepare and trafficking complex proteins are well expressed in this tissue. However, cytosol from other cell types can be used for more specific studies (e.g., epithelial cells). Cytosol must be prepared in order to reach a concentration of 5–10 mg/ml of total protein in the binding assays. A typical protocol is described below.

1. 30 mice (CD1, 12–14 weeks old females) are decapitated and the brains are dissected.
2. The brains are homogenized with a 7-ml Dounce homogenizer (Gerresheimer, Germany) in 4 ml of recruitment buffer (20 mM HEPES/KOH pH 7.2, 125 mM K-acetate, 2.5 mM Mg-acetate) supplied with complete protease inhibitor cocktail (Roche, Germany).
3. The homogenate is cleared from cell debris by centrifugation for 60 min, $370{,}000 \times g$ at 4 °C and another centrifugation of the supernatant for 60 min, $170{,}000 \times g$ at 4 °C.
4. The supernatant is collected and should be immediately used for recruitment assay. Take an aliquot for estimating the protein concentration.

2.2. Expression and purification of 6xHis-MBP-Tev-Cys-tagged proteins

We use a modified pET28-derived expression plasmid to express tagged versions of cytoplasmic domains of transmembrane proteins. The cytosolic domains of type I membrane proteins are flanked at their amino-terminus by a 6xHis-maltose-binding protein (MBP) dual tag and a Tev protease cleavage site followed by a cysteine (Fig. 13.1). This allows us to purify the recombinant protein, to remove the double His–MBP tag with a high

```
MGSSHHHHHHSSGLVPRGSHMASGKIEEGKLVIWNGDKGYNGLAEVGKKFEKDTGIKVTVEHPDKL
EEKFPQVAATGDGPDIIFWAHDRFGGYAQSGLLAEITPDKAFQDKLYPFTWDAVRYNGKLIAYPIAVEA
LSLIYNKDLLPNPPKTWEEIPALDKELKAKGKSALMFNLQEPYFTWPLIAADGGYAFKYENGKYDIKDV
GVDNAGAKAGLTFLVDLIKNKHMNADTDYSIAEAAFNKGETAMTINGPWAWSNIDTSKVNYGVTVLPT
FKGQPSKPFVGVLSAGINAASPNKELAKEFLENYLLTDEGLEAVNKDKPLGAVALKSYEEELAKDPRIA
ATMENAQKGEIMPNIPQMSAFWYAVRTAVINAASGRQTVDEALKDAQTRITKGENLYFQGSCGRK
RRQSEGTYSPSQQEVAGARLEMDSVLKVPPEERLI
```

Figure 13.1 Sequence of the recombinant, His- and MBP-tagged Crumbs 2. Black: peptide after Tev-cleavage that is coupled to the liposomes via an introduced cysteine (C), representing the cytoplasmic domain of the type I transmembrane protein crumbs 2.

sequence-specific protease, and to use a cysteine for coupling peptides to the maleimide lipid anchor. The cysteine is placed at the carboxy-terminus of the cytoplasmic domain of type II membrane proteins.

1. An *E. coli* expression strain (BL21) is transformed with a pET28-derived expression plasmid of interest. A 5-ml overnight culture is prepared from the respective expression strain in LB medium with the appropriate antibiotic (kanamycin, 30 mg/l) at 37 °C.
2. The culture is then transferred into a 2.5-l flask containing 500 ml of LB medium with kanamycin and cultivated at 37 °C until the optical density at 600 nm reaches a value of 0.6 (1-cm optical path).
3. The culture is cooled to 15 °C and expression of the recombinant protein of interest is induced with 0.1 mM IPTG.
4. The expression is performed overnight at 15 °C.
5. Bacteria are harvested by centrifugation at $2500 \times g$ for 20 min at 4 °C.
6. The pellets are resuspended in 15 ml of ice-cold cell lysis buffer: 50 mM Tris–HCl pH 7.4, 100 mM NaCl, protein inhibitor mix M (Serva, Germany), 20 mM imidazole.
7. 250 U of benzonase (Merck, Germany) are added.
8. The bacteria are lysed using three rounds of homogenization with an EmulsiFlex-C5 Homogenizer (Avestin, Canada).
9. Cell debris are pelleted by centrifugation at $10{,}000 \times g$ for 30 min at 4 °C.
10. The supernatant is passed through a 0.45-μm filter and kept on ice.
11. An empty 10 ml Poly-Prep column (BIO-RAD, CA, USA) is filled with 1 ml IMAC sepharose (GE Healthcare, Sweden).
12. The sepharose is washed three times with 10 ml H_2O, once with 10 ml NiCl, and then three times with 10 ml H_2O.
13. The sepharose is washed one time with cell lysis buffer.
14. The bacterial extract is loaded onto the column, which is then washed three times with 10 ml cell lysis buffer.

15. A first elution step is performed by three washes with 1.8 ml of cell lysis buffer containing 250 mM imidazole. The eluates are kept separately on ice.
16. A second elution step is performed by three washes with 1.8 ml cell lysis buffer containing 500 mM imidazole. The eluates are kept separately on ice.
17. The sepharose column is then washed once with 100 mM EDTA, three times with 10 ml H$_2$O, then filled with 20% ethanol, and kept at 4 °C until reuse for the purification of the same recombinant protein.
18. An aliquot of each of the different eluates is analyzed by SDS-PAGE.
19. The eluates of interest are dialyzed overnight against 20 mM HEPES/KOH pH 7.2, 125 mM K-acetate, 1 mM EDTA.
20. The protein concentration of the eluates is determined, adjusted to 10 mg/ml, and aliquots of 50 μl are kept on −80 °C.

2.3. Preparation of liposomes

Liposomes are prepared using the following molar ratios of lipids: 42% PC, 11% PS, 33% PE, 11% C, 1% PI, 2% anchor (PE) (see Table 13.1). The use of an extruder to control the size of liposomes is optional when binding assays are followed by protein identification using mass spectrometry-based proteomic approaches. Usually, triplicate samples for each condition are needed for label-free quantitative proteomics.

1. For each triplicate, lipids are mixed as follows: 18.0 μl PC, 33.2 μl PE, 12.1 μl PS, 5.75 μl C, 32 μl of the respective phosphoinositides and 6.6 μl PE-MCC in a siliconized 1.5 ml Eppendorf tube.
2. The mixture is thoroughly vortexed and dried down under a stream of nitrogen to yield a thin and homogeneous film of lipids.
3. Lipids are resuspended in 330 μl of coupling buffer (20 mM HEPES/KOH pH 7.2, 125 mM K-acetate, 1 mM EDTA).
4. Six cycles of 4-min-long freezing in liquid nitrogen and 4-min-long thawing in a thermoshaker at 25 °C with thorough vortexing are applied. After these six cycles, the suspension should be homogeneous.
5. For immediate coupling of peptides, three aliquots of the liposome suspension are prepared in new siliconized Eppendorf tubes (100 μl/sample) and kept at room temperature to avoid phase separation for immediate coupling of peptides.
6. An optional extrusion step can be applied before aliquoting. A LipoFast handheld extruder (Avestin, Canada) is assembled according to the

Table 13.1 Lipids used to prepare liposomes

Lipid	Name	Vendor (#order)	Stem solution
PI3P	1,2-Dioleoyl-sn-glycero-3-phospho-(1'-myo-inositol-3'-phosphate)	Avanti (850150)	0.2 mg/ml in $CHCl_3/MEOH/H_2O$ (65:35:8)
PI4P	L-α-Phosphatidylinositol-4-phosphate	Avanti (840045)	0.2 mg/ml in $CHCl_3/MEOH/H_2O$ (65:35:8)
PC	1,2-Dioleoyl-sn-glycero-3-phosphocholine	Avanti (850375)	25 mg/ml in $CHCl_3$
PS	L-α-Phosphatidylserine	Avanti (840032)	10 mg/ml in $CHCl_3$
PE	1,2-Dioleoyl-sn-glycero-3-phosphoethanolamine	Avanti (850725)	10 mg/ml in $CHCl_3$
C	Cholesterol	Avanti (700000)	25 mg/ml in $CHCl_3$
PE-MCC	1,2-Dioleoyl-sn-glycero-3-phosphoethanolamine-N-[4-(p-maleimidomethyl)cyclohexane-carboxamide]	Avanti (780201)	0.5 mg/ml in $CHCl_3$
PE-CHO	Di-O-hexadecyl-rac-glyceraldehyde was synthesized as described[a]		10.0 mg/ml in $CHCl_3$

[a] Bourel-Bonnet et al. (2005).

manufacturer's manual using a polycarbonate membrane with a pore diameter of 400 nm. The liposome preparation is passed 21 times through the filter and harvested with the opposite syringe.

2.4. Coupling of peptides to liposomes via cysteine

1. The chosen His-MBP-TEV-cytoplasmic domain is defrosted (for each sample, 50 μl in 20 mM HEPES/KOH pH 7.2, 125 mM K-acetate, 2.5 mM Mg-acetate). Three samples are routinely used for label-free quantitative proteomics.
2. An aliquot of 2.5 μl is removed from each sample and mixed with 2.5 μl of 2×-Laemmli sample buffer for subsequent analysis.
3. 2.5 μl of dialyzed TEV protease (1 mg/ml in 50 mM Tris–HCl pH 7.0, 150 mM NaCl, 20% glycerol, 1 mM TCEP) is added to

the remaining 47.5 µl of His-MBP-TEV-cytoplasmic domain. TCEP is adjusted to a final concentration of 1 mM.
4. Digestion is allowed to take place overnight at 25 °C.
5. An aliquot of 2.5 µl is removed and mixed with 2.5 µl of 2×-Laemmli sample buffer. The remaining 47.5 µl is kept on ice, until the liposome preparation is ready.
6. Cysteine controls (triplicate samples) are prepared as follows: for each sample, 47.5 µl of cysteine (1 mg /ml) in coupling buffer and 1 mM TCEP are added to a 1.5 ml newly siliconized Eppendorf tubes.
7. The 47.5 µl of digested recombinant protein or 47.5 µl of cysteine is added to each 100 µl-liposome suspension. Thus, six samples are prepared, three with the recombinant cytoplasmic domain, and three with cysteine.
8. The recombinant cytoplasmic domain or cysteine is allowed to couple to liposomes for 60 min at room temperature.
9. To saturate the remaining maleimido groups, beta-mercaptoethanol is added to the suspension to a final concentration of 1 mM and incubated at room temperature for 15 min.
10. Liposomes are pelleted in an Eppendorf table-top centrifuge (20,000 × g for 5 min at room temperature).
11. The supernatants (~150 µl) are removed and an aliquot of 7.5 µl is kept and mixed with 7.5 µl of 2×-Laemmli sample buffer for subsequent SDS-PAGE analysis.
12. The liposome pellets are washed once with 150 µl of recruitment buffer (20 mM HEPES/KOH pH 7.2, 125 mM K-acetate, 2.5 mM Mg-acetate) and centrifuged at 20,000 × g for 5 min at room temperature. The supernatants are discarded.
13. The liposome pellets are resuspended in 200 µl of recruitment buffer.
14. An aliquot of 10 µl is removed from each suspension and mixed with 10 µl 2×-Laemmli sample buffer for subsequent SDS-PAGE analysis.
15. The collected samples are analyzed by SDS-PAGE (2.5 µl of pre-TEV digest, 2.5 µl of post-TEV digest, 7.5 µl of postcoupling supernatant, and 10 µl of resuspended pellet). After electrophoresis, the gel is stained with Coomassie Blue to assess the efficiencies of the TEV digestion and of the coupling reaction (Fig. 13.2).

2.5. Coupling of peptides to liposomes via an alpha-hydrazino acetyl group

This procedure was initially used to couple synthetic, 50 amino acid long peptides to liposomes (Baust et al., 2006; Bourel-Bonnet et al., 2005). It

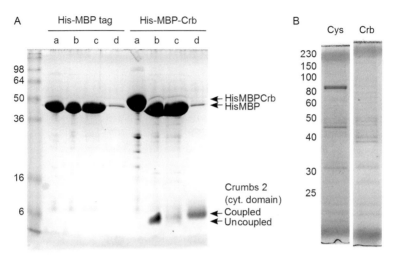

Figure 13.2 SDS-PAGE analysis of the different steps in the production of a recombinant, tagged cytoplasmic domain of a transmembrane protein, its proteolytic cleavage by the TEV protease and the coupling of the cleaved cytoplasmic domain to liposomes. The recombinant crumb cytoplasmic domain (His-MBP-Crbs) is taken as an example and compared to the recombinant protein without the Crumb cytoplasmic domain (His-MPP tag). (Aa) The expressed recombinant protein and control after purification. (Ab) The recombinant protein and control after Tev cleavage. (Ac) Supernatant collected after coupling to liposomes. (Ad) Cytoplasmic domains and controls coupled to liposome after covalent linkage to liposomes. Arrows indicate coupling efficiencies. (B) Coomassie Blue staining of SDS-PAGE gel lanes of cytosolic proteins recruited onto PIP3P and Crumb cytoplasmic containing liposomes (Crbs) and onto the same liposomes with cysteine (Cys).

requires the specific synthesis of the reactive group at the N-terminus of synthetic peptides. This procedure can only be used for chemically synthesized cytoplasmic domains of type I transmembrane proteins.

1. For three samples, lipids are mixed as follows: 18.0 μl PC, 33.2 μl PE, 12.1 μl, 5.75 μl C, 32 μl of the respective phosphoinositide, and 8.5 μl of CHO-anchor in a siliconized 1.5-ml Eppendorf tube.
2. The mixture is thoroughly vortexed and dried down under a stream of nitrogen to yield a thin and homogeneous film of lipids.
3. Lipids are resuspended in 330 μl of hydrazino-coupling buffer (15.4 mM citric acid, 69.2 mM Na$_2$HPO$_4$, pH 6.4).
4. Ten cycles of 4 min of freezing in liquid nitrogen and 4 min of thawing in a thermoshaker at 25 °C with thorough vortexing are applied.
5. The liposomes are mixed with the chemically synthesized and modified peptides overnight at room temperature.

6. A LipoFast handheld extruder (Avestin, Canada) is assembled according to the manufacturer's manual with a polycarbonate membrane with a pore diameter of 400 nm. The liposome preparation is passed 21 times through the filter and harvested with the opposite syringe.
7. Optional gel filtration to remove an excess of uncoupled peptides can be performed. For this, a NAP-5 column (GE Healthcare, Sweden) is equilibrated with 10 ml recruitment buffer (see Section 2.1).
8. The liposome suspension is adjusted to 500 µl and loaded onto the column.
9. When the liposome suspension completely entered the gel bed, a new Eppendorf tube is placed under the column and liposomes are collected by elution with 1 ml recruitment buffer.
10. The liposomes should be immediately used for binding assays.

2.6. Binding of cytosolic proteins to functionalized liposomes

The binding of cytosolic proteins to liposomes can be assayed and multiple parameters can be tested, that is, the concentration of cytosolic proteins, the lipid composition of liposomes (phosphoinositides or others or lipid anchor), the presence of GTP or GTP-γ-S, which influences the activity of GTPases or the effect reagents affecting actin polymerization (ATP, ATP-regenerative system, or the actin depolymerization drug latrunculin B). A given reaction is described below.

1. 500 µl of mouse brain cytosol is supplemented with GTPγS (final concentration 0.21 mM).
2. The recruitment of cytosolic proteins is initiated by the addition of 200 µl liposome suspension (final protein concentration of 3–5 mg/ml) and the incubation is carried out at 37 °C for various period of time (the binding reaction is complete after 20 min).

2.7. Purification of coated liposomes

The procedure to isolate coated liposomes from unbound cytosolic components essentially depends on the type of the subsequent analyses. Conventional centrifugation can be used for subsequent analysis by Western blotting. Mass spectrometry-based analyses need a better purification scheme as described below.

1. The binding of cytosolic proteins to liposomes is stopped by adding 1.4 ml of ice-cold 60% sucrose in recruitment buffer (see

Section 2.1) and rapid mixing. The mixture is transferred into a 13-ml ultracentrifugation tube.
2. The resulting mixture is overlaid with 30% sucrose in recruitment buffer up to 1 cm below the top of the tube.
3. The 30% sucrose solution is overlaid with recruitment buffer up to 2 mm below the top of the tube.
4. The tubes are centrifugated at $200.000 \times g$ for 14 h at 4 °C.
5. The liposomes floating to the top of the 30% sucrose fraction are collected into a 4.5-ml ultracentrifugation tube, diluted with recruitment buffer, and filled up to 2 mm below the top of the tube.
6. Liposomes are pelleted by centrifugation at $370.000 \times g$ for 1 h at 4 °C.
7. Supernatants are removed. The tube walls are dried with a dust-free wipe avoiding touching the pellet.
8. 30 μl of 2× Laemmli sample buffer is added, scratching the pellet with a pipette tip.
9. The tubes are placed into an ultrasonic bath and sonicated for 10 min at 65 °C.
10. The samples are transferred into 0.5-ml Eppendorf tubes and spun down.
11. The sample volume is adjusted to 30 μl with 2× Laemmli sample buffer.

3. IDENTIFICATION OF CORE MACHINERIES BY MASS SPECTROMETRY-BASED, LABEL-FREE QUANTITATIVE PROTEOMICS

3.1. SDS-PAGE and in-gel digestion

1. SDS-PAGEs are performed with a separation lane length of 4 cm. Cysteine- and cytoplasmic tail-samples should be run on separated gels to avoid cross contamination by diffusion (Knaust & Shevchenko, 2012).
2. After electrophoresis, the lanes are cut into 10 slices, and each slice is cut into cubes of 1 mm and transferred into a 0.5-ml Eppendorf tube.
3. 400 μl H$_2$O is added. The solution is incubated for 10 min, vortexed, spun down, and the supernatant removed.
4. 150 μl of 25 mM ammonium bicarbonate in acetonitrile is added. The solution is incubated for 5 min, vortexed, spun down, and the supernatant removed.

5. 150 μl of acetonitrile is added. The solution is incubated for 5 min, vortexed, spun down, and the supernatant removed.
6. 150 μl of 10 mM dithiothreitol/100 mM ammonium bicarbonate is added. The solution is incubated for 45 min at 56 °C, vortexed, spun down, and removed.
7. 150 μl of 55 mM iodoacetamide/100 mM ammonium bicarbonate is added, incubated for 20 min in the dark, vortexed, spun down, and the supernatant removed.
8. 150 μl of 25 mM ammonium bicarbonate in acetonitrile is added. The solution is incubated for 5 min, vortexed, spun down, and the supernatant removed.
9. 150 μl of acetonitrile is added. The solution is incubated for 5 min, vortexed, spun down, and the supernatant is removed.
10. The resulting gel pieces are dried in a speed vacuum system for 15 min.
11. 0.15 μg of trypsin in 5 mM ammonium bicarbonate is added to the gel pieces and incubated for 5 min.
12. 30 μl of 5 mM ammonium bicarbonate is added. The tubes are closed and incubated for 16 h at 37 °C.
13. 60 μl of 0.5% trifluoroacetic acid in acetonitrile is added and tubes are sonicated in an ultrasonic bath for 10 min and spun down.
14. The peptide eluates are transferred into empty glass vials used for in MS analyses.
15. 60 μl of acetonitrile is added to the gel pieces and tubes are sonicated in an ultrasonic bath for 10 min and spun down.
16. The eluates are added to the previous ones.
17. The peptides are dried in a speed vacuum system.
18. 20 μl 0.1% formic acid (LC-buffer A) is added.
19. The glass vials are closed and sonicated for 10 min in an ultrasonic bath and spun down.

3.2. LC–MS acquisition

We use an UltiMate 3000 HPLC-System (Thermo-Scientific, Germany) equipped with a 15-cm analytical C18-column with an inner diameter of 75 mm and particle size 2 mm, 100 Å (Thermo-Scientific, Germany). The linear gradient 5–45% acetonitrile in 0.1% formic acid was applied during 80 min to elute peptides.

Mass spectrometry is performed on an LTQ-Orbitrap XL system in a Top8-procedure. MS1 is hereby accomplished in the orbitrap with a resolution of 60,000 (at 400 Th) triggering parallel LTQ acquisition (resolution of 30,000) of MS2 of the eight most intensive MS1 features. Dynamic exclusion is enabled with a repeat count of 2, a repeat duration of 5 s, an exclusion duration of 15 s with an exclusion list size of 500. Lock mass is optionally set to a polysiloxane adduct of 445.12 (Keller et al., 2008). Collision-activated dissociation is executed with an activation-Q of 0.25 and a normalized collision energy of 35 for 30 ms.

3.3. Postacquisition workflow (MaxQuant, Perseus, STRING)

All raw files from XCalibur are loaded into MaxQuant (V1.2.2.5) (Cox and Mann, 2008) and an experimental table is created in a way that all slices of one lane are assigned to the same experiment. Carbamidomethylation of cysteines is considered as a fixed modification and oxidation of methionine, acetylation of the N-terminus, and deamidation of asparagine or glutamine as variable modifications. Other parameters are multiplicity of 1, trypsin as enzyme, maximum number of modifications per peptide of 3, maximal missed cleavages of 2, maximum charge of 4 with individual peptide mass tolerances allowed. A suitable FASTA database is assigned (UNIPROT-TrEMBL 2012-01). Parameters in the protein identification tab are as follows: peptide and site FDR of 0.01, maximum peptide posterior error probability of 1, all minimum peptides of 1, filter-labeled amino acids disabled, second peptides, requantify, label-free quantification, and match between runs (2 min) enabled.

The proteinGroups.txt-file is loaded into Perseus (Hubner et al., 2010) with all LFQ intensities as "expression." All proteins matching any of the three categorical annotations are deleted, LFQ intensities are logarithmized, experiments are grouped to sample or control, and proteins that are not detected in all three replicates of at least one group are deleted. Missing values are imputed by normal distribution with parameters that should be adapted using histogram plots. A two-sided t-test is performed with a false discovery rate in between 0.1% and 1%, and a slope value of 0.2–1, the control should hereby be assigned to "group 2." A scatterplot is created with "−log t-test p-value" versus "t-test difference" showing the volcano plot. Selecting the "t-test significant" row in the right table labels all proteins that passed the t-test with proteins that are recruited on the right side and liposomal membrane binding proteins that are displaced during recruitment are on the left side of the plot (Figs. 13.3 and 13.4) (Table 13.2).

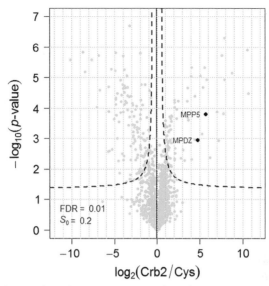

Figure 13.3 Volcano plot of proteins recruited on liposomes containing PI3P and cumbs 2 versus those recruited on liposomes with PI3P and cysteine. The two known direct crumbs 2-interactors (MPP5 and MPDZ) emerge on the upper right side among the proteins significantly recruited in the presence of crumbs 2.

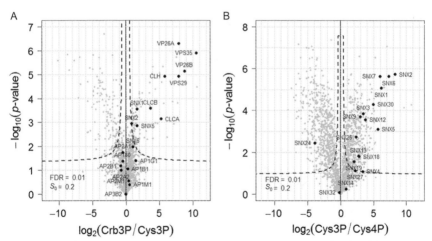

Figure 13.4 (A) Volcano plot of cytosolic proteins recruited on liposomes with crumbs and PI3P versus those with cysteine and PI3P. The cargo-recognizing part of the retromer complex (VPS26A, VPS26B, VPS29, VPS35) is recruited due to the presence of the cytosolic domain of crumbs 2. (B) Volcano plot of cytosolic proteins recruited on liposomes containing cysteine and either PI3P or PI4P. The phosphoinositide-recognizing components of the retromer complex SNX1, SNX2, SNX5, SNX6 are recruited due to the presence of PI3P. Recruitment of those sorting nexins is further enhanced by the presence of the tail, indicating coincidence detection of PI3P and VPS-proteins.

Table 13.2 Recruited proteins in a Crumbs 2- and PI3P-dependent manner with a Gene Ontology-assignment to signaling (GO:0023052)

Protein	UniProt	Protein name
Crumbs interactors		
Mpdz	Q8VBX6	Multiple PDZ domain protein
Mpp5	Q9JLB2	MAGUK p55 subfamily member 5
Proteins involved in trafficking		
Snx17	Q8BVL3	Sorting nexin-17
Snx27	Q3UHD6	Sorting nexin-27
Sv2a	Q9JIS5	Synaptic vesicle glycoprotein 2A
Rab21	P35282	Ras-related protein Rab-21
Rab2a	P53994	Ras-related protein Rab-2A
Rab5c	P35278	Ras-related protein Rab-5C
Zfyve9	A2A8R0	Zinc finger, FYVE domain containing 9
Others		
Dclk1	Q9JLM8	Serine/threonine-protein kinase DCLK1
Plcb1	Q9Z1B3	1-Phosphatidylinositol 4,5-bisphosphate phosphodiesterase beta-1
Prkci	Q62074	Protein kinase C iota type
Sel1l	Q9Z2G6	Protein sel-1 homolog 1

4. VISUALIZATION OF PROTEIN DYNAMICS ON GIANT UNILAMELLAR VESICLES BY FLUORESCENCE MICROSCOPY

Two types of artificial membranes systems, giant unilamellar vesicles (GUVs) and supported lipid bilayers, have been extensively used during the last twenty years to study the behavior of proteins according to changes in their biochemical composition and physical properties like fluidity, rigidity lipid packing (liquid-ordered and disorder phases), and curvature. GUVs are spherical membranes with diameters of 1–50 μm and represent the most convenient system when compared to supported lipid bilayers (Chan & Boxer, 2007). Due to their size, GUVs can be observed by conventional microscopy and studied using various microscopic and spectrometric methods (Kahya, Scherfeld, Bacia, & Schwille, 2004). Their mechanical strength ensures easy handling and even allows micromanipulation when combined with optical traps (Sens, Johannes, & Bassereau, 2008). We describe here how GUVs can be used and the methods to visualize the binding and the behavior of cytosolic components on GUVs mimicking the protein and lipid composition of intracellular compartments of the secretory and the endocytic pathways. We take here as an example the clathrin/AP-1-dependent post-Golgi transport to endosome (Anitei et al., 2010).

4.1. Preparation of cytosol from cells stably expressing fluorescently tagged proteins

1. BSC-1 cells stably expressing GFP-AP-1σ or GFP-clathrin light chain (CLC) or tomato-CLC are cultured to confluency in DMEM containing 10% fetal calf serum on five 16-cm dishes.
2. Cells are washed with cold PBS and gently detached using a cell lifter.
3. Cells are pelleted by a 5-min-long centrifugation at $450 \times g$ at 4 °C and resuspended in 600 μl of recruitment buffer containing complete protease inhibitor cocktail.
4. Cells are homogenized on ice by repeatedly passing the solution 10 times through a 22-Gauge needle of a syringe and then 15 times through a 27-Gauge needle of a syringe.
5. The homogenate is then centrifuged at $10,000 \times g$ for 15 min.
6. The supernatant is recovered and centrifuged at $150,000 \times g$ for 45 min at 4°CC. The resulting supernatant is aliquoted, snap frozen, and stored at -80 °C. Cytosols are prepared in order to obtain a final protein concentration of 5–10 mg/ml.

4.2. Preparation of GUVs from liposomes

The shape of membranes can be manipulated *in vitro* using different means (Sens et al., 2008). To visualize membranes, the lipid dyes DiI-C16 or DiD-C16 (Molecular Probes/Life Technologies, UK) are added at molar ratio of 0.01–0.05% to the lipid mixture (as described earlier) prior to liposome preparation. The liposomes with covalently linked cytoplasmic domains of transmembrane proteins are prepared as described previously. GUVs are prepared from these proteoliposomes by electro-swelling using an alternating electric field. For this reason, salts and charged solutes must be removed from the proteoliposome suspension. Therefore, liposomes are washed and resuspended in deionized water after peptide coupling. Alternatively, the proteoliposomes suspension can be desalted using a small gel filtration column. Proteoliposomes can be stored at $-20\ ^\circ C$ for several months and then be used for GUV preparation.

1. 10–20 μl of proteoliposomes are applied in small droplets onto two indium tin oxide (ITO)-covered glass slides (Präzisions Glas & Optik, Germany). A voltmeter is used to determine the conductive side of the ITO slide.
2. The proteoliposomes are dried for 20 min in a dessicator under reduced pressure.
3. After drying, the slides are assembled, separated by a rubber ring spacer, and fixed using clamps (Fig. 13.5).
4. 600 μl of 330 mM sucrose is added in between the slides. The concentration of the sucrose solution is chosen to match the osmolality of the solution.
5. GUVs are formed by applying an alternating voltage (10 Hz, 1.8 V) for 2 h using a Voltcraft 8202 1-channel-function generator (Conrad Electronic, Hirschau, Germany).
6. The GUV containing solution is removed from the chamber with a pipette and diluted in 2 volumes of recruitment buffer. The osmolality of this buffer must match the osmolality inside the GUVs.

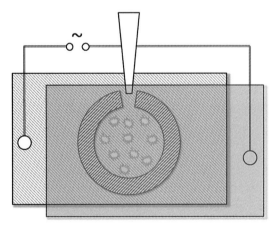

Figure 13.5 A chamber for GUV formation: two ITO slides containing dried liposomes are assembled, separated by a rubber ring. The sucrose solution is pipetted inside the ring. The ring is closed and each slide is connected via crocodile clips to a function generator.

7. GUVs are allowed to sediment for 30 min and 200 µl from the bottom of the tube is recovered as a GUV suspension. GUVs should be immediately used for experiments.

4.3. Imaging of protein dynamics

1. Cytosols of cells stably expressing proteins of interest tagged with a given fluorescent protein are defrosted. They are centrifuged at $150,000 \times g$ for 30 min at 4 °C to sediment protein aggregates. When two different proteins or protein complexes are studied simultaneously, then two cytosols are mixed in equal amounts. One reaction is carried out in a final volume of 100 µl keeping the concentration of cytosolic proteins around 5–10 mg of total protein/ml.
2. To prevent GUV spreading, a eight-well Lab-Tek glass bottom chamber (Nunc, Germany) is coated for 30 min with 10 mg/ml BSA in water, washed with water, and then dried.
3. 10 µl of GUV suspension is added to 100 µl of cytosol in the presence of 150 µM GTP-γ-S. The solution is mixed by up and down pipetting. The reaction is incubated at 37 °C for various periods of time as needed.
4. GUVs sediment at the bottom of the chamber due to their higher density and can be analyzed by microscopy.

Examples of such studies are shown in Fig. 13.6.

5. SUMMARY

Transport carriers mediate the vectorial transport of proteins and lipids between the compartments of the secretory and endocytic pathways. The formation of these transport carriers involved the interplay of a number of cytosolic machineries whose interactions with membranes are timely regulated in order to concentrate cargos into a nascent carrier, to bend and elongate membranes, and to trigger the release of a mature carrier. A number of biological methods have been used during the last two decades to identify the key molecules regulating carrier budding from a donor compartment and their fusion with an acceptor compartment. More recently, synthetic membrane biology has emerged as a key approach not only for visualizing the dynamic behavior of key molecules during membrane deformation and scission but also for identifying the complete protein networks participating in these events by label-free quantitative proteomics.

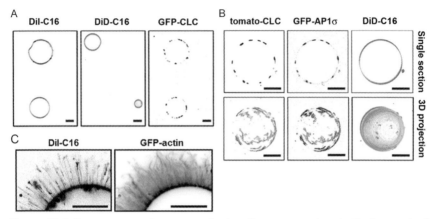

Figure 13.6 Fluorescence microscopy of cytosolic components selectively recruited onto GUV surfaces. (A) GUVs with an intact cytoplasmic domain of the Varicella Zoster Virus glycoprotein I (gpI) are labeled with the red lipid dye dil-C16 contain. GUVs with gpI lacking its sorting motif are labeled with the far-red lipid dye DiD-C16. After incubation with cytosol prepared from cells expressing GFP-CLC, these GUVs were imaged using confocal microscopy. (B) GUVs containing an intact gpI cytoplasmic domain and DiD-C16 were incubated in a mixture of cytosol from cells expressing GFP-AP-1σ or tomato-CLC. GUVs were imaged using confocal microscopy. Both proteins colocalized as best seen in 3D projections of z-stacks. (C) GUVs containing a wild-type gpI cytoplasmic domain and Dil-C16 were incubated with a cytosol of cells expressing GFP-actin in the presence of an ATP-regenerating system promoting actin polymerization. Bar: 10 μm.

ACKNOWLEDGMENTS

This work was supported by grants from DFG (TRR 83/1-2010, HO 2584/1-1, HO 2584/2-1, HO 2584/6-1, HO 2584/8-1, HO 2584/9-1) and TU-Dresden.

REFERENCES

Anitei, M., & Hoflack, B. (2012). Bridging membrane and cytoskeleton dynamics in the secretory and endocytic pathways. *Nature Cell Biology, 14*, 11–19.

Anitei, M., Stange, C., Parshina, I., Baust, T., Schenck, A., Raposo, G., et al. (2010). Protein complexes containing CYFIP/Sra/PIR121 coordinate Arf1 and Rac1 signalling during clathrin-AP-1-coated carrier biogenesis at the TGN. *Nature Cell Biology, 12*, 330–340.

Baust, T., Czupalla, C., Krause, E., Bourel-Bonnet, L., & Hoflack, B. (2006). Proteomic analysis of adaptor protein 1A coats selectively assembled on liposomes. *Proceedings of the National Academy of Sciences of the United States of America, 103*, 3159–3164.

Bonifacino, J. S., & Glick, B. S. (2004). The mechanisms of vesicle budding and fusion. *Cell, 116*, 153–166.

Bonifacino, J. S., & Hurley, J. H. (2008). Retromer. *Current Opinion in Cell Biology, 20*, 427–436.

Bonifacino, J. S., & Traub, L. M. (2003). Signals for sorting of transmembrane proteins to endosomes and lysosomes. *Annual Review of Biochemistry, 72*, 395–447.

Bourel-Bonnet, L., Pecheur, E. I., Grandjean, C., Blanpain, A., Baust, T., Melnyk, O., et al. (2005). Anchorage of synthetic peptides onto liposomes via hydrazone and alpha-oxo hydrazone bonds. Preliminary functional investigations. *Bioconjugate Chemistry, 16*, 450–457.

Campelo, F., & Malhotra, V. (2012). Membrane fission: The biogenesis of transport carriers. *Annual Review of Biochemistry*, *81*, 407–427.

Chan, Y. H., & Boxer, S. G. (2007). Model membrane systems and their applications. *Current Opinion in Chemical Biology*, *11*, 581–587.

Cox, J., & Mann, M. (2008). MaxQuant enables high peptide identification rates, individualized p.p.b.-range mass accuracies and proteome-wide protein quantification. *Nature Biotechnology*, *26*, 1367–1372. http://dx.doi.org/10.1038/nbt.1511. Epub 2008 Nov 30.

De Matteis, M. A., & Godi, A. (2004). PI-loting membrane traffic. *Nature Cell Biology*, *6*, 487–492.

Di Paolo, G., & De Camilli, P. (2006). Phosphoinositides in cell regulation and membrane dynamics. *Nature*, *443*, 651–657.

Duden, R. (2003). ER-to-Golgi transport: COP I and COP II function (Review). *Molecular Membrane Biology*, *20*, 197–207.

Henne, W. M., Buchkovich, N. J., & Emr, S. D. (2011). The ESCRT pathway. *Developmental Cell*, *21*, 77–91.

Hirst, J., Barlow, L. D., Francisco, G. C., Sahlender, D. A., Seaman, M. N., Dacks, J. B., et al. (2011). The fifth adaptor protein complex. *PLoS Biology*, *9*, e1001170.

Hubner, N. C., Bird, A. W., Cox, J., Splettstoesser, B., Bandilla, P., Poser, I., et al. (2010). Quantitative proteomics combined with BAC TransgeneOmics reveals in vivo protein interactions. *Journal of Cell Biology*, *189*, 739–754.

Itoh, T., & De Camilli, P. (2006). BAR, F-BAR (EFC) and ENTH/ANTH domains in the regulation of membrane-cytosol interfaces and membrane curvature. *Biochimica et Biophysica Acta*, *1761*, 897–912.

Kahya, N., Scherfeld, D., Bacia, K., & Schwille, P. (2004). Lipid domain formation and dynamics in giant unilamellar vesicles explored by fluorescence correlation spectroscopy. *Journal of Structural Biology*, *147*, 77–89.

Keller, B. O., Sui, J., Young, A. B., & Whittal, R. M. (2008). Interferences and contaminants encountered in modern mass spectrometry. *Analytica Chimica Acta*, *627*, 71–81. http://dx.doi.org/10.1016/j.aca.2008.04.043. Epub 2008 Apr 25.

Knaust, A., & Shevchenko, A. (2012). Horizontal carryover of proteins on one-dimensional polyacrylamide gels may jeopardize gel-enhanced liquid chromatography mass spectrometry proteomic interpretations. *Analytical Biochemistry*, *421*, 779–781.

Krauss, M., & Haucke, V. (2007a). Phosphoinositide-metabolizing enzymes at the interface between membrane traffic and cell signalling. *EMBO Reports*, *8*, 241–246.

Krauss, M., & Haucke, V. (2007b). Phosphoinositides: Regulators of membrane traffic and protein function. *FEBS Letters*, *581*, 2105–2111.

McMahon, H. T., & Gallop, J. L. (2005). Membrane curvature and mechanisms of dynamic cell membrane remodelling. *Nature*, *438*, 590–596.

Pocha, S. M., Wassmer, T., Niehage, C., Hoflack, B., & Knust, E. (2011). Retromer controls epithelial cell polarity by trafficking the apical determinant Crumbs. *Current Biology*, *21*, 1111–1117.

Robinson, M. S., & Bonifacino, J. S. (2001). Adaptor-related proteins. *Current Opinion in Cell Biology*, *13*, 444–453.

Rothman, J. E., & Wieland, F. T. (1996). Protein sorting by transport vesicles. *Science*, *272*, 227–234.

Sens, P., Johannes, L., & Bassereau, P. (2008). Biophysical approaches to protein-induced membrane deformations in trafficking. *Current Opinion in Cell Biology*, *20*, 476–482.

Spang, A. (2002). ARF1 regulatory factors and COPI vesicle formation. *Current Opinion in Cell Biology*, *14*, 423–427.

Stagg, S. M., LaPointe, P., & Balch, W. E. (2007). Structural design of cage and coat scaffolds that direct membrane traffic. *Current Opinion in Structural Biology*, *17*, 221–228.

CHAPTER FOURTEEN

Mouse Models of PI(3,5)P$_2$ Deficiency with Impaired Lysosome Function

Guy M. Lenk, Miriam H. Meisler[1]

Department of Human Genetics, University of Michigan, Ann Arbor, Michigan, USA
[1]Corresponding author: e-mail address: meislerm@umich.edu

Contents

1. Introduction	246
2. Design of Mouse Models	247
3. A Spontaneous Null Mutation of *Fig4*: The *Pale Tremor* Mouse	249
4. Tissue-Specific *Fig4* Transgenes: Neurons Versus Astrocytes	251
5. Conditional Knockout of *Fig4* in Neurons	251
6. The Human Disease Mutation FIG4-I41T in Transgenic Mice	252
7. A Spontaneous Missense Mutation of *Vac14* in the *ingls* Mouse	252
8. A Null Gene-Trap Allele of *Vac14*	253
9. A Hypomorphic Gene-Trap Allele of *Pikfyve* (*Fab1*)	253
10. A Conditional Knockout of Pikfyve	254
11. Genetic Interactions: *Fig4*, *Vac14*, and *Mtmr2*	254
12. Genetic Effects of Strain Background	255
13. Future Applications of Mouse Models of PI(3,5)P$_2$ Deficiency	255
Acknowledgments	257
References	257

Abstract

The endolysosomal system and autophagy are essential components of macromolecular turnover in eukaryotic cells. The low-abundance signaling lipid PI(3,5)P$_2$ is a key regulator of this pathway. Analysis of mouse models with defects in PI(3,5)P$_2$ biosynthesis has revealed the unique dependence of the mammalian nervous system on this signaling pathway. This insight led to the discovery of the molecular basis for several human neurological disorders, including Charcot–Marie–Tooth disease and Yunis–Varon syndrome. Spontaneous mutants, conditional knockouts, transgenic lines, and gene-trap alleles of *Fig4*, *Vac14*, and *Pikfyve* (*Fab1*) in the mouse have provided novel information regarding the role of PI(3,5)P$_2$ *in vivo*. This review summarizes what has been learned from mouse models and highlights the utility of manipulating complex signaling pathways *in vivo*.

1. INTRODUCTION

Low-molecular-weight effectors regulate many aspects of cell biology through specific interactions with protein recognition domains. Phosphatidyl inositols are amphipathic lipids that are associated with cell membranes and interact with cytoplasmic proteins through their polar domains. PI(3,5)P_2 (phosphatidyl-inositol (3,5) bisphosphate) is a low-abundance member of this family that regulates vesicle maturation and trafficking in the endosomal/lysosomal pathway in eukaryotic cells (Volpicelli-Daley & De Camilli, 2007; McCartney et al., 2013). PI(3,5)P_2 was first identified in 1997 as a component of the yeast vacuole and mammalian cells (Dove et al., 1997; Whiteford, Brearley, & Ulug, 1997). Reduction of PI(3,5)P_2 in yeast results in an enlarged vacuole due to defects in vacuole fission and retrograde traffic to the Golgi (Bonangelino et al., 2002; Gary et al., 2002; Wiradjaja et al., 2007). Hyperosmotic shock in yeast results in elevation of PI(3,5)P_2 within 10 min, suggesting a role in environmental adaptation (Duex et al., 2006). The signals regulating PI(3,5)P_2 in mammalian cells have not been identified.

In yeast, PI(3,5)P_2 is generated by a protein complex that includes the lipid 5-kinase *Pikfyve*, the 5-phosphatase *Fig4*, and the scaffolding protein *Vac14*, and these three proteins are highly conserved in mammalian genomes (Botelho et al., 2008; Jin et al., 2008; Sbrissa et al., 2007). Initial studies in mammalian cells utilized reduction of PI(3,5)P_2 by transfection of COS7 cells with a dominant negative mutant of the PIKfyve kinase (Ikonomov et al., 2002) and transfection of HEK cells with an siRNA to VAC14 (Sbrissa et al., 2004). Both treatments resulted in cell vacuolization reminiscent of the enlarged vacuole in yeast mutants, indicating that the role of PI(3,5)P_2 is conserved in eukaryotic cells.

The discovery of spontaneous mouse mutations affecting *Fig4* and *Vac14*, and the generation of targeted mutations of all three proteins in the mouse germ line, have produced insight into the role of this pathway in tissues of the intact animal. Although these genes are expressed in all mammalian cells, the central and peripheral nervous system are particularly sensitive to deficiency of PI(3,5)P_2, resulting in extensive spongiform neurodegeneration (Chow et al., 2007; Jin et al., 2008; Zhang et al., 2007; Zolov et al., 2012). This insight from mouse models led to the identification of patient mutations in human neurological disorders including Charcot–Marie–Tooth disease (Chow et al., 2007; Nicholson et al., 2011),

Table 14.1 Mouse models with altered metabolism of PI(3,5)P$_2$

Gene	Genetic mouse model	References
Fig4	Spontaneous null (*pale tremor*)	Chow et al. (2007)
	Neuron-specific transgene	Ferguson et al. (2012)
	Astrocyte-specific transgene	Ferguson et al. (2012)
	Human disease variant transgene	Lenk et al. (2011)
	Floxed allele	Ferguson et al. (2012)
Vac14	Spontaneous missense allele (*ingls*)	Jin et al. (2008)
	Gene-trap null allele	Zhang et al. (2007)
Pikfyve	Gene-trap hypomorphic allele	Zolov et al. (2012)
	Floxed allele	Ikonomov et al. (2011)
Mtmr2	Mtmr2 null	Bolino et al. (2004)

Yunis–Varon syndrome (Campeau et al., 2013), amyotrophic lateral sclerosis (ALS) (Chow et al., 2009) and polymicrogyria (Baulac et al., 2013). Another important outcome from mouse models was elucidation of the pathogenic mechanism of the common human pathogenic variant *FIG4-*I41T found in patients with CMT4J (Lenk et al., 2011). Overexpression of the I41T variant rescued lethality and neurodegeneration of *Fig4* null mice indicating that increased expression of this allele could be therapeutic. A third important observation was the discovery that loss of *Fig4* in neurons has a secondary effect on myelination (Winters et al., 2012). Analysis of mouse models has also provided evidence for genetic interaction between genes regulating PI(3,5)P$_2$ biosynthesis (Vaccari et al., 2011). This review will focus on mouse models of PI(3,5)P$_2$ deficiency caused by the mutations of the genes *Fig4*, *Vac14*, *Pikfyve* (*Fab1*), and *Mtmr2* (Table 14.1).

2. DESIGN OF MOUSE MODELS

The earliest mouse models of human disorders were spontaneous mutants that were detected by their visible phenotypes (Paigen, 2003). The spontaneous mutations of *Fig4* and *Vac14* were identified by their visible neurological dysfunction and early lethality. These mutants exhibit global expression of the mutated gene in all tissues, as is the case for patients with human inherited disorders. In addition to the global mutants, the design of tissue-specific mutations can provide unique biological information, especially when the global mutant causes early lethality. Tissue-specific models including transgenic, conditional null, and gene-trap alleles have been used to study the genes regulating the $PI(3,5)P_2$ pathway. Choices among these alternative technologies are dictated by both practical and theoretical considerations.

Classically, "transgenic" mice are generated by the addition of a cloned transgene to the germline of a wild-type mouse via microinjection of fertilized eggs followed by random chromosomal insertion of multiple copies of the transgene. The transgene typically contains a previously characterized tissue-specific promoter fragment, between a few hundred base pairs and a few kilobases in length, fused upstream of a mutant or wild-type cDNA, to achieve tissue-specific expression. A broad range of tissue-specific promoter fragments have been characterized for this purpose (Donahue et al., 2012). It is important to characterize at least two independent transgenic lines to control for the unanticipated effects of chromosomal insertion site on transgene expression and to compare the effects of different quantitative levels of transgene expression. Other practical issues include leaky expression of tissue-specific promoters in nontargeted tissues and incomplete expression of the transgene in the population of targeted cells.

Gene-trap mice are generated by infection of ES cells with transposons that insert at random sites in the genome (Nord et al., 2006). The precise location of the insertion in each ES cell clone is then determined by PCR. Libraries of ES cells carrying gene-trap alleles at known positions are available (Bradley et al., 2012). Many gene-trap insertions cause "hypomorphic alleles" with reduced gene expression, while some completely abolish expression and generate null alleles. The level of residual gene expression must be determined for each gene-trap allele after the mouse is generated from the mutated ES cells via chimeric embryos.

Targeted knockout lines are produced by inserting CRE recombinase recognition sites (loxP sites) flanking an exon using homologous recombination in ES cells. Mice carrying the "floxed" allele are crossed with transgenic mice that express the CRE recombinase under the regulation of a global or tissue-specific promoter (Murray et al., 2012). The use of floxed alleles requires access to CRE transgenic lines with the desired tissue specificity. The international mouse knockout project, KOMP, is generating targeted alleles of every gene in the mouse genome for distribution to investigators on request (Ayadi et al., 2012; Saunders, 2010). Potential limitations of this approach include off-target expression of the CRE recombinase and incomplete deletion of the floxed allele in the targeted tissue. Combining one null allele with one floxed allele is a popular approach toward increasing the extent of deletion in the targeted tissue.

The use of targeted mutations often involves crosses between mice with different strain backgrounds. For example, floxed alleles generated by targeting in an ES cell line from strain 129 may be bred with CRE alleles maintained on strain C57BL/6 J. The resulting segregation of genetic variation from the two background strains can produce phenotypic variation among the mutant mice. Other issues include variable susceptibility of different floxed alleles to the CRE recombinase, depending upon chromosomal location, and unanticipated expression of transgenes in the male or female germ line, which can subvert the intended tissue specificity (e.g., Rempe et al., 2006).

3. A SPONTANEOUS NULL MUTATION OF *FIG4*: THE *PALE TREMOR* MOUSE

The first PI(3,5)P$_2$-deficient mouse to be identified was the spontaneous mutant *pale tremor*. The recessive mutant was recognized in our mouse colony by its diluted pigmentation and resting tremor. The loss-of-function mutation was caused by insertion of a 5.5 kb ETn2b retrotransposon into intron 18 of *Fig4*, preventing processing of the full-length mRNA (Chow et al., 2007) and resulting in lack of FIG4 protein (Lenk et al., 2011). Homozygous *Fig4$^{plt/plt}$* mice do not survive beyond 6 weeks of age. The cellular phenotype of PI(3,5)$_2$ deficiency is evident in cultured fibroblasts, which accumulate large empty vacuoles, as shown in Fig. 14.1A. The vacuole membranes contain the lysosomal proteins LAMP1 and LAMP2 (Chow et al., 2007; Ferguson et al., 2009). These vacuoles do not stain for lipid or carbohydrate and also appear empty by electron microscopy. Measurement of PI(3,5)P$_2$ level by HPLC detected an overall twofold reduction compared with wild-type cells (Chow et al., 2007). Vacuolization is also visible in tissue sections from brain and spleen, and primary cultures of neurons, consistent with a basic housekeeping role for PI(3,5)P$_2$. The appearance of vacuolated cells in *Fig4* null tissues is reminiscent of the enlargement of the primary vacuole in PI(3,5)P$_2$-deficient yeast strains (Bonangelino et al., 2002; Gary et al., 2002; Wiradjaja et al., 2007). It has recently been demonstrated that PI(3,5)P$_2$ is an activator of several cation channels in the lysosomal membrane that permit the exit of calcium and

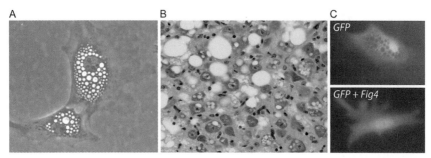

Figure 14.1 Phenotypes of the spontaneous *Fig4* null mutant "pale tremor." (A) Accumulation of cytoplasmic vacuoles in primary cultured fibroblasts (MEFs). (B) Spongiform degeneration of the brain stem at P21, with neuronal vacuolization. (C) The cytoplasmic vacuoles in primary fibroblasts cultured from *Fig4* null mice are rescued by transfection with wild-type *Fig4* cDNA plus GFP, but not by GFP alone.(For color version of this figure, the reader is referred to the online version of this chapter.)

other cations (Dong et al., 2010; Wang et al., 2012). Thus, the enlarged vesicles may result from increased osmotic pressure within the lysosome, and the osmotic swelling may trap the lysosomes in an inactive form and prevent the normal regeneration process.

Backup in the autophagy pathway that feeds into lysosomal degradation was first identified in tissue slices of *Fig4* null brain, which contain numerous inclusion bodies containing p62 and ubiquitinated proteins (Ferguson, Lenk, & Meisler, 2009). Two types of vesicles accumulate in mutant brain: small electron dense bodies with the appearance of autolysosomes, and large empty vesicles similar to those shown in cultured fibroblasts earlier (Chow et al., 2007; Ferguson et al., 2009; Katona et al., 2011; Zhang et al., 2008; GML and MHM, unpublished observations). Both types of vesicles appear to be derived from the late endolysosomal system since they contain lysosomal but not endosomal markers. Costaining with GFAP demonstrated that the dense inclusion bodies contain p62 and ubiquitinated proteins are localized to astrocytes rather than neurons in the mutant brain.

Another pathological characteristic discovered in the null mouse is defective myelination in CNS and PNS, resulting in major reduction of myelin proteins in the brain, dysplasia of the corpus callosum and white tracts of the cerebellum, thinning of the myelin sheath of optic nerve and sciatic nerve, and reduction in nerve conduction velocity (Chow et al., 2007; Winters et al., 2012). The spongiform degeneration of *Fig4*-deficient brain thus results in abnormalities of neurons, astrocytes, and oligodendrocytes. To elucidate the relationships between these cellular phenotypes, we generated cell-type-specific transgenic lines and a floxed allele of *Fig4*, as described in the following sections.

4. TISSUE-SPECIFIC *FIG4* TRANSGENES: NEURONS VERSUS ASTROCYTES

Both neurons and astrocytes are morphologically abnormal in *Fig4* null mice (Ferguson et al., 2009). We evaluated the relative contributions of these two cell types to the neurological disorder by directing *Fig4* cDNA expression specifically in neurons or astrocytes, using a neuron-specific promoter (NSE) or an astrocyte-specific promoter (GFAP) (Ferguson et al., 2012). The tissue-specific transgenes were bred to the *Fig4* null background and the phenotypes of the transgene-positive, *Fig4* null mice were determined. Remarkably, the neuron-specific transgene was sufficient to rescue virtually all of the pathogenic effects of *Fig4* deficiency, including juvenile lethality, size, tremor, spongiform degeneration, and accumulation of inclusion bodies in astrocytes. Although expression of *Fig4* in the oligodendrocytes is absent in the NSE-Tg mice, the myelination deficit was rescued in CNS (Winters et al., 2012) and PNS (Ferguson et al., 2012). In contrast, *Fig4* expression in astrocytes corrected the accumulation of inclusion bodies but did not extend the lifetime of the Fig4 null mice (Ferguson et al., 2012). This result demonstrated that *Fig4* deficiency is primarily a neuronal disorder, and it will be necessary to treat the neuronal defect in order to alleviate neurodegeneration in patients.

5. CONDITIONAL KNOCKOUT OF *FIG4* IN NEURONS

To investigate the requirement for *Fig4* expression in specific tissues, we generated a floxed allele of *Fig4* with loxP sites flanking exon 4 (Ferguson et al., 2012). The floxed allele can be combined with a variety of CRE recombinase transgenic mice to examine the effects of *Fig4* deletion at various developmental points as well as in different cell types. In crosses with a neuron-specific synapsin CRE line, mice lacking *Fig4* expression in neurons developed spongiform degeneration, a movement disorder, and tremor. However, these mice survive several months longer than the global null mice, indicating that expression of *Fig4* in other tissues can moderate some of the effects of neuronal *Fig4* deficiency (Ferguson et al., 2012). The neuron-specific knockout mouse demonstrates that *Fig4* expression in neurons is *necessary* for survival, while the neuron-specific transgenic rescue demonstrates that expression of *Fig4* in neurons is *sufficient* for survival.

6. THE HUMAN DISEASE MUTATION FIG4-I41T IN TRANSGENIC MICE

In order to examine the mechanism of pathogenesis of the human mutant I41T found in patients with Charcot–Marie–Tooth neuropathy type 4J, we generated a transgene with the mutant *Fig4* cDNA driven by the globally expressed chicken β-actin promoter (Lenk et al., 2011). The transgene was crossed on to the *Fig4* null background, in order to generate a mouse that produced only mutant FIG4-I41T protein, like CMT4J patients. An interesting effect of expression level of the I41T mutant protein was observed. In mice with 5× overexpression of the mutant transcript, there was complete rescue of the *Fig4* null phenotype, restoration of normal brain morphology, and survival beyond 2 years of age. This rescue was obtained despite the very low level of mutant protein in the transgenic mouse tissues, 10% of normal protein level. However, in a second line of transgenic mice with 2× transcript expression, survival was much shorter, only 3–4 months. It thus appears that the minimum requirement for the *Fig4* enzyme is approximately one tenth of wild-type expression level. This work indicates that upregulation of the I41T allele in CMT4J patients could offer a therapeutic strategy. Further analysis demonstrated that the instability of the I41T protein is a consequence of impaired binding to the VAC14 scaffold protein. The level of FIG4-I41T is also very low in patient fibroblasts and can be increased by inhibition of degradation by the proteasome pathway (Ikonomov et al., 2010; Lenk et al., 2011).

7. A SPONTANEOUS MISSENSE MUTATION OF *VAC14* IN THE *INGLS* MOUSE

The spontaneous *ingls* mutation (infantile gliosis) was identified at the Jackson Laboratory in 1991 by its juvenile lethality, and mapped to a chromosomal location close to *Vac14*, the scaffold protein in the PI(3,5)P_2 biosynthetic protein complex (Bronson et al., 2003). Homozygous *ingls* mutants exhibit neurodegeneration and gliosis that closely resemble the phenotype of the *Fig4* null mutant. We therefore tested *Vac14* as a positional candidate gene for *ingls*. Sequencing the exons of *Vac14* identified the mutation Leu156Arg in the fourth heat repeat domain of the protein (Jin et al., 2008). The mutation prevents binding of VAC14 to the kinase protein PIKFYVE, resulting in a 50% reduction in the level of PI(3,5)P$_2$ in cultured fibroblasts. Tissues of homozygous *ingls* mutants contain normal levels of FIG4, VAC14, and PIKFYVE proteins, but the impaired formation of the protein complex leads to deficiency of PI(3,5)P$_2$. The similarity in pathology between the *ingls* mutant and the *Fig4* null mouse, including vacuolization of fibroblasts, neurodegeneration, astrocytosis, and formation of inclusion bodies in astrocytes containing p62 and ubiquitinated protein, supports the conclusion that reduced PI(3,5)P$_2$ levels are responsible for the pathology in both mutants (Ferguson et al., 2009; Jin et al., 2008).

8. A NULL GENE-TRAP ALLELE OF *VAC14*

Mice null for *Vac14*, the scaffold protein in the biosynthetic complex for PI(3,5)P$_2$, were generated in the KOMP project by random insertion of the β-geo gene-trap vector into intron 1 of *Vac14*. This insertion ablated detectable protein from mice homozygous for the "trapped" allele, resulting in earlier lethality than the Leu156Pro mutation in the ingls mouse (Zhang et al., 2007). Similar to the *Fig4* null mice described above, the *Vac14* null mice exhibit spongiform degeneration of the brain, extensive vacuolization of primary cultured fibroblasts, and a 50% reduction of PI(3,5)P$_2$ in fibroblasts (Zhang et al., 2007). Mislocalization of the cation-independent mannose-6-phosphate receptor (CI-MPR) and CI-MPR cargo cathepsin D in this mutant implicates PI(3,5)P$_2$ in retrograde transport from the endo-lysosomal system to the trans-Golgi network (Zhang et al., 2007). Synaptic localization of VAC14 in hippocampal neurons and impaired turnover of the postsynaptic AMPA receptor have also been observed in *Vac14* null mice (Zhang et al., 2012). The similarity of the pathological changes in mutant mice suggests that mutations of human *VAC14*, like *FIG4*, could cause peripheral neuropathy or Yunis–Varon syndrome, but to date patient mutations have not been identified.

9. A HYPOMORPHIC GENE-TRAP ALLELE OF *PIKFYVE* (*FAB1*)

A global null mutant of *Pikfyve*, the 5-kinase of the PI(3,5)P$_2$ synthetic complex, was recently described. Deletion of floxed exon 6 by CRE recombinase resulted in preimplantation lethality, possibly as early as E3.5

(Ikonomov et al., 2011). This is much earlier than the lethality caused by lack of *Fig4* or *Vac14*, suggesting that loss of *Pikfyve* may result in complete deficiency of PI(3,5)P$_2$. Transfection of CRE recombinase into homozygous *Pikfyve*$^{flox/flox}$ fibroblasts also resulted in vacuolization and arrested cell division, demonstrating the cell autonomy of the defect in fibroblasts (Ikonomov et al., 2011).

10. A CONDITIONAL KNOCKOUT OF PIKFYVE

A gene-trap allele of *Pikfyve* generated from a KOMP ES cell line resulted in 85% reduction of transcript level (Zolov et al., 2012). The residual 15% expression was sufficient for completion of embryonic development leading to postnatal lethality during the first 3 weeks after birth. The PI(3,5)P$_2$ levels in cultured fibroblasts of these mice were approximately 50% of normal, comparable to the *Fig4* and *Vac14* mutants with similar postnatal survival (Zolov et al., 2012). Spongiform degeneration of the brain and vacuolization of several other tissues including lung and heart were observed (Zolov et al., 2012).

11. GENETIC INTERACTIONS: *FIG4*, *VAC14*, AND *MTMR2*

The availability of multiple mouse mutants in a pathway of interest makes it possible to examine genetic interactions using crosses between mutant lines. Myotubularin-related 2 (MTMR2) is a phosphatase that removes the 3-phosphate from PI(3,5)P$_2$, while FIG4 removes the 5-phosphate. Mice lacking MTMR2 are models of peripheral neuropathy CMT4B1 (Bolino et al., 2004) and are thought to accumulate excess PI(3,5)P$_2$. To test the hypothesis that *Mtmr2* and *Fig4* act on the same subcellular pool of PI(3,5)P$_2$, mice with mutations at both loci were generated. Interestingly, heterozygosity for the null allele of *Fig4* reduced the severity of neurodegeneration and myelin outfolding in *Mtmr2* null mice, demonstrating that the two phosphatases can access the same substrate pool and that MTMR2 does hydrolyze PI(3,5)P$_2$ *in vivo* (Vaccari et al., 2011).

Null heterozygotes for either *Fig4* or *Vac14* do not exhibit visible abnormalities. To evaluate the possibility that double heterozygosity for *Fig4* and *Vac14* might cause a visible defect, we generated *Fig4*$^{+/-}$, *Vac14*$^{+/-}$ mice. In this case, no interaction was observed and the mice were viable and fertile, with normal life span.

The stability of the FIG4 protein is reduced by the human pathogenic mutation I41T mutation that impairs binding to VAC14 (Lenk et al.,

2011). To evaluate the dependence of the wild-type FIG4 protein on interaction with VAC14, we carried out Western blotting of VAC14 null tissues. We observed a complete absence of FIG4 protein in the VAC14 mutant, clearly demonstrating that the stability of wild-type FIG4 is dependent on interaction with VAC14 (Lenk et al., 2011).

These examples demonstrate the utility of testing genetic interactions using mutant mice. The availability of both global and conditional alleles for the major components of the PI(3,5)P$_2$ biosynthetic pathway will be useful for further analysis of *in vivo* gene interactions in this pathway.

12. GENETIC EFFECTS OF STRAIN BACKGROUND

During the generation of transgenic and conditional knockout mice, it is often difficult to avoid mixing the genetic backgrounds of different inbred strains of mice. While the segregation of modifier variants in different inbred backgrounds can complicate the characterization of mutant phenotypes, the positive aspect of interstrain variation is the potential to identify the critical differences between strains and to better understand the underlying pathogenic mechanisms. For example, analysis of strain differences in the phenotypes of sodium channel mutations led to identification of the Scnm1 splice factor affecting the *Scn8a* transcript (Buchner et al., 2003) and the modifying effect of potassium channel *Kcnv2* on seizures caused by mutation of sodium channel *Scn2a* (Jorge et al., 2011). To examine the effect of inbred strain backgrounds on *Fig4* null lethality, we crossed the spontaneous null allele plt onto strains C57BL/6J and C3H by repeated backcrossing of *Fig4*$^{+/-}$ heterozygotes to wild-type mice of each strain for more than ten generations. A significant difference in survival was observed, with neonatal lethality on the congenic strain B6.plt/plt (Fig. 14.2). Interestingly, null homozygotes on the hybrid F1 background survive longer than either inbred strain, suggesting that there is interaction between multiple loci affecting dependence on *Fig4*. Identification of these modifier loci could provide novel targets for treatment of *Fig4* deficiency.

13. FUTURE APPLICATIONS OF MOUSE MODELS OF PI(3,5)P$_2$ DEFICIENCY

Mouse models provide tools for evaluation of the rare variants that are being rapidly discovered in patient populations by exome sequencing. For example, primary fibroblasts cultured from *Fig4* null mice exhibit extensive

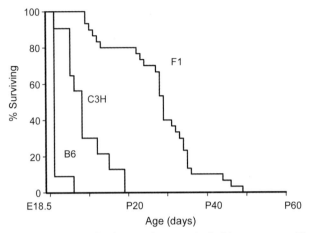

Figure 14.2 Effects of genetic background on survival of homozygous *Fig4* null mice. The congenic lines B6.plt and C3H.plt carrying the *Fig4*^{plt} null mutation were generated by repeated backcrossing of *Fig4*$^{+/-}$ heterozygotes to inbred strains C57BL/6J and C3H, for more than ten generations. Crosses between heterozygotes within each line generated the congenic homozygotes. Crosses between B6 and C3H congenic heterozygotes generated the homozygous *Fig4*$^{-/-}$ F1 mutants. The difference in survival of *Fig4* null homozygotes on each genetic background is indicated. B6.plt/plt, $n=50$; C3H.plt/plt, $n=18$; F1.plt/plt, $n=30$.

cytoplasmic vacuolization that can be rescued by transfection of wild-type *Fig4* cDNA (Fig. 14.1C). The functional effects of human variants can be tested by transfection of mutated *Fig4* cDNAs into the null fibroblasts and comparison of rescue efficiency with the wild-type cDNA (e.g., Campeau et al., 2013; Baulac et al., 2013). Pathogenic mechanisms can also be investigated by expression of the mutated cDNA in transgenic mice (e.g., Lenk et al., 2011).

Mice provide a valuable system for testing therapies for inherited disorders *in vivo*. For example, reduced turnover of the *Fig4* I41T variant by proteasome inhibitors could be evaluated in the I41T transgenic mouse. It is an attractive hypothesis that the disrupted vesicle trafficking in PI(3,5)P$_2$-deficient cells may be secondary to reduced activity of lysosomal ion channels, since the cation channels *MCOLN1* (TRPML1), *TPC1*, and *TPC2* are directly activated by PI(3,5)P$_2$ (Dong et al., 2010; Wang et al., 2012). Therapeutic activation of these channels to treat neurodegeneration is an exciting possibility that can be tested in the *Fig4* null mice.

Long-lived mutants like the neuron-specific *Fig4* transgenic mouse will be useful for analysis of nonneuronal pathology, such as the recently recognized dysmyelination and bone dysplasia in *Fig4* null mice (Campeau et al.,

2013; Ferguson et al., 2012). Careful analysis of these mice may reveal additional effects of PI(3,5)P$_2$ deficiency and thereby suggest new patient populations for screening.

The variable age of onset and clinical diversity in CMT4J patients suggest that genetic modifiers may influence disease severity (Nicholson et al., 2011). Mouse models offer a unique resource for identification of genetic modifiers segregating in crosses between inbred strains.

Several human diseases are known to result from defects in the PI(3,5)P$_2$ pathway and more are likely to be identified in the coming years through the application of high-throughput genome sequencing. A variety of mouse models will continue to contribute to characterizing the physiological roles of PI(3,5)P$_2$ signaling and testing therapies for PI(3,5)P$_2$-deficiency disorders.

ACKNOWLEDGMENTS

This work was supported by NIH grant GM24872 (M. H. M.). G. M. L. is a fellow of the Postdoctoral Translational Scholars Program of the Michigan CTSA (UL1 TR000433).

REFERENCES

Ayadi, A., Birling, M. C., Bottomley, J., Bussell, J., Fuchs, H., Fray, M., et al. (2012). Mouse large-scale phenotyping initiatives: Overview of the European Mouse Disease Clinic (EUMODIC) and the Welcome Trust Sanger Institute Mouse Genetics Project. *Mammalian Genome, 23*, 9–10.

Baulac, S., Lenk, G. M., Defresnois, B., Bencheikh, B. O. A., Couarch, P., Renard, J., et al. (2013). Role of phosphoinositide phosphatase FIG4 gene in familial epilepsy with polymicrogyria. *Neurology*, in press.

Bolino, A., Bolis, A., Previtali, S. C., Dina, G., Bussini, S., Dati, G., et al. (2004). Disruption of Mtmr2 produces CMT4B1-like neuropathy with myelin outfolding and impaired spermatogenesis. *Journal of Cell Biology, 22*, 711–721.

Bonangelino, C. J., Nau, J. J., Duex, J. E., Brinkman, M., Wurmser, A. E., Gary, J. D., et al. (2002). Osmotic stress-induced increase of phosphatidylinositol 3,5-bisphosphate requires Vac14p, an activator of the lipid kinase Fab1p. *Journal of Cell Biology, 18*, 1015–1028.

Botelho, R. J., Efe, J. A., Teis, D., & Emr, S. D. (2008). Assembly of a Fab1 phosphoinositide kinase signaling complex requires the Fig4 phosphoinositide phosphatase. *Molecular Biology of the Cell, 10*, 4273–4286.

Bradley, A., Anastassiadis, K., Ayadi, A., Battey, J. F., Bell, C., Birling, M. C., et al. (2012). The mammalian gene function resource: The International Knockout Mouse Consortium. *Mammalian Genome, 23*, 580–586.

Bronson, R., Perides, G., Sweet, H., Ward-Bailey, P. F., & Davisson, M. (2003). Infantile gliosis (ingls) a new spontaneous mutation in the mouse. MGI Direct Data Submission http://mousemutant.jax.org/gliosis_paper.html.

Buchner, D. A., Trudeau, M., & Meisler, M. H. (2003). SCNM1, a putative RNA splicing factor that modifies disease severity in mice. *Science, 301*, 967–969.

Campeau, P. M., Lenk, G. M., Lu, J. T., Bae, Y., Burrage, L., Turnpenny, P., et al. (2013). Yunis-Varón syndrome is caused by mutations in FIG4, encoding a phosphoinositide phosphatase. *American Journal of Human Genetics, 92*, 782–791.

Chow, C. Y., Landers, J. E., Bergren, S. K., Sapp, P. C., Grant, A. E., Jones, J. M., et al. (2009). Deleterious variants of FIG4, a phosphoinositide phosphatase, in patients with ALS. *American Journal of Human Genetics, 84*, 85–88.

Chow, C. Y., Zhang, Y., Dowling, J. J., Jin, N., Adamska, M., Shiga, K., et al. (2007). Mutation of FIG4 causes neurodegeneration in the pale tremor mouse and patients with CMT4J. *Nature, 5*, 68–72.

Donahue, L. R., Hrabe de Angelis, M., Hagn, M., Franklin, C., Lloyd, K. C., Magnuson, T., et al. (2012). Centralized mouse repositories. *Mammalian Genome, 23*, 559–571.

Dong, X. P., Shen, D., Wang, X., Dawson, T., Li, X., Zhang, Q., et al. (2010). controls membrane trafficking by direct activation of mucolipin Ca(2+) release channels in the endolysosome. *Nature Communications, 13*, 1–38.

Dove, S. K., Cooke, F. T., Douglas, M. R., Sayers, L. G., Parker, P. J., & Michell, R. H. (1997). Osmotic stress activates phosphatidylinositol-3,5-bisphosphate synthesis. *Nature, 13*, 187–192.

Duex, J. E., Nau, J. J., Kauffman, E. J., & Weisman, L. S. (2006). Phosphoinositide 5-phosphatase Fig 4p is required for both acute rise and subsequent fall in stress-induced phosphatidylinositol 3,5-bisphosphate levels. *Eukaryotic Cell, 5*, 723–731.

Ferguson, C. J., Lenk, G. M., Jones, J. M., Grant, A. E., Winters, J. J., Dowling, J. J., et al. (2012). Neuronal expression of *Fig4* is both necessary and sufficient to prevent spongiform neurodegeneration. *Human Molecular Genetics, 21*, 3525–3534.

Ferguson, C. J., Lenk, G. M., & Meisler, M. H. (2009). Defective autophagy in neurons and astrocytes from mice deficient in PI(3,5)P2. *Human Molecular Genetics, 15*, 4868–4878.

Gary, J. D., Sato, T. K., Stefan, C. J., Bonangelino, C. J., Weisman, L. S., & Emr, S. D. (2002). Regulation of Fab1 phosphatidylinositol 3-phosphate 5-kinase pathway by Vac7 protein and Fig4, a polyphosphoinositide phosphatase family member. *Molecular Biology of the Cell, 13*, 1238–1251.

Ikonomov, O. C., Sbrissa, D., Delvecchio, K., Xie, Y., Jin, J. P., Rappolee, D., et al. (2011). The phosphoinositide kinase PIKfyve is vital in early embryonic development: Preimplantation lethality of PIKfyve-/- embryos but normality of PIKfyve +/- mice. *Journal of Biological Chemistry, 15*, 13404–13413.

Ikonomov, O. C., Sbrissa, D., Fligger, J., Delvecchio, K., & Shisheva, A. (2010). ArPIKfyve regulates Sac3 protein abundance and turnover: Disruption of the mechanism by Sac3I41T mutation causing Charcot-Marie-Tooth 4J disorder. *Journal of Biological Chemistry, 285*, 26760–26764.

Ikonomov, O. C., Sbrissa, D., Mlak, K., Kanzaki, M., Pessin, J., & Shisheva, A. (2002). Functional dissection of lipid and protein kinase signals of PIKfyve reveals the role of PtdIns 3,5-P2 production for endomembrane integrity. *Journal of Biological Chemistry, 15*, 9206–9211.

Jin, N., Chow, C. Y., Liu, L., Zolov, S. N., Bronson, R., Davisson, M., et al. (2008). VAC14 nucleates a protein complex essential for the acute interconversion of PI3P and PI(3,5)P (2) in yeast and mouse. *EMBO Journal, 17*, 3221–3234.

Jorge, B. S., Campbell, C. M., Miller, A. R., Rutter, E. D., Gurnett, C. A., Vanoye, C. G., et al. (2011). Voltage-gated potassium channel KCNV2 (Kv8.2) contributes to epilepsy susceptibility. *Proceedings of the National Academy of Sciences of the United States of America, 108*, 5443–5448.

Katona, I., Zhang, X., Bai, Y., Shy, M. E., Guo, J., Yan, Q., et al. (2011). Distinct pathogenic processes between Fig4-deficient motor and sensory neurons. *European Journal of Neuroscience, 33*, 1401–1410.

Lenk, G. M., Ferguson, C. J., Chow, C. Y., Jin, N., Jones, J. M., Grant, A. E., et al. (2011). Pathogenic mechanism of the FIG4 mutation responsible for Charcot-Marie-Tooth disease CMT4J. *PLoS Genetics*, *7*, e1002104.

McCartney, A. J., Zhang, Y., & Weisman, L. S. (2013). Phosphatidylinositol 3,5-bis phosphate: Low abundance. High significance. *BioEssays*, in press.

Murray, S. A., Eppig, J. T., Smedley, D., Simpson, E. M., & Rosenthal, N. (2012). Beyond knockouts: Cre resources for conditional mutagenesis. *Mammalian Genome*, *23*, 587–599.

Nicholson, G., Lenk, G. M., Reddel, S. W., Grant, A. E., Towne, C. F., Ferguson, C. J., et al. (2011). Distinctive genetic and clinical features of CMT4J: A severe neuropathy caused by mutations in the PI(3,5)P₂ phosphatase FIG4. *Brain*, *134*, 1959–1971.

Nord, A. S., Chang, P. J., Conklin, B. R., Cox, A. V., Harper, C. A., Hicks, G. G., et al. (2006). The International Gene Trap Consortium Website: a portal to all publicly available gene trap cell lines in mouse. *Nucleic Acids Research*, *34*, D642–648.

Paigen, K. (2003). One hundred years of mouse genetics: an intellectual history. I. The classical period (1902–1980). *Genetics*, *163*, 1–7.

Rempe, D., Vangeison, G., Hamilton, J., Li, Y., Jepson, M., & Federoff, H. J. (2006). Synapsin I Cre Transgene expression in male mice produces germline recombination in progeny. *Genesis*, *44*, 44–49.

Saunders, T. L. (2010). A survey of internet resources for mouse development. *Methods in Enzymology*, *476*, 3–21.

Sbrissa, D., Ikonomov, O. C., Fu, Z., Ijuin, T., Gruenberg, J., Takenawa, T., et al. (2007). Core protein machinery for mammalian phosphatidylinositol 3,5-bisphosphate synthesis and turnover that regulates the progression of endosomal transport. Novel Sac phosphatase joins the ArPIKfyve-PIKfyve complex. *The Journal of Biological Chemistry*, *282*, 23878–23891.

Sbrissa, D., Ikonomov, O. C., Strakova, J., Dondapati, R., Mlak, K., Deeb, R., et al. (2004). A mammalian ortholog of Saccharomyces cerevisiae Vac14 that associates with and up-regulates PIKfyve phosphoinositide 5-kinase activity. *Molecular and Cellular Biology*, *24*, 10437–10447.

Vaccari, I., Dina, G., Tronchère, H., Kaufman, E., Chicanne, G., Cerri, F., et al. (2011). Genetic interaction between MTMR2 and FIG4 phospholipid phosphatases involved in Charcot-Marie-Tooth neuropathies. *PLoS Genetics*, *7*, e1002319.

Volpicelli-Daley, L., & De Camilli, P. (2007). Phosphoinositides' link to neurodegeneration. *Nature Medicine*, *7*, 784–786.

Wang, X., Zhang, X., Dong, X. P., Samie, M., Li, X., Cheng, X., et al. (2012). TPC proteins are phosphoinositide-activated sodium-selective ion channels in endosomes and lysosomes. *Cell*, *12*, 372–383.

Whiteford, C. C., Brearley, C. A., & Ulug, E. T. (1997). Phosphatidylinositol 3,5-bisphosphate defines a novel PI 3-kinase pathway in resting mouse fibroblasts. *Biochemical Journal*, *1*, 597–601.

Winters, J. J., Ferguson, C. J., Lenk, G. M., Giger-Mateeva, V. I., Shrager, P., Meisler, M. H., et al. (2012). Congenital CNS hypomyelination in the Fig4 null mouse is rescued by neuronal expression of the PI(3,5)P(2) phosphatase Fig4. *Journal of Neuroscience*, *30*, 17736–17751.

Wiradjaja, F., Ooms, L. M., Tahirovic, S., Kuhne, E., Devenish, R. J., Munn, A. L., et al. (2007). Inactivation of the phosphoinositide phosphatases Sac1p and Inp54p leads to accumulation of phosphatidylinositol 4,5-bisphosphate on vacuole membranes and vacuolar fusion defects. *The Journal of Biological Chemistry*, *282*, 16295–16307.

Zhang, X., Chow, C. Y., Sahenk, Z., Shy, M. E., Meisler, M. H., & Li, J. (2008). Mutation of FIG4 causes a rapidly progressive, asymmetric neuronal degeneration. *Brain*, *131*, 1990–2001.

Zhang, Y., Zolov, S. N., Chow, C. Y., Slutsky, S. G., Richardson, S. C., Piper, R. C., et al. (2007). Loss of Vac14, a regulator of the signaling lipid phosphatidylinositol 3,5-bisphosphate, results in neurodegeneration in mice. *Proceedings of the National Academy of Sciences of the United States of America, 30*, 17518–17523.

Zhang, Y., McCartney, A. J., Zolov, S. N., Ferguson, C. J., Meisler, M. H., Sutton, M. A., et al. (2012). Modulation of synaptic function by VAC14, a protein that regulates the phosphoinositides PI(3,5)P$_2$ and PI(5)P. *The EMBO Journal, 31*, 3442–3456.

Zolov, S. N., Bridges, D., Zhang, Y., Lee, W. W., Riehle, E., Verma, R., et al. (2012). In vivo, Pikfyve generates PI(3,5)P2, which serves as both a signaling lipid and the major precursor for PI5P. *Proceedings of the National Academy of Sciences of the United States of America, 23*, 17472–17477.

CHAPTER FIFTEEN

Monitoring Endosomal Trafficking of the G Protein-Coupled Receptor Somatostatin Receptor 3

Cristy Tower-Gilchrist*, Melanie L. Styers[†], Bradley K. Yoder[‡], Nicolas F. Berbari[‡], Elizabeth Sztul[‡,1]

*Department of Cell Biology, Emory University, Atlanta, Georgia, USA
[†]Department of Biology, Birmingham-Southern College, Birmingham, Alabama, USA
[‡]Department of Cell, Developmental and Integrative Biology, University of Alabama at Birmingham, Birmingham, Alabama, USA
[1]Corresponding author: e-mail address: esztul@uab.edu

Contents

1. Introduction	262
2. Development of Cell Lines Stably Expressing SSTR3	264
2.1 Generation of a tagged SSTR3 construct	265
2.2 Transfection of IMCD-3 cells via electroporation	266
2.3 Selection of IMCD-3 cells stably expressing SSTR3	267
3. Live Imaging of SSTR3 and RABS in Mammalian Kidney Cells	269
3.1 Transfection of Rabs into IMCD-3 cells	269
3.2 Microscope setup	270
3.3 Confocal microscopy colocalization of SSTR3 and Rabs	271
4. Dynamics of SSTR3 Transit Relative to RABS	272
4.1 Time-lapse dual color imaging of SSTR3 and Rabs	272
4.2 Analysis of half-time residency and clearance times	274
5. Effects of Dominant Negative Rabs on SSTR3 Trafficking	276
5.1 Generating mutant forms of endosomal Rabs	276
5.2 Dynamics of SSTR3 traffic in cells expressing mutant Rabs	277
6. Summary	278
Acknowledgments	278
References	278

Abstract

Endocytic trafficking of G protein-coupled receptors (GPCRs) regulates the number of cell surface receptors available for activation by agonists and serves as one mechanism that controls the intensity and duration of signaling. Deregulation of GPCR-mediated signaling pathways results in a multitude of diseases, and thus extensive efforts have been directed toward understanding the pathways and molecular events that regulate endocytic trafficking of these receptors. The general paradigms associated with internalization and recycling, as well as many of the key regulators involved in endosomal trafficking of GPCRs have been identified. This knowledge provides goalposts to facilitate the analysis of endosomal pathways traversed by previously uncharacterized GPCRs. Some of the most informative markers associated with GPCR transit are the Rab members of the Ras-related family of small GTPases. Individual Rabs show high selectivity for distinct endosomal compartments, and thus colocalization of a GPCR with a particular Rab informs on the internalization pathway traversed by the receptor. Progress in our knowledge of endosomal trafficking of GPCRs has been achieved through advances in our ability to tag GPCRs and Rabs with fluorescent proteins and perform live cell imaging of multiple fluorophores, allowing real-time observation of receptor trafficking between subcellular compartments in a cell culture model.

1. INTRODUCTION

G protein-coupled receptors (GPCRs) are comprised of a large class of seven transmembrane-spanning receptors with key roles in regulation of cellular physiological responses. Disruption of GPCR function or trafficking can lead to perturbation in the activity of downstream signaling pathways and can result in a multitude of diseases, including cancer, congestive heart failure, nephrogenic diabetes insipidus, retinitis pigmentosa, hypo- and hyperthyroidism, and fertility disorders (Lappano & Maggiolini, 2012; Schöneberg et al., 2004; Sorkin & von Zastrow, 2009).

GPCRs bind ligands at the cell surface and signal by coupling to different types of heterotrimeric $G\alpha\beta\gamma$ proteins (Lappano & Maggiolini, 2012). This coupling event catalyzes the exchange of GDP for GTP on the $G\alpha$ subunit, leading to dissociation of $G\alpha$ from $G\beta\gamma$, and the subsequent activation of downstream effectors by the $G\alpha$ and the $G\beta\gamma$ moieties (reviewed in Lappano & Maggiolini, 2012). GPCRs can exist in different active conformations that are preferentially stabilized by specific agonists, and thus, binding of different ligands can result in the same GPCR coupling to different signaling pathways. In addition, GPCR pairs have been shown to interact and form heterocomplexes *in vitro* and *in vivo*, which may contribute to functional specificity of GPCR signaling (Gurevich & Gurevich, 2008; Hubbard & Hepler, 2006; Milligan, 2010).

Following ligand binding and signaling, many GPCRs are rapidly phosphorylated by GPCR kinases (GRKs) that selectively phosphorylate agonist-activated receptors. Phosphorylation promotes the binding of β-arrestin, which prevents the interaction of GPCRs with G proteins and terminates signaling. Arrestin binding also promotes endocytosis of GPCRs, leading to the elimination of the GPCR from cell surface. The pathways taken by GPCRs after endocytosis differ; some are targeted for lysosomal degradation, while others are recycled back to the cell surface through either rapid or slow traffic from endosomes (Jean-Alphonse & Hanyaloglu, 2011; Sorkin & Von Zastrow, 2002). Thus, the plasma membrane and the endosomal pathway together represent a sophisticated communication and homeostatic device used by the cell to regulate GPCR-mediated signaling. Considering the extreme clinical importance of GPCRs, it is not surprising that elucidating the intracellular pathways traversed by diverse GPCRs and the molecular machinery regulating passage of GPCRs through these compartments is under active investigation.

Rab GTPases belong to a large family of Ras-related small GTPases and have been shown to regulate membrane trafficking by influencing multiple steps, including vesicle formation, tethering, and fusion (Stenmark, 2009). Importantly, Rabs have been shown to localize to specific vesicular compartments and can be used as spatial markers to monitor passage of GPCRs through the endosomal pathway (reviewed in Jean-Alphonse & Hanyaloglu, 2011; Stenmark, 2009). Within the endocytic pathway, Rab4 and Rab5 localize to early endosomes, Rab7 localizes to late endosomes, and Rab11 localizes to recycling endosomes. Even within single compartments, early endosomal Rabs show spatial and temporal separation, as Rab4 and Rab5 show extensive colocalization, but also areas of separation (Sönnichsen, De Renzis, Nielsen, Rietdorf, & Zerial, 2000). Therefore, Rabs represent useful molecular markers to monitor endosomal trafficking of GPCRs by measuring the dynamics of GPCR movement through specific compartments and subcompartments defined by each endosomal Rab.

In this chapter, we use the somatostatin receptor 3 (SSTR3), a GPCR involved in signal transmission initiated by the neuropeptide somatostatin, as a model protein to describe the methodology to monitor endosomal trafficking of GPCRs. SSTR3 is one of five somatostatin receptors (somatostatin receptors 1–5; SSTR1–5) found in many different tissues that exhibit different expression patterns during development (Barnett, 2003; Handel et al., 1999). Previous histological and immunofluorescence studies localized SSTR3 to many regions of the rodent brain and demonstrated its localization to the primary cilium in hippocampal neurons (Berbari, Bishop, Askwith, Lewis, & Mykytyn, 2007, Berbari, Johnson, Lewis, Askwith, & Mykytyn, 2008; Handel et al., 1999). Biochemical and molecular studies document that the binding of somatostatin induces SSTR3 internalization from the cell membrane via clathrin-coated vesicles in a process that requires β-arrestin 2 and adaptor protein 2 (Jacobs & Schulz, 2008; Tulipano et al., 2004). Internalized SSTR3 has been detected in recycling endosomes, as evidenced by colocalization with Rab11 and endocytosed transferrin, two markers of recycling endosomes (Kreuzer, Krisch, Déry, Bunnett, & Meyerhof, 2001). In addition, internalized SSTR3 also localizes to early endosomal compartments lacking transferrin. However, the exact pathway traversed by SSTR3 upon internalization and the dynamics of its passage through individual endosomal compartments can only be assessed by dynamic live cell imaging.

To investigate the endosomal trafficking of GPCRs in live cells, a number of key parameters must be considered. First, it is essential that the receptor be tagged with a fluorescent moiety (usually the green fluorescent protein (GFP) or a spectral derivative of GFP) and that the tagged receptor is expressed at levels sufficient for direct imaging. Second, the tag must not affect the ability of the receptor to bind ligand or signal at a level comparable to that of the endogenous receptor. Third, the tagged receptor must traffic to the correct cellular compartment based on current knowledge regarding the distribution and/or function of the endogenous untagged receptor. Fourth, fluorescently tagged markers must be available to label the organelle through which transport will be monitored. Fifth, the expression levels of the fluorescently tagged receptor and the organellar marker within the cell should be in the low-to-medium range to allow the detection of small trafficking intermediates formed from highly labeled endosomal compartments. These five criteria should be assessed prior to live cell imaging.

In this chapter, we discuss approaches and provide examples of methodology used to express epitope-tagged SSTR3 and Rabs in mammalian cells, assess levels of colocalization between the SSTR3 and different Rabs, and measure dynamic passage of SSTR3 relative to different Rabs by real-time microscopy. Techniques to be discussed include tagging a GPCR with a fluorescent protein, mammalian tissue culture on glass and plastic, transfection of DNA constructs into mammalian cells by electroporation, selection of stable cell lines expressing the tagged GPCR, and methods for live imaging and image processing to define parameters associated with endosomal transport of the GFP-tagged GPCR relative to red fluorescent protein (RFP)- and mCherry-tagged organellar markers.

2. DEVELOPMENT OF CELL LINES STABLY EXPRESSING SSTR3

A variety of methods can be used to generate cell lines stably expressing fluorescent fusion proteins. Here we describe a method of selection of stable clones expressing the exogenous gene after transient transfection. We describe in detail the generation and characterization of the enhanced green fluorescent protein (EGFP)-SSTR3 cell line that has previously been used to analyze the endosomal trafficking of SSTR3 (Tower et al., 2011). The major steps we will cover include generation of the tagged SSTR3 construct, transfection of target cells, and clonal selection and characterization. This entire process requires approximately 4–8 weeks.

2.1. Generation of a tagged SSTR3 construct

The goal of this procedure is to generate a construct that drives expression of SSTR3 fused to EGFP in mammalian cells. The pEGFP-N3 plasmid (Clontech, Mountain View, CA) was selected as the expression vector, as it allows for fusion of EGFP at the C-terminus of SSTR3. Although there are commercially available vectors that can be used to fuse EGFP to the N-terminus of GPCRs (pEGFP-C series, Clontech), we recommend the use of the C-terminal EGFP tag when possible. Use of the N-terminal EGFP tag requires additional steps be taken to ensure proper membrane threading of the seven transmembrane domain GPCR, as previously described (Huang & Willars, 2011).

1. Isolate total RNA from postnatal day 30 mouse whole brain using Trizol reagent according to the manufacturer's protocol (#15596-026, Life Technologies, Carlsbad, CA).
2. Synthesize complementary DNA (cDNA) using Superscript II reverse transcriptase according to the manufacturer's recommended instructions (#18064-022, Life Technologies). We recommend 5 µg of total RNA be used as a template for the reaction.
3. Design primers that bind sequences located in the untranslated regions of SSTR3 to ensure successful amplification of the entire coding region. The National Center for Biotechnology Information has a useful Web site for primer design (http://www.ncbi.nlm.nih.gov/tools/primer-blast/).
4. Amplify the coding sequence of mouse SSTR3 via the polymerase chain reaction (PCR) using mouse cDNA as a template. The Platinum® *Taq* DNA Polymerase High Fidelity kit (Invitrogen, Carlsbad, CA), or a similar high fidelity polymerase kit, should be used according to manufacturer instructions. Genomic DNA may also be used as a template for the reaction, as the SSTR3 gene, like many GPCRs, does not contain introns.
5. Subclone the resulting PCR product, containing the entire open reading frame of SSTR3, into a TA cloning vector, such as pSTBlue-1 (Novagen, San Diego, CA), according to manufacturer instructions.

6. Confirm that the sequence of SSTR3 is correct by DNA sequencing. This open reading frame will serve as a template for further subcloning.
7. Amplify SSTR3 by PCR for insertion into the pEGFP-N3 vector. Primers for the PCR reaction must add restriction enzyme sites compatible with the vector (XhoI and KpnI). Because pEGFP-N3 contains a C-terminal-EGFP tag, the reverse primer must also remove the SSTR3 stop codon in order to allow fusion with the tag in the vector.
8. The resulting PCR product can be digested with both XhoI and KpnI and ligated into the pEGFP-N3 vector digested with the same enzymes. Detailed experimental procedures for subcloning have been described and can be found at http://www.scribd.com/doc/23261720/Molecular-Cloning-A-Laboratory-Manual-On-The-Web-Maniatis.

2.2. Transfection of IMCD-3 cells via electroporation

IMCD-3 cells, derived from mouse inner medullary collecting duct cells, were chosen to generate a stable line expressing SSTR3-EGFP because they retain many characteristics of the original kidney epithelial tissue (Rauchman, Nigam, Delpire, & Gullans, 1993). These cells can be grown in monolayers on plastic and glass or as monolayers on filters, where they show basolateral polarity (Goel, Sinkins, Zuo, Estacion, & Schilling, 2006). IMCD-3 cells are refractory to some common methods of transfection, such as the use of Lipofectamine 2000 (Invitrogen) and FuGENE (Promega, Madison, WI). For this reason, we have found that electroporation is the preferred and most efficient method.

1. IMCD-3 cells (ATCC, Manassas, VA) should be cultured in Dulbecco's Modified Eagle Medium/Ham's F12 50/50 (DMEM:F12) media supplemented with 10% fetal bovine serum (FBS), 1.2 g/L sodium bicarbonate, 0.5 mM sodium pyruvate, 2.5 mM L-glutamine, and 100 units/mL penicillin/streptomycin (Invitrogen). Cells should be maintained at 37 °C in a humidified incubator containing 5% CO_2.
2. To prepare for transfection, split the cells in a 1:3 dilution into a 100-mm culture dish.
3. When confluent (2–3 days later), collect the IMCD-3 cells from the culture dish by first washing cells with 10 mL sterile phosphate-buffered saline (PBS; Invitrogen), and then incubating cells in 1 mL 0.25% trypsin/EDTA (Invitrogen) for approximately 5 min at room temperature. Rinse cells off of the plate using 10 mL DMEM/F12, and pipette up and down (5 ×). Collect cells by centrifugation for 2 min at 200 × g in a table-top centrifuge.

4. Remove the media, and resuspend 5×10^6 cells in 800 μL of cytomix (120 mM KCl, 0.15 mM CaCl$_2$, 10 mM K$_2$HPO$_4$, 10 mM KH$_2$PO$_4$, 2.5 mM HEPES, 2 mM EGTA, 4 mM MgCl$_2$, pH adjusted to 7.6 with KOH), which mimics the cytoplasmic cellular environment (van den Hoff, Moorman, & Lamers, 1992), supplemented with fresh 2 mM ATP and 5 mM glutathione. Gently mix the cell suspension by pipetting up and down three times.
5. To a 1.5-mL sterile, nuclease-free microcentrifuge tube, add 10 μg of the pEGFP-N3-SSTR3 DNA construct described above. The DNA-containing tubes, as well as the electroporation cuvettes (4 mm gap cuvette, Fisher, Freemont, CA), should be maintained on ice.
6. To each DNA sample, add 400 μL of the cell suspension prepared in step 3. Gently mix the DNA with the cells by briefly pipetting up and down. Transfer the DNA-cell suspension mixture to a labeled 4 mm electroporation cuvette, and place on ice.
7. Electroporate the cells using BioRad Gene Pulser II (Hercules, CA) using the following parameters: 3200 V, 950 μF.
8. Remove the cuvette from the ice, and quickly add 0.5 mL of fresh culture media. Resuspend the cells using a Pasteur pipette. The cells should be gently mixed three times before addition to culture dishes.
9. Transfer the transfected cells to a 100-mm plate containing 10 mL IMCD-3 media (described above), and incubate for 24–48 h.

2.3. Selection of IMCD-3 cells stably expressing SSTR3

This procedure describes the isolation of single clones of IMCD-3 cells stably expressing SSTR3-EGFP. Use of cells derived from a single clone is preferable for imaging studies because it alleviates the need for cotransfection of multiple constructs and allows for better consistency in expression levels of the tagged protein. An alternative approach to what is described here would be to use flow cytometry to sort the stably transfected cells and select cells expressing the desired levels of the EGFP-tagged protein before subculturing.

1. After allowing electroporated cells to recover for 24–48 h, replace the standard media with media containing 400 μg/μL G418 (Invitrogen). This step will select cells that have stably incorporated SSTR3-EGFP into the genomic DNA.
2. Carefully change the media in the plate daily, replacing the media with media containing 400 μg/μL G418. Massive cell death should be expected, and the vast majority of the cells will detach from the plate, leaving colonies of stably transfected cells behind.

3. After approximately 2 weeks of selection, detach all cells by digesting with 1 mL 0.25% trypsin/EDTA (Invitrogen) for approximately 5 min at 37 °C. Add the cells to 9 mL media containing 400 μg/μL G418, and plate on a 15 cm tissue culture dish. Grow for 2–5 days until visible colonies of cells form.
4. Isolate individual colonies using cloning rings attached to the dish with autoclaved vacuum grease. To select a colony, aspirate medium within the ring, and then detach cells within a cloning ring by digesting with 1 mL 0.25% trypsin/EDTA for approximately 5 min at 37 °C. Transfer cells from a single ring to a 6-well plate containing media with 400 μg/μL G418. Multiple colonies should be selected to ensure that a colony with the desired expression level of the tagged protein is obtained.
5. When the cells reach confluence, wash with sterile PBS, and then detach with 100 μL of trypsin-EDTA for approximately 5 min at room temperature. Plate 50 μL of the cells into one well of a 6-well plate and 50 μL into one well of a 12-well plate containing a 12 mm glass coverslip (Fisher Scientific, Pittsburgh, PA).
6. After 24–48 h, process the coverslips using fluorescence microscopy to identify clones that express properly localized SSTR3-EGFP. SSTR3 has previously been localized to cilia (Berbari et al., 2008; Handel et al., 1999). We show that EGFP-SSTR3 localizes to cilia in IMCD-3 cells (Fig. 15.1).

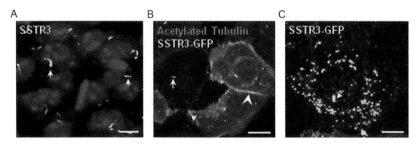

Figure 15.1 Localization of SSTR3 in IMCD cells. (A) A frozen section of wild-type adult mouse brain was processed for IF with anti-SSTR3 antibodies (sc-11617, 1:500; Santa Cruz, Dallas, TX). The CA3 region of the hippocampus shows cilia-specific localization of SSTR3. Draq5 nuclear stain is in blue. (B, C) IMCD-3 cells stably expressing SSTR3-GFP were imaged directly (C), or processed for immunofluorescence using an anti-GFP polyclonal antibody (ab290-50, 1:500; Abcam, Cambridge, MA) (to detect SSTR3) and anti-acetylated tubulin monoclonal antibody (T6793, 1:2000; Sigma, Saint Louis, Missouri Sigma) (to detect cilia) (B). SSTR3 localizes to cilia (arrows) and is also detected on the plasma membrane (arrowheads) and in internal endosomes. Scale bars, 10 μm (A, B) and 19 μm (C). *Adapted from Figure 1 of Tower et al. (2011) with permission from Elsevier.* (See color plate.)

7. To ensure that full-length SSTR3-GFP is expressed in the selected clonal lines, an α-GFP antibody (ab290, Abcam, Cambridge, MA) can be used to perform a Western immunoblot on lysates from the selected cells.
8. Passage confirmed clones from the 6-well plate into a minimum of two dishes: one for freezing as Passage 0, and one for further passaging (if the line is already clonal).
9. Once a clone(s) of interest has been identified, a lower concentration of G418 (200 μg/mL) may be used for maintaining the cells.

3. LIVE IMAGING OF SSTR3 AND RABS IN MAMMALIAN KIDNEY CELLS

3.1. Transfection of Rabs into IMCD-3 cells

This procedure describes the transfection of IMCD-3 cells stably expressing EGFP-SSTR3 (SSTR3#1 IMCD-3) with expression vectors for markers of endosomal compartments, including Cherry-Rab4, mRFP-Rab5, Ds-Red-Rab7, or Ds-Red Rab11. SSTR3#1 IMCD-3 cells were generated as described in Section 2. Use of a clonal cell line stably expressing SSTR3 is preferable because this alleviates the need to electroporate two DNA constructs at once to achieve coexpression of both SSTR3 and Rab within a single cell. Also, the use of a stable cell line allows for better control of SSTR3 expression levels across experiments for more accurate comparisons.

1. Grow SSTR3#1 IMCD-3 cells in plastic dishes in DMEM/F12 culture media supplemented with 200 μg/mL G418.
2. Coat glass-bottom plates (Warner Instruments, Hamden, CT) by adding 200 μL filtered 0.1% gelatin diluted in molecular grade or sterile distilled water (a volume sufficient to cover only the glass) to the bottom of each dish. Aspirate any excess solution. This step helps cells to adhere to the glass, but not to the sides of the dish, following electroporation.
3. Transfect SSTR3#1 IMCD-3 cells with each DNA construct (empty vector and vector containing the epitope-tagged wild-type or mutant Rab; mutant constructs described below) by electroporation as described in Section 2.2.
4. Add each transfected cell suspension to a single 0.1% gelatin-coated, 35 mm glass-bottom dish containing 2.5 mL fresh prewarmed media.
5. Place cells in an incubator in 5% CO_2 at 37 °C. After 4–6 h, change the media to fresh DMEM/F12 containing 10% FBS, 100 units/mL penicillin/streptomycin, 1.2 g/L sodium bicarbonate, 0.5 mM sodium

pyruvate, and 200 μg/mL G418. This step appears to decrease the toxicity associated with electroporation.

6. Change the media after 24 h. It is normal to have ~25% of cells detached and floating after 24 h. Live imaging can be performed 24–48 h after transfection, depending upon the expression levels of the transfected Rab.

3.2. Microscope setup

Imaging can be performed on any microscope equipped to acquire rapid fluorescence images. A warming stage and a CO_2 chamber are beneficial but not essential. Although multiple microscope setups can be utilized, here we describe our imaging studies performed using the PerkinElmer Ultra-VIEWERS 6FE-US spinning disk confocal attached to a Nikon TE2000-U with either a CFI Plan APO $63\times$ or $100\times$ oil immersion objective, equipped with laser and filter sets for GFP, TRITC, Cy5, and DAPI fluorescence. In addition, this microscope has a temperature-adjustable stage-top (for our studies the temperature was set to 37 °C) incubation system equipped with a CO_2 gas mixer. A Hamamatsu C9100-50 camera was used for image acquisition.

1. Turn on the power to the microscope a minimum of 15–20 min before beginning the experiment to allow the laser and CCD camera to stabilize.
2. Add a drop of immersion oil (Cargille Labs, Cedar Grove, NJ) to the $100\times$-objective, and place the glass-bottom plate of transfected cells in the center of the heated stage above the objective. Cover the heated stage with the CO_2 chamber. Raise the objective to the bottom of the plate.
3. Start up the image acquisition software on the computer, and input parameters appropriate for the microscope used. We utilized Volocity software (Volocity 5.2, Perkin Elmer, Shelton, CT) for the experiments described. However, newer software (version 6.1) is now available. The software package selected should be appropriate for the microscope and camera and should allow rapid acquisition of live images.
4. To acquire images, set up the light pathways by selecting appropriate laser/filter combinations for GFP and RFP for dual color imaging. Filters should be managed at maximum speed for live imaging. Set the z step size to 0.3 mm.
5. Set the exposure time for each channel so that the image intensity fits well within the dynamic range of the camera (maximum intensity of

around 5000–10,000 for this setup). Utilize a fast sequential time-lapse imaging setting. We do not recommend the use of auto contrast and recommend a sensitivity of approximately 155 for image acquisition.
6. Adjust the intensity of each laser according to the cell selected for imaging. We recommend that the laser intensity be set to a lower level (<55% for this setup). Use of lower laser intensities and shorter exposure times while acquiring images helps to minimize photobleaching of the fluorophores in the cell. The ideal cell is one in which the GFP signal and RFP or mCherry signal are bright enough to be detected with similar laser settings.
7. For each image, set the top and bottom of the z-stack after finding an appropriate cotransfected cell and before initiating time-lapse image acquisition.
8. Acquire images every 5 s for 10–20 min.

3.3. Confocal microscopy colocalization of SSTR3 and Rabs

1. Prewarm DMEM/F12 culture medium in a 37 °C water bath for 10 min. Remove the 35 mm glass-bottom plates containing SSTR#3 IMCD-3 cells transfected with Rabs from the incubator, and replace the media with 2 mL prewarmed fresh media.
2. Add 50 µL 1 M HEPES buffer, pH 7.4, to each 35 mm glass-bottom dish. This step helps in maintaining the pH of the media to ensure longer cell survival during imaging.
3. Place the plate on the heated stage of the microscope inside the humidifying chamber and a CO_2 chamber. Add a single drop of immersion oil to the objective and bring the objective up to the bottom of the plate.
4. To prevent a shift of focal plane during recording caused by thermal drift, after setting up the plate on the microscope, it is best to wait for approximately 15 min before continuous imaging. During this time, scan the field of cells to assess the transfection efficiency and the expression levels of the transfected proteins in the cells.
5. Ensure that the selected cells exhibit moderate expression and correct localization of SSTR3 and Rabs. SSTR3 localizes to cilia (Berbari et al., 2008; Handel et al., 1999), and thus SSTR3#1 IMCD-3 cells showing ciliary localization should be selected for subsequent imaging. The correct distribution of Rabs can be assessed by similarity to published patterns (see Choudhury et al., 2002; Daro, van der Sluijs, Galli, & Mellman, 1996; Vonderheit & Helenius, 2005; Ward et al., 2005).

6. Focus on a region of interest (ROI) within the cell. Acquire multiple focal planes for each fluorophore in each ROI within a cell. Acquire images from at least 10 different cells from a minimum of three independent experiments.
7. Merge the images of the fluorophores from a single ROI to obtain a visual representation of the colocalization of SSTR3 and Rabs (Fig. 15.2A–D).
8. Quantify the level of SSTR3/Rab colocalization in all collected ROIs using Volocity or similar software. Calculate the Pearson's correlation coefficients for the z-stacked images. Different endosomes can be selected within ROIs to determine colocalization of each object through time. Colocalization is based on a scale of -1 to 1, where values close to 0 indicate that two fluorophores do not show a linear colocalization relationship. A value of -1.0 indicates a complete negative correlation, and a value of 1 indicates complete colocalization. For our experiments, the no colocalization threshold was set at/below 0.100 (Fig. 15.2E).

4. DYNAMICS OF SSTR3 TRANSIT RELATIVE TO RABS

4.1. Time-lapse dual color imaging of SSTR3 and Rabs

1. Place the plate of SSTR3#1 IMCD-3 cells transfected with a specific Rab on the heated stage of the microscope inside the CO_2 chamber. Add a single drop of immersion oil to the $100\times$ objective, and bring the objective up to the bottom of the plate.
2. Select a cell expressing moderate levels of SSTR3 and Rab, and focus on an area of the cell containing a representative distribution of Rab. In order to minimize photobleaching of dual-labeled cells, the laser power should be set as low as possible. In our case, the 488 nm laser intensity was set to 15% and the 561 nm laser to 15%.
3. Acquire z-stack images with a spacing of 0.3 mm every 5 s for 10 min.
4. Merge the images of the fluorophores at each time to obtain a visual representation of the global localization of SSTR3 relative to each Rab at each time point.
5. To analyze entry and exit of SSTR3 from endosomal compartments, use $100\times$ TIRF oil objective and focus on a single endosome to observe fusion and budding events. Image as described above for 10 min (Fig. 15.3 A and B).

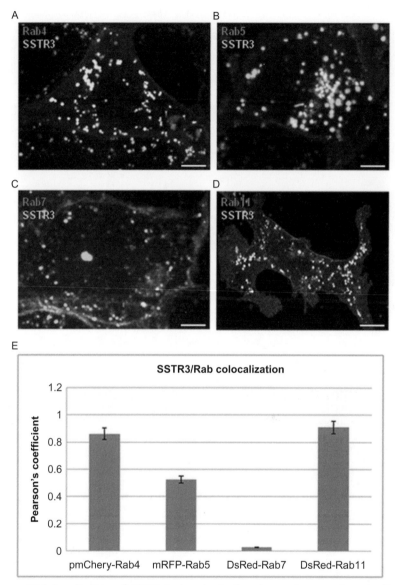

Figure 15.2 Localization of SSTR3 to a subset of endosomal compartments. (A–D) IMCD-3 cells stably expressing EGFP-SSTR3 were transfected with mCherry-Rab4 (A), mRFP-Rab5 (B), Ds-Red-Rab7 (C), or Ds-Red Rab11 (D). Cells were imaged at 48 h post-transfection. SSTR3 localizes to Rab4 and Rab5-positive early endosomes, as well as Rab11-positive recycling endosomes, but not to Rab7-positive endosomes. The level of SSTR3 colocalization with each subset of endosomes was measured using Pearson's correlation coefficient (E). Scale bars, 19 μm (A–D). *Adapted from Tower et al. (2011) with permission from Elsevier.* (See color plate.)

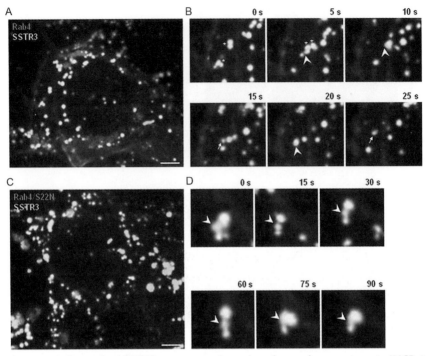

Figure 15.3 Dynamics of SSTR3 movement through endosomal compartments. IMCD-3 cells stably expressing SSTR3 were transfected with mCherry-Rab4 wild-type (A and B) or with mCherry-Rab4/S22N (C and D). Live imaging was carried out, and still images from movies were selected. (B) Arrowheads point to SSTR3 (green) rapidly budding from Rab4-containing (yellow) endosomes. (D) Arrowheads point to SSTR3-only (green) elements that remain in association with Rab4-containing (yellow) elements for an extended time period. Scale bars 19 μm (A and C). (See color plate.)

4.2. Analysis of half-time residency and clearance times

For each of these assays, imaging should be performed for at least 10 different cells and at least 20–25 different endosomes in each of three independent experiments to obtain average transport rates of SSTR3 through a Rab-marked compartment.

1. To assess half-time residency ($t_{1/2}$), collect images over 10 min and use them to assess colocalization coefficients between SSTR3 and Rabs at each time point. Based on these values generate $t_{1/2}$ times.
   a. To quantify fusion and fission events, select and analyze 21–25 fluorescent objects from different fields of the cell.
   b. Analyze each structure using the Measure Colocalization Application in Volocity (or a similar application) and empirically determine the threshold values. These values can be generated by using

automated thresholding (see Volocity 5.2 user manual available at http://cellularimaging.perkinelmer.com/pdfs/manuals/ VolocityUserGuide.pdf).

- c. Record the Pearson's correlation coefficient for each object throughout the duration of the time-lapse sequences. As measured by Pearson's correlation coefficient, the changes in colocalization of SSTR3 with endosomes can be used to estimate the time the proteins remain in the same organelle. In our study, the colocalization coefficient was recorded until it decreased to zero, indicating no overlap between the SSTR3-positive structure (green) and the red-labeled endosome.
- d. Plot the percentage of colocalized objects on the y-axis versus the time in seconds on the x-axis. Objects were considered to be colocalized until the Pearson's correlation coefficient reached a value less than 0.1.
- e. Calculate the $t_{1/2}$ based on the graph (Fig. 15.4). The $t_{1/2}$ value is defined as the time at which 50% of the objects are no longer colocalized.

2. To assess clearance, collect images over 10–20 min, and use these images to assess colocalization correlation coefficients between SSTR3 and Rab-positive endosomes at each time point.
   - a. Randomly select 20–25 individual Rab/SSTR3 positive endosomes and assess the Pearson's coefficient for each structure over time. When

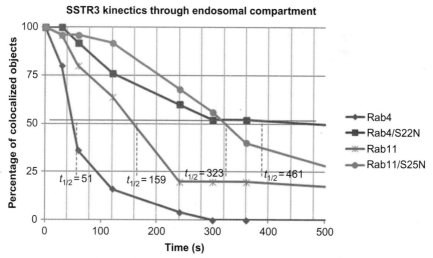

Figure 15.4 Effects of mutant Rabs on SSTR3 dynamics. The Pearson's colocalization coefficients of SSTR3 with wild-type Rab4 and Rab11 or with dominant negative Rab4/S22N and Rab11/S25N were extracted from different times during imaging as in Fig. 15.3 and used to calculate the percentage of colocalized objects at each time point. (See color plate.)

SSTR3 is associated with a Rab-labeled endosome, the structure should be yellow in color and have a high (0.7–1.0) Pearson's coefficient. Over time, as SSTR3 is sorted out of the Rab-positive endosome, the yellow structure begins to separate into a red Rab-positive endosome and a green SSTR3-positive structure. The measured Pearson's coefficient of these independent structures should approach 0.

  b. Repeat this analysis for each endosome. Record the time point at which SSTR3 is no longer colocalized with the Rab (Pearson's correlation coefficient <0.1). This time is considered to be the clearance time because it represents the point at which SSTR3 is no longer associated with the Rab-positive endosome.

5. EFFECTS OF DOMINANT NEGATIVE RABS ON SSTR3 TRAFFICKING

The role of Rabs in trafficking of cargo proteins through the endosomal pathway can be probed based on the availability of mutant forms of Rabs that arrest trafficking within a specific endosomal compartment. Rab GTPases cycle between two nucleotide-bound states: (1) a GDP-bound inactive state and (2) a GTP-bound active state. The inactive form of each Rab can be generated by mutating a single amino acid within the switch region (reviewed in Fukuda, 2010; Hutagalung & Novick, 2011). For the key human endosomal Rabs, the following substitutions generate dominant inactive forms: mutation of serine 22 of Rab4 to asparagine (Rab4/S22N) and of serine 25 of Rab11 to asparagine (Rab11/S25N). When expressed in cells, these proteins inhibit trafficking at early endosomes (Rab4/S22N) or at recycling endosomes (Rab11/S25N) (Bucci et al., 1992, 1994; Choudhury et al., 2002; Daro, Sheff, Gomez, Kreis, & Mellman, 1997; Pellinen et al., 2006; Simpson et al., 2004; van der Sluijs et al., 1992; Ward et al., 2005). We first generated a red fluorescently tagged, dominant inactive mutant of Rab4. We then expressed dominant inactive mutants of Rab4 and Rab11 in SSTR3#1 IMCD-3 cells to determine whether the mutant Rab affected specific SSTR3 trafficking events.

5.1. Generating mutant forms of endosomal Rabs

1. Clone or obtain constructs in which the wild-type Rab is tagged with an RFP, such as mCherry or RFP. These constructs will serve as a template for site-directed mutagenesis. We obtained the pmCherry-C2 Rab4 wild-type construct from Dr. James Goldenring (Vanderbilt University Medical Center, Nashville, TN).

2. Perform site-directed mutagenesis using the PCR to substitute conserved Ser for Asn in the α1 helix of wild-type Rab4. There are a number of commercially available mutagenesis kits. We used the QuickChange XL-Site directed mutagenesis kit (# 200516 and 200517, Stratagene, La Jolla, CA) to generate pmCherry-C2 Rab4/S22N. The experimental details associated with this procedure are described in http://www.scribd.com/doc/23261720/Molecular-Cloning-A-Laboratory-Manual-On-The-Web-Maniatis.

3. If wild-type and dominant inactive forms of the Rab are not available with a red tag but are available in other vectors, they can be subcloned into an appropriate vector containing a red tag. Verify the correct (in frame) cloning and mutagenesis of the new constructs by DNA sequencing.

5.2. Dynamics of SSTR3 traffic in cells expressing mutant Rabs

1. Transfect SSTR3#1 IMCD-3 cells with the inactive (GDP-locked) Rab4/S22N or Rab11/S25N (Addgene, Cambridge, MA, USA) mutants by electroporation (see Section 3.1 for method). Change culture media after 24 h, and add fresh prewarmed media supplemented with 200 μg/mL G418.

2. Prior to imaging, add fresh media containing 2 μM HEPES buffer, pH 7.4, and place plate on the heated stage of the humidifying CO_2 chamber.

3. Select the 100× oil immersion objective. Allow the objective to warm before adding drop of oil.

4. Acquire images as described in Section 3. Record the images in the green and red channels for each time point (Fig. 15.3C and D).

5. Process images using Volocity 5.2 or similar image acquisition and processing software. The program calculates the average fluorescence for each endosome (with background subtracted) in every image of the z-series.

6. Export the time series into a QuickTime movie format that plays at five frames per second (Adobe Systems, Inc., San Jose, CA). Monitor individual endosomes containing SSTR3 (yellow) over time to determine the kinetics of SSTR3 (green) clearance from the red-labeled endosome.

This data can be analyzed as described in Section 4.2 to assess the length of time SSTR3 remains in endosomes expressing the wild-type or mutant Rab.

6. SUMMARY

Advances in the field of somatostatin receptor trafficking have been impressive; however, further studies need to be conducted to identify additional molecular components that regulate endosomal transport of SSTR3 and to analyze the precise roles of these proteins in distinct steps of endocytosis and recycling of these GPCRs. Importantly, these studies will provide a better understanding of somatostatin physiology and aid in design of therapeutic strategies to target subclasses of GPCRs.

Live cell imaging has proven to be a useful tool to study intracellular trafficking. Over the years, the advancement of technology has led to the development of better means of tagging proteins for live imaging and led to significant improvements in microscopes and image analysis software, allowing for improved visualization of transport events taking place in real time in a single cell or organelle. These techniques can provide spatial and temporal information regarding the trafficking events and have proven useful for addressing key questions concerning the endosomal transit of SSTR3. In addition, this method permits transport kinetics to be quantified, as well as assessing the effects of specific proteins on kinetic parameters. Instead of simply providing a fixed single snapshot of cellular trafficking events, this procedure provides a window into continuous processes occurring in cells over time, allowing for improved characterization of the coordinated membrane trafficking within the cell.

ACKNOWLEDGMENTS

This work was supported by a Pilot and Feasibility Award (to E. S.) from the UAB Hepatorenal Fibrocystic Disease Core Center NIH P30 DK074038 (To B. K. Y.), NSF grant MCB0744471 (to E. S.), NIH RO1 grant DK075996 (to B. K. Y.), and an F32 postdoctoral award F32 DK088404 (to N. F. B.).

REFERENCES

Barnett, P. (2003). Somatostatin and somatostatin receptor physiology. *Endocrine*, *20*(3), 255–264.

Berbari, N. F., Bishop, G. A., Askwith, C. C., Lewis, J. S., & Mykytyn, K. (2007). Hippocampal neurons possess primary cilia in culture. *Journal of Neuroscience Research*, *85*(5), 1095–1100.

Berbari, N. F., Johnson, A. D., Lewis, J. S., Askwith, C. C., & Mykytyn, K. (2008). Identification of ciliary localization sequences within the third intracellular loop of G protein-coupled receptors. *Molecular Biology of the Cell, 19*(4), 1540–1547.

Bucci, C., Parton, R. G., Mather, I. H., Stunnenberg, H., Simons, K., Hoflack, B., et al. (1992). The small GTPase rab5 functions as a regulatory factor in the early endocytic pathway. *Cell, 70*(5), 715–728.

Bucci, C., Wandinger-Ness, A., Lütcke, A., Chiariello, M., Bruni, C. B., & Zerial, M. (1994). Rab5a is a common component of the apical and basolateral endocytic machinery in polarized epithelial cells. *Proceedings of the National Academy of Sciences of the United States of America, 91*(11), 5061–5065.

Choudhury, A., Dominguez, M., Puri, V., Sharma, D. K., Narita, K., Wheatley, C. L., et al. (2002). Rab proteins mediate Golgi transport of caveola-internalized glycosphingolipids and correct lipid trafficking in Niemann–Pick C cells. *Journal of Clinical Investigation, 109*(12), 1541–1550.

Daro, E., Sheff, D., Gomez, M., Kreis, T., & Mellman, I. (1997). Inhibition of endosome function in CHO cells bearing a temperature-sensitive defect in the coatomer (COPI) component epsilon-COP. *Journal of Cell Biology, 139*(7), 1747–1759.

Daro, E., van der Sluijs, P., Galli, T., & Mellman, I. (1996). Rab4 and cellubrevin define different early endosome populations on the pathway of transferrin receptor recycling. *Proceedings of the National Academy of Sciences of the United States of America, 93*(18), 9559–9564.

Fukuda, M. (2010). How can mammalian Rab small GTPases be comprehensively analyzed?: Development of new tools to comprehensively analyze mammalian Rabs in membrane traffic. *Histology and Histopathology, 25*(11), 1473–1480.

Goel, M., Sinkins, W. G., Zuo, C. D., Estacion, M., & Schilling, W. P. (2006). Identification and localization of TRPC channels in the rat kidney. *American Journal of Physiology Renal Physiology, 290*(5), F1241–F1252.

Gurevich, V. V., & Gurevich, E. V. (2008). How and why do GPCRs dimerize? *Trends in Pharmacological Sciences, 29*(5), 234–240.

Handel, M., Schulz, S., Stanarius, A., Schreff, M., Erdtmann-Vourliotis, M., Schmidt, H., et al. (1999). Selective targeting of somatostatin receptor 3 to neuronal cilia. *Neuroscience, 89*(3), 909–926.

Huang, Y., & Willars, G. B. (2011). Generation of epitope-tagged GPCRs. *Methods in Molecular Biology, 746*, 53–84.

Hubbard, K. B., & Hepler, J. R. (2006). Cell signalling diversity of the Gqalpha family of heterotrimeric G proteins. *Cellular Signalling, 18*(2), 135–150.

Hutagalung, A. H., & Novick, P. J. (2011). Role of Rab GTPases in membrane traffic and cell physiology. *Physiological Reviews, 91*(1), 119–149.

Jacobs, S., & Schulz, S. (2008). Intracellular trafficking of somatostatin receptors. *Molecular and Cellular Endocrinology, 286*(1–2), 58–62.

Jean-Alphonse, F., & Hanyaloglu, A. C. (2011). Regulation of GPCR signal networks via membrane trafficking. *Molecular and Cellular Endocrinology, 331*(2), 205–214.

Kreuzer, O. J., Krisch, B., Déry, O., Bunnett, N. W., & Meyerhof, W. (2001). Agonist-mediated endocytosis of rat somatostatin receptor subtype 3 involves beta-arrestin and clathrin coated vesicles. *Journal of Neuroendocrinology, 13*(3), 279–287.

Lappano, R., & Maggiolini, M. (2012). GPCRs and cancer [Review]. *Acta Pharmacologica Sinica, 33*(3), 351–362.

Milligan, G. (2010). The role of dimerisation in the cellular trafficking of G-protein-coupled receptors. *Current Opinion in Pharmacology, 10*(1), 23–29.

Pellinen, T., Arjonen, A., Vuoriluoto, K., Kallio, K., Fransen, J. A., & Ivaska, J. (2006). Small GTPase Rab21 regulates cell adhesion and controls endosomal traffic of beta1-integrins. *Journal of Cell Biology, 173*(5), 767–780.

Rauchman, M. I., Nigam, S. K., Delpire, E., & Gullans, S. R. (1993). An osmotically tolerant inner medullary collecting duct cell line from an SV40 transgenic mouse. *American Journal of Physiology*, *265*(3 Pt 2), F416–F424.

Schöneberg, T., Schulz, A., Biebermann, H., Hermsdorf, T., Römpler, H., & Sangkuhl, K. (2004). Mutant G-protein-coupled receptors as a cause of human diseases. *Pharmacology & Therapeutics*, *104*(3), 173–206.

Simpson, J. C., Griffiths, G., Wessling-Resnick, M., Fransen, J. A., Bennett, H., & Jones, A. T. (2004). A role for the small GTPase Rab21 in the early endocytic pathway. *Journal of Cell Science*, *117*(Pt 26), 6297–6311.

Sönnichsen, B., De Renzis, S., Nielsen, E., Rietdorf, J., & Zerial, M. (2000). Distinct membrane domains on endosomes in the recycling pathway visualized by multicolor imaging of Rab4, Rab5, and Rab11. *Journal of Cell Biology*, *149*(4), 901–914.

Sorkin, A., & Von Zastrow, M. (2002). Signal transduction and endocytosis: Close encounters of many kinds. *Nature Reviews Molecular Cell Biology*, *3*(8), 600–614.

Sorkin, A., & von Zastrow, M. (2009). Endocytosis and signalling: Intertwining molecular networks. *Nature Reviews Molecular Cell Biology*, *10*(9), 609–622.

Stenmark, H. (2009). Rab GTPases as coordinators of vesicle traffic [Review]. *Nature Reviews Molecular Cell Biology*, *10*(8), 513–525.

Tower, C., Fu, L., Gill, R., Prichard, M., Lesort, M., & Sztul, E. (2011). Human cytomegalovirus UL97 kinase prevents the deposition of mutant protein aggregates in cellular models of Huntington's disease and ataxia. *Neurobiology of Disease*, *41*(1), 11–22.

Tulipano, G., Stumm, R., Pfeiffer, M., Kreienkamp, H. J., Höllt, V., & Schulz, S. (2004). Differential beta-arrestin trafficking and endosomal sorting of somatostatin receptor subtypes. *Journal of Biological Chemistry*, *279*(20), 21374–21382.

van den Hoff, M. J., Moorman, A. F., & Lamers, W. H. (1992). Electroporation in 'intracellular' buffer increases cell survival. *Nucleic Acids Research*, *20*(11), 2902.

van der Sluijs, P., Hull, M., Webster, P., Mâle, P., Goud, B., & Mellman, I. (1992). The small GTP-binding protein rab4 controls an early sorting event on the endocytic pathway. *Cell*, *70*(5), 729–740.

Vonderheit, A., & Helenius, A. (2005). Rab7 associates with early endosomes to mediate sorting and transport of Semliki forest virus to late endosomes. *PLoS Biology*, *3*(7), e233.

Ward, E. S., Martinez, C., Vaccaro, C., Zhou, J., Tang, Q., & Ober, R. J. (2005). From sorting endosomes to exocytosis: Association of Rab4 and Rab11 GTPases with the Fc receptor, FcRn, during recycling. *Molecular Biology of the Cell*, *16*(4), 2028–2038.

SECTION III

Proteins

CHAPTER SIXTEEN

Genetic Circuitry Modulating Notch Signals Through Endosomal Trafficking

Kazuya Hori[1], Anindya Sen[1], Spyros Artavanis-Tsakonas[2]
Department of Cell Biology, Harvard Medical School, Boston, Massachusetts, USA
[1]These authors contributed equally to this work
[2]Corresponding author: e-mail address: artavanis@hms.harvard.edu

Contents

1. Introduction	284
2. Genetic Screen Using the Exelixis Collection	285
2.1 The Exelixis collection	285
2.2 Materials	286
2.3 Genetic screen for modifiers of Dx–Krz action	287
3. Notch Localization in Endosomes	289
3.1 Immunostaining in cultured cells and imaginal discs	289
3.2 Endocytosis assay in cultured cells and imaginal discs	291
4. Optical Approaches	292
4.1 Construction of fluorescently tagged Notch and ligand molecules	293
4.2 Generation of stable cell lines	294
4.3 Rescue experiments	295
5. Ubiquitination Status of Notch	295
5.1 Materials	295
5.2 Ubiquitination assay	296
6. Conclusion	297
Acknowledgments	297
References	297

Abstract

Genetic modifier screens offer a powerful, indeed a uniquely powerful tool for the analysis and identification of elements capable of modulating specific cellular functions in development. Here, we describe the methodology that allowed us to explore the genetic circuitry that affects a Notch mutant phenotype caused by the abnormal endosomal trafficking of the Notch receptor. Endosomal trafficking events are increasingly appreciated to play a major role in controlling Notch signaling in development.

1. INTRODUCTION

The Notch pathway is used throughout development to couple the fate choices of a particular cell to those of neighboring cells, ultimately affecting proliferation, apoptosis, and differentiation (Artavanis-Tsakonas, Rand, & Lake, 1999; Bray, 2006; Schweisguth, 2004; Hori et al., 2013). Notch malfunction, which may result either in the up- or downregulation of the signal, has been associated with abnormal development in all metazoans examined and with diseases in humans including cancer (Louvi & Artavanis-Tsakonas, 2012). *Notch* encodes a single-pass transmembrane receptor and the developmental logic of this signaling pathway relies on the interaction of the receptor expressed on one cell with membrane-bound ligands expressed on its neighboring cell. The canonical signaling model has the Notch receptor being activated through a series of proteolytic events after it interacts with the ligands, Delta or Serrate (Bray, 2006). The crucial cleavage event for signaling depends on γ-secretase and results in releasing the intracellular domain of Notch from the membrane, allowing it to translocate to the nucleus, where it participates directly in a core transcriptional complex together with the DNA-binding protein Suppressor of Hairless and the nuclear effector Mastermind, to activate the transcription of target genes (Bray, 2006; Schweisguth, 2004).

The role of endocytic trafficking in the regulation of Notch signaling has been increasingly appreciated (Fortini, 2009; Yamamoto, Charng, & Bellen, 2010). Several endocytic factors modulating the degradation of the Notch receptor and consequently the negative attenuation of signaling have been identified (Fortini, 2009; Yamamoto et al., 2010). However, sorting of the receptor through the endocytic compartments has also been shown to result in the activation of the receptor. Such intracellular events have been associated with ligand-dependent (Coumailleau, Fürthauer, Knoblich, & González-Gaitán, 2009) as well as ligand-independent, that is, non-canonical, activation of the receptor (Childress, Acar, Tao, & Halder, 2006; Hori et al., 2004; Hori, Fuwa, Seki, & Matsuno, 2005; Hori, Sen, Kirchhausen, & Artavanis-Tsakonas, 2011; Thompson et al., 2005; Vaccari & Bilder, 2005; Vaccari, Lu, Kanwar, Fortini, & Bilder, 2008; Vaccari et al., 2009; Wilkin et al., 2004, 2008). Mutations in elements of the endosomal and multivesicular bodies sorting machinery have been shown to be sufficient to trigger ligand independent signaling of the receptor, but the mechanisms underlying these events and the genetic circuitry capable of modulating such intracellular Notch signaling remain opaque.

We previously showed that Kurtz (Krz), the single nonvisual β-arrestin homolog in *Drosophila*, together with the ubiquitin ligase Deltex (Dx), affects trafficking of the Notch receptor and regulates Notch signaling by modulating the turnover of the receptor (Mukherjee et al., 2005). To gain further insight into how Krz and Dx regulate the trafficking of the receptor, we carried out genetic screens for modifiers of the Krz and Dx-dependent synergy, which is manifested *in vivo* as a typical Notch loss of function as evidenced by the notched wing phenotype. Here, we describe the genetic approach that allowed us to search for genetic modifiers of a double mutant *krz/dx* combination. This specific approach, originally designed to unveil elements of the endocytic machinery, also serves as an experimental paradigm for exploring a genetic circuitry that can affect any phenotype of choice.

2. GENETIC SCREEN USING THE EXELIXIS COLLECTION

The advent of genome wide, molecularly characterized, mutant collections [e.g., Bloomington *Drosophila* Stock Center (BDSC), http://flystocks.bio.indiana.edu/), Vienna *Drosophila* RNAi Center (VDRC, http://www.vdrc.at), and *Drosophila* Genetic Resource Center at Kyoto (DGRC, http://www.dgrc.kit.ac.jp/en/)] over the past few years has revolutionized the way we can elucidate genetic circuitries that affect specific phenotypic parameters. We can now carry out genetic screens with the highest degree of saturation of *Drosophila* genome, coupled with essentially instantaneous identification of the modifier genes. We have been extensively using the Exelixis collection (http://drosophila.med.harvard.edu/; Artavanis-Tsakonas, 2004) for probing genetic circuitries.

2.1. The Exelixis collection

The collection is composed of 15,500 transposon-induced gene disruptions, resulting in mutations in ~53% of the *Drosophila* genome (Artavanis-Tsakonas, 2004; Parks et al., 2004; Thibault et al., 2004). Each insertion is derived from one of four vector types, three piggyBac-derived (PB, RB, and WH) and a fourth, a P-element variant (XP) (Parks et al., 2004; Thibault et al., 2004). Currently, the collection is 22% PB, 20% RB, 35% WH, and 23% XP. There are two classes of disruption events, those leading to inactivation of loci or those driving expression of downstream genes when combined with GAL4, due to the presence of Upstream Activation Sequence (UAS) sequences within the insertional transposon

(WH and XP elements) (Parks et al., 2004; Thibault et al., 2004). In the presence of GAL4, UAS-containing insertions could theoretically inactivate loci even if oriented to drive expression, for example, by generating antisense RNA products. This can be investigated genetically by screening the collection in a background containing a loss-of-function *mam* or *Notch* allele without any GAL4 driver. WH insertions also contain splice acceptor sites (splice traps), permitting normal transcription of tagged genes but are designed such that WH, rather than endogenous, splice acceptors are used, allowing for a piece of the piggyBac transposon to be spliced into the final transcript, thereby disrupting translation. In the presence of GAL4, UAS-containing WH and XP insertions may represent hypomorphic, hypermorphic, neomorphic, or antisense alleles. In contrast, PB and RB insertions lack UAS sequences and are likely to represent null or hypomorphic alleles. Therefore, screening in a genetic background containing a GAL4-dependent phenotype allows one to exploit the full potential of the collection and to recover interactors representing both classes of insertional events. In the screen we describe here, we used the Exelixis collection to search for dominant modifiers of the wing-notching phenotype elicited by coexpression of Dx and Krz under *C96-GAL4* control (Fig. 16.1B) (Hori et al., 2011). Given that this synergistic phenotype is a result of degradation of the Notch receptor via a ubiquitin-related mechanism within the endocytic pathway, we expected to identify elements of the endocytic machinery that affect this phenotypic parameter.

2.2. Materials

1. The Exelixis collection and the Exelixis deficiency kits were obtained from Harvard Medical School (https://drosophila.med.harvard.edu).
2. A deficiency kit was obtained from the Bloomington Stock Center (http://flystocks.bio.indiana.edu/).
3. The following mutant alleles were used: N^{54l9} (Lindsley & Zimm, 1992), dx^{152} (Fuwa et al., 2006), and krz^1 (Roman, He, & Davis, 2000).
4. The UAS lines used were *UAS-Flag:Dx* (Mukherjee et al., 2005), *UAS-HA:Krz* (Mukherjee et al., 2005), and $UAS\text{-}N^{FL}$ (Hori et al., 2004). The UAS constructs are driven by *C96-Gal4* (Gustafson & Boulianne, 1996). All crosses were carried out at 25 °C.
5. The genotypes of fly strains used in the screen were *UAS-Flag:Dx*/+; *C96-Gal4, UAS-HA:Krz*/+ (C96-Dx+Krz), *UAS-Flag:Dx*/+; *C96-Gal4*/+ (C96-Dx), *C96-Gal4, UAS-HA:Krz*/+ (C96-Krz), dx^{152}; *C96-Gal4*/+ (dx^{152}, C96), and krz^1, *C96-Gal4*/+ (krz^1, C96).

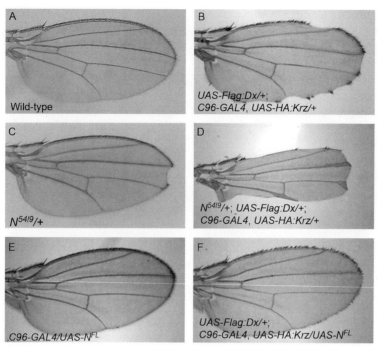

Figure 16.1 Interaction of the screening stock with *Notch*. (A) Wild-type adult wing. (B) Coexpression of Dx and Krz driven by *C96-Gal4* shows wing-nicking phenotype. (C) Heterozygous Notch null allele ($N^{54l9}/+$) is associated with the typical wing nicking. (D) $N^{54l9}/+$ enhances Dx- and Krz-mediated wing-nicking phenotype. (E) Expression of full length Notch driven by *C96-Gal4* does not affect wing morphology under our experimental conditions. (F) The wing-nicking phenotype associated with coexpression of Dx and Krz is rescued by expressing a transgene encoding wild-type Notch.

2.3. Genetic screen for modifiers of Dx–Krz action

1. *Validation of the screening phenotype*: In order to explore the feasibility of the screen for modifiers of the synergistic action of Dx and Krz, we needed to ensure that the phenotype was adequate for screening. We first determined that the genotype we generated for the screen (C96-Dx+Krz) is suited for screening given the health of the stock and its sensitivity to Notch pathway modulation, using known Notch pathway mutations. Figure 16.1 shows that the screening phenotype can be either enhanced by reducing the gene dosage of Notch (Fig. 16.1D) as N^{54l9} is a null allele of *Notch* or suppressed by expressing a wild-type copy of Notch transgene (Fig. 16.1F).

2. *Validation of the screening strategy*: To explore whether modifiers of the Dx–Krz phenotype can be identified in an unbiased fashion, we first used the Bloomington deficiency kit (Bloomington Stock Center at Indiana University) as well as the Exelixis deficiency kit which is composed of smaller deletions (Harvard Medical School). C96-Dx+Krz virgin females were crossed with males carrying autosomal deficiency. For deficiency on X chromosome, C96-Dx+Krz males were crossed with virgin females carrying deficiency, and the F1 progeny was scored for phenotypic modifications. Both enhancer and suppressor deficiencies were thus identified, ensuring the feasibility of such a screen. Furthermore, the number of modifiers identified was not inordinately large, something that could render the significance of the results from the designed screens questionable. It is also worth pointing out that the deficiency screens revealed regions harboring known components of the Notch pathway.
3. *Primary screen*: In the primary screen, C96-Dx+Krz virgin females were crossed with males carrying autosomal or viable X-linked insertions. C96-Dx+Krz males were crossed with virgin females carrying lethal insertions on the X chromosome and the F1 progeny was scored for phenotypic modifications (Fig. 16.2). Modifying transposons were

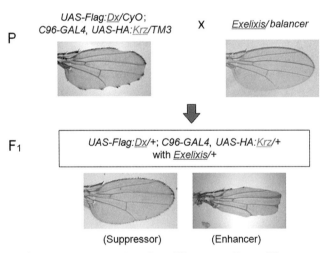

Figure 16.2 Schematic representation of modifier screen. F1 modifier screen to identify genetic modifiers of Dx- and Krz-mediated wing phenotype using the Exelixis collection, composed of 15,500 transposon-induced gene disruptions. Individual Exelixis fly stocks harboring a unique transposon insertion was crossed to *UAS-Flag:Dx/CyO; C96-Gal4, UAS-HA:Krz/TM3* and the F1 generation was screened for modifiers. (For color version of this figure, the reader is referred to the online version of this chapter.)

categorized as enhancers or suppressors of weak, moderate, or strong intensity based on the observed wing-nicking phenotype.
4. *Secondary screens*: Mutations identified in the primary screen were crossed to flies carrying *C96-Gal4* alone, to identify genes that affected the wing unilaterally and hence would be scored as positives (albeit false-positives) in the primary screen. Moreover, positive secondary tests were performed to examine the interaction with either *dx* or *krz*, alone, using wing phenotypes that result from the expression of Dx (C96-Dx), or the expression of Krz (C96-Krz) as well as mutant versions of *dx* mutant (dx^{152}, C96), or *krz* (krz^1, C96). Modifying transposons were again categorized as enhancers or suppressors of weak, moderate, or strong intensity in these independent screens.

3. NOTCH LOCALIZATION IN ENDOSOMES

The genetic screen provided a roster of genes that are involved in different aspects of Notch activation mediated by Dx and Krz. To identify novel molecular players involved in endosomal trafficking of Notch, we probed the Notch localization and signaling output in either Schneider 2 (S2) cells, or *Drosophila* imaginal discs. S2 cells are derived from a primary culture of late stage (20–24 h old) *Drosophila melanogaster* embryo, from a macrophage-like lineage (Schneider, 1972). Analysis of Notch localization in endosomes involved colocalization of Notch (antibodies against Notch ECD and ICD) with different endosomal markers—antibodies against Rab5 (Abcam), Hook (Krämer & Phistry, 1996), Hrs (Lloyd et al., 2002), Sara (Coumailleau et al., 2009), Rab7 (Chinchore, Mitra, & Dolph, 2009), LAMP1 (Abcam), Lysotracker (Molecular Probes), and so on. In addition, the signaling output of Notch activity was measured using antibodies against the Notch targets cut and wingless, in wing imaginal discs (Hori et al., 2011).

3.1. Immunostaining in cultured cells and imaginal discs
3.1.1 Materials
1. *Drosophila* S2 cells are cultured in Schneider's *Drosophila* medium (Gibco) with 10% fetal bovine serum (Gibco) and penicillin–streptomycin (Gibco) at 25 °C.
2. Wing imaginal discs dissected from third instar *Drosophila* larvae reared at 25 °C.
3. Glass-bottom Petri dishes (MatTek Corporation).

4. Concanavalin A (ConA, Sigma) treatment: To ConA coat glass-bottom dishes, spread 200 μl of the ConA solution (a 0.5 mg/ml in sterile water) onto the lid-facing side of the glass slide at the bottom of the Petri dish. After 30 min incubation of ConA, rinse the surface of the Petri dish with sterile water. The ConA-coated dishes are stored at room temperature (RT) and can be used months after preparation.
5. Fixative: $10 \times$ PBS, pH 7.0 (200 mM KPO$_4$, 140 mM NaCl) 500 mM EGTA pH 6.4, 37% formaldehyde (Sigma).
6. Alternative fixative: 0.5 M PIPES pH 6.9, 500 mM EGTA pH 7.4, 1 M MgSO$_4$, 37% formaldehyde (Sigma).
7. Blocking solution PBT: $10 \times$ PBS, pH 7.0 (200 mM KPO$_4$, 140 mM NaCl) 500 mM EGTA pH 7.4, Triton X-100 BSA.
8. Wash buffer PT: $10 \times$ PBS, pH 7.0 (200 mM KPO$_4$, 140 mM NaCl) 500 mM EGTA pH 7.4, Triton X-100.
9. Mounting medium: Vectashield (Vector Laboratories).

3.1.2 Methods
1. *Drosophila* S2 cells were cultured in glass-bottomed Petri dishes treated with ConA.
2. Fixative: Make a stock solution that is $1.33 \times$ PBS and 67 mM EGTA ("*in situ* fix") or one that is 100 mM PIPES, 2 mM EGTA, and 1 mM MgSO$_4$ ("PEMFA buffer"). These fixatives can be filter sterilized and stored for ~1 year.
3. Make PT ($1 \times$ PBS pH 7.0, 0.1% Triton X-100) and PBT ($1 \times$ PBS pH 7.0, 0.1% Triton X-100, 2% BSA). Make PT in 1 l volumes, filter sterilize it, and store it at RT. To make PBT, use 50 ml of PT in a 50 ml conical and add 1 g of BSA. PBT spoils very quickly; store it at 4 °C for 2–3 days.
4. (a) *Drosophila* S2 cells were fixed in 4% formaldehyde for 20 min at RT. (b) *Drosophila* third instar larval wing discs were dissected in cold $1 \times$ PBS and fixed in 4% formaldehyde for 40 min. The imaginal discs are processed in 1.5 ml microfuge tubes.
5. The fixative was poured off and the fixed cells or wing imaginal discs washed with four washes of PT, 15 min each.
6. Following this, 1 ml of blocking solution (PBT) was added for 1 h at RT.
7. The blocking solution was removed and primary antibody (diluted in blocking solution) was added directly onto the cells or wing imaginal discs. A total of 200 μl of antibody solution should completely cover

the cells on a plate. The Petri lid was placed back on dish to prevent evaporation of the antibody solution. The antibody was incubated with the cells overnight at 4 °C.

8. The antibody solution was removed and three PT washes were performed. A transfer pipette was used to add 2 ml of PBT to the dish; each wash was 15 min for a total time of 1 h.
9. After the last wash, 200 μl of fluorescent secondary antibody (diluted in blocking solution PBT) was added directly onto the cells or wing imaginal discs. The Petri dish (or 1.5 ml microfuge tubes) was covered with aluminum foil to prevent bleaching of fluorescent secondary antibody. The antibody was incubated with the cells for 2 h at RT.
10. The secondary antibody was removed, followed by four 15-min washes with PT.
11. After the last wash, a few drops of Vectashield were added and the dish was stored in a dark environment at 4 °C, ready for light microscopy. Wing imaginal discs were mounted on a glass slide in a drop of Vectashield, covered with a glass cover slip and stored at 4 °C.

3.2. Endocytosis assay in cultured cells and imaginal discs

Either Notch–ligand interaction or ligand-independent activation of Notch in endosomes requires Notch trafficking through different endosomal compartments. To track such trafficking of the Notch receptor, imaginal discs or S2 cells are used to assay antibody uptake.

3.2.1 Materials

1. *Drosophila* S2 cells are cultured in Schneider's *Drosophila* medium (Gibco) with 10% fetal bovine serum (Gibco) and penicillin–streptomycin (Gibco) at 25 °C.
2. Wing imaginal discs from third instar *Drosophila* larvae reared at 25 °C.
3. Glass-bottom Petri dishes (MatTek Corporation).
4. 0.5 mg/ml solution of ConA (in sterile water).
5. (1) Fixative: $10\times$ PBS, pH 7.0 (200 mM KPO$_4$, 140 mM NaCl) 500 mM EGTA pH 6.4, 37% formaldehyde (Sigma). (2) Alternative fixative: 0.5 M PIPES pH 6.9, 500 mM EGTA pH 7.4, 1 M MgSO$_4$ 37% formaldehyde (Sigma).
6. Blocking solution PBT: $10\times$ PBS, pH 7.0 (200 mM KPO$_4$, 140 mM NaCl) 500 mM EGTA pH 7.4, Triton X-100, BSA.
7. Wash buffer PT: $10\times$ PBS, pH 7.0 (200 mM KPO$_4$, 140 mM NaCl) 500 mM EGTA pH 7.4, Triton X-100.

8. Mounting medium: Vectashield (Vector Laboratories).
9. Primary antibodies: anti-Notch ECD, an antibody for Notch extracellular domain (2H, Developmental Studies Hybridoma Bank).

3.2.2 Methods
1. (a) *Drosophila* S2 cells were cultured in glass-bottomed Petri dishes, treated with ConA. (b) Imaginal discs were dissected in standard M3 medium.
2. Pulse: Imaginal discs (or S2 cells) were incubated for either 15 or 40 min (depending on the experimental approach) with anti-Notch ECD (1:500) antibody at 25 °C in M3 medium (Gibco).
3. Chase: The anti-Notch ECD antibody was washed off the cells (or wing imaginal discs in 1.5 ml microfuge tubes) with three washes of 10 min each of standard M3 medium at RT.
4. *Drosophila* S2 cells (or wing imaginal discs in 1.5 ml microfuge tubes) were fixed in 4% formaldehyde for 20 min at RT.
5. The fixative was poured off and fixed cells or wing imaginal discs were washed with four 15 min washes with PT.
6. 1 ml of blocking solution (PBT) was added to the cells (or wing imaginal discs in 1.5 ml microfuge tubes) for 1 h.
7. PBT was removed and 200 µl of fluorescent secondary antibody (diluted in blocking solution PBT) was added directly onto the cells or wing imaginal discs. The Petri dish or 1.5 ml microfuge tubes were covered with aluminum foil to prevent bleaching of fluorescent secondary antibody. The antibody was incubated with the cells for 2 h at RT.
8. The secondary antibody was removed, followed by four 15-min washes with PT.
9. After the last wash, a few drops of Vectashield were added and the dish was stored in a dark environment at 4 °C, ready for light microscopy. For wing imaginal disc, the discs were mounted on a glass slide in a drop of Vectashield, covered with a glass cover slip and stored at 4 °C.

4. OPTICAL APPROACHES

To follow the trafficking of the Notch and Delta in live cells, we constructed and utilized fluorescently tagged Notch and Delta probes. Crucial to this is not only the construction of the probes but also a demonstration that they are functional and display the proper subcellular localization. We thus expressed each transgene in cultured cells to evaluate the subcellular localization of the corresponding tagged protein compared to the wild type and probed functionality by examining their ability to rescue loss-of-function phenotypes.

4.1. Construction of fluorescently tagged Notch and ligand molecules

1. We have tagged the Notch receptor and the ligands in multiple sites to ensure that we can generate functional molecules. A summary of the successful constructs is depicted in Fig. 16.3A.

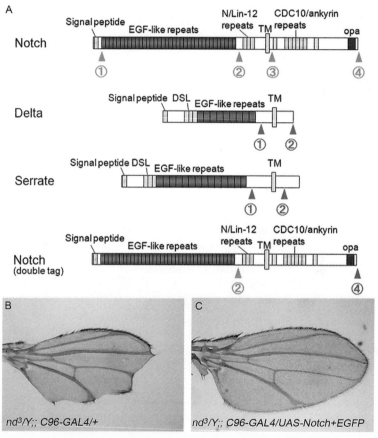

Figure 16.3 Construction and validation of the Notch and ligands transgenes. (A) Schematic representation of EGFP-tag (green) in Notch and tdTomato-tag (red) in Delta and Serrate. (B) nd^3/Y; C96-GAL4/+ males show characteristic notching at the wing tip. nd^3 is a hypomorphic allele of *Notch*. (C) The nd^3 phenotype was rescued by *UAS-Notch-EGFP* transgene expressed in the wing margin under the control of the *C96-GAL4* driver. (For interpretation of the references to color in this figure legend, the reader is referred to the online version of this chapter.)

2. All constructs were expressed in cultured cells via transgenes under a metallothionein promoter (Invitrogen) or *in vivo* via transgenes under a *UAS* promoter.

4.2. Generation of stable cell lines
4.2.1 Materials
1. *Drosophila* S2R+ cells are cultured in Schneider's *Drosophila* medium (Gibco) with 10% fetal bovine serum (Gibco) and penicillin–streptomycin (Gibco) at 25 °C.
2. DNA constructs: pMK33-Notch+EGFP, pMK33-Dl+tdTomato, pMK33-Ser+tdTomato. The pMK33-derived vector contains a Hygromycin resistance gene that can be used for selection. As a negative control for selection, any vector that does not contain a Hygromycin resistant gene can be used.
3. Transfection reagent: TransIT-2020 (Mirus). We have evaluated several transfection reagents and found that the TransIT-2020 shows the highest efficiency of transfection in our assay.
4. Hygromycin (Invitrogen).

4.2.2 Methods
1. S2R+ cells were seeded in 6-well plate at 0.5×10^6 cells/ml.
2. DNA constructs were transfected using TransIT-2020 (Mirus) following manufacturer's recommended protocol.
3. After 48 h of transfection, cells were collected by gently pipetting, transferred to 15 ml falcon tube, and centrifuged at 500 rpm for 5 min.
4. The cells were resuspended in 6 ml medium with Hygromycin (100 μg/ml) and transferred into a T25 flask.
5. Every 2–3 days, cells were changed into fresh medium containing Hygromycin (100 μg/ml).
6. For 2–3 weeks, continuous cell death was visible. Controls (e.g., transfected with any empty vector) completely die out after 3–4 weeks.
7. After about a month, the stable line was established and they were expanded as necessary.
8. Expression of transgene was induced by 0.35 mM CuSO$_4$ for 16–24 h. The subcellular localization of the receptors and ligands was confirmed to be on the membrane as well as in some intracellular endosomes.
9. The cells were passaged at 1:5 dilutions every 5 days into fresh medium containing Hygromycin (100 μg/ml).

4.3. Rescue experiments

1. To confirm the functionality of the fluorescently tagged Notch and ligands, we performed rescue experiments. For this purpose, we established transgenic flies carrying *pUAST-Notch+EGFP*, *pUAST-Delta+tdTomato*, and *pUAST-Serrate+tdTomato*.
2. The *UAS-Notch+EGFP* construct was driven by *C96-Gal4* in the mutant background of *Notch*. Figure 16.3B and C shows an example showing rescue of the mutant phenotype with fluorescence-tagged Notch construct.
3. We have also demonstrated the biological activity of Delta+tdTomato and Serrate+tdTomato transgenes, by eliciting gain-of-function phenotypes in the wing.

5. UBIQUITINATION STATUS OF NOTCH

Membrane trafficking and ubiquitination are generally intimately linked processes and a gene that affects Notch by modulating ubiquitylation is encoded by *shrub*. *shrub* encodes a core component of the ESCRT-III complex and was identified as a modifier of the *dx–krz* synergistic wing phenotype (Hori et al., 2011). To understand the role of Shrub in Notch signaling, we examined the subcellular localization of Notch by modulating the expression level of Shrub. We found that expression of Shrub leads to the accumulation of Notch in endosomes, which are positive for FK1 (Biomol), an antibody that recognizes poly-ubiquitinated proteins (Hori et al., 2011). This result suggested that Shrub regulates the trafficking of Notch though a ubiquitinylation process. Furthermore, we showed that Dx can also influence the ubiquitination status of Notch, which is a phenotype paralleled by an upregulation of Notch signaling (Hori et al., 2011).

Utilizing antibody staining *in vivo*, we understood that Dx and Shrub regulate the trafficking of Notch and its activity. However, we still need to directly assess the relative roles of Dx and Shrub in the ubiquitinylation of Notch. For the purpose, we relied on S2R+ cultured cells, which do not express Notch endogenously.

5.1. Materials

1. DNA constructs: pMT-Notch^{FL}, pMT-Dx, and pMT-Flag-Ubiquitin^{WT} (Ub^{WT}), pMT-Flag-Ub^{Mono}. Flag-Ub^{MONO} is a Flag-tagged mutated version of ubiquitin that can only participate in monoubiquitination events.

2. Lysis buffer: 50 mM Tris pH 7.5, 125 mM NaCl, 1.5 mM MgCl$_2$, 5% glycerol, 0.2% NP-40, 1 mM DTT, 25 mM NaF, 1 mM Na$_3$VO$_4$, and complete protease inhibitor (Roche).
3. Proteasome inhibitor: MG132 (Calbiochem).

5.2. Ubiquitination assay

1. S2R+ cells were seeded at 1×10^7 cells/ml into 100 mm dishes.
2. pMT-Notch^{FL}, pMT-Dx, and pMT-Flag-Ub^{WT} (or pMT-Flag-Ub^{Mono}) are transfected by using the TransIT-2020 reagent (Mirus). Equal amounts of DNA constructs are transfected, and the total amount of DNA is kept constant by adding empty vector (15 μg of DNA constructs are transfected in total). One day after transfection, plasmid expression was induced with 0.35 mM CuSO$_4$ overnight.
3. Cells were treated for 4 h at 25 °C with a 50 μM concentration of the proteasome inhibitor MG132 (Calbiochem).
4. The cell culture dishes were placed on ice. The medium was drained, and cells were collected by pipetting in cold PBS.
5. The cells were washed using cold PBS twice and then gently transferred into precooled 1.5 ml microfuge tubes.
6. The cells were lysed in ice-cold lysis buffer. The lysates were incubated on ice for 20 min.
7. The lysates were cleared by centrifugation at 2000 rpm for 10 min at 4 °C.
8. The supernatant was collected and placed in a fresh tube kept on ice.
9. On ice, ~1000 μg cell lysate was added to the anti-Notch ICD (9C6, Developmental Studies Hybridoma Bank) (1:1000).
10. The sample was incubated with the antibody at 4 °C under rotary agitation overnight.
11. The protein G-agarose beads were prepared (Roche) (20 μl of the slurry) by washing twice in 1 ml cold lysis buffer.
12. The beads were added to each sample and incubated at 4 °C under rotary agitation for 2–3 h.
13. The precipitates were washed in 1 ml cold lysis buffer five times.
14. Protein complexes were eluted with 30 μl of 2× LDS sample buffer (Invitrogen) and heated at 70 °C for 10 min.
15. The samples were run on a 3–7% Tris–Acetate gel (Invitrogen) and transferred onto PVDF membrane (Invitrogen).
16. The ubiquitinated Notch is detected by western blot using rabbit anti-FLAG (Sigma, 1:1000).

6. CONCLUSION

Recent genetic and molecular studies have led to the increased appreciation of how important and indeed diverse, trafficking events related to the receptor and the ligands are for the developmental control of Notch signals. Here we described paradigmatic methodology to obtain an informative picture regarding the complexity and the nature of the genetic circuitry that affects Notch trafficking controlled by Dx and Krz. Such genetic screens while focused on a specific phenotype affecting one particular tissue, the value of the screen is more general. This particular screen recovered more than 250 modifiers of the wing phenotype we screened with only less than 20 modifiers associated with membrane trafficking as judged by Gene Ontology analysis (Flybase, http://flybase.org/) with the rest falling into diverse functional categories including, for example, gene regulations, metabolism, cytoskeleton, as well as several genes of unknown function (unpublished data). Linking molecularly each category or each gene with Notch is a nontrivial task (see e.g., Hori et al., 2011), notwithstanding the value of genetic, that is, functional links unveiled by genetic analyses.

ACKNOWLEDGMENTS

We would like to thank Robert A. Obar and K. G. Guruharsha for critically reading the manuscript. This work was supported by NIH Grants NS26084 and CA98402 (S. A. -T.), a JSPS Postdoctoral Fellowship for Research Abroad (K. H.), and a Postdoctoral Fellowship from the FSMA (A. S.).

REFERENCES

Artavanis-Tsakonas, S. (2004). Accessing the Exelixis collection. *Nature Genetics*, 36, 207.
Artavanis-Tsakonas, S., Rand, M. D., & Lake, R. J. (1999). Notch signaling: Cell fate control and signal integration in development. *Science*, 284, 770–776.
Bray, S. J. (2006). Notch signalling: A simple pathway becomes complex. *Nature Reviews Molecular Cell Biology*, 7, 678–689.
Childress, J. L., Acar, M., Tao, C., & Halder, G. (2006). Lethal giant discs, a novel C2-domain protein, restricts notch activation during endocytosis. *Current Biology*, 16, 2228–2233.
Chinchore, Y., Mitra, A., & Dolph, P. J. (2009). Accumulation of rhodopsin in late endosomes triggers photoreceptor cell degeneration. *PLoS Genetics*, 5, e1000377.
Coumailleau, F., Fürthauer, M., Knoblich, J. A., & González-Gaitán, M. (2009). Directional Delta and Notch trafficking in Sara endosomes during asymmetric cell division. *Nature*, 458, 1051–1055.
Fortini, M. E. (2009). Notch signaling: The core pathway and its posttranslational regulation. *Developmental Cell*, 16, 633–647.

Fuwa, T. J., Hori, K., Sasamura, T., Higgs, J., Baron, M., & Matsuno, K. (2006). The first deltex null mutant indicates tissue-specific deltex-dependent Notch signaling in Drosophila. *Molecular Genetics & Genomics, 275,* 251–263.

Gustafson, K., & Boulianne, G. L. (1996). Distinct expression patterns detected within individual tissues by the GAL4 enhancer trap technique. *Genome, 39,* 174–182.

Hori, K., Fostier, M., Ito, M., Fuwa, T. J., Go, M. J., Okano, H., et al. (2004). Drosophila deltex mediates suppressor of Hairless-independent and late-endosomal activation of Notch signaling. *Development, 131,* 5527–5537.

Hori, K., Fuwa, T. J., Seki, T., & Matsuno, K. (2005). Genetic regions that interact with loss- and gain-of-function phenotypes of deltex implicate novel genes in Drosophila Notch signaling. *Molecular Genetics & Genomics, 272,* 627–638.

Hori, K., Sen, A., Kirchhausen, T., & Artavanis-Tsakonas, S. (2011). Synergy between the ESCRT-III complex and Deltex defines a ligand-independent Notch signal. *Journal of Cell Biology, 195,* 1005–1015.

Hori, K., Sen, A., & Artavanis-Tsakonas, S. (2013). Notch signaling at a glance. *Journal of Cell Science, 126,* 2135–2140.

Krämer, H., & Phistry, M. (1996). Mutations in the Drosophila hook gene inhibit endocytosis of the boss transmembrane ligand into multivesicular bodies. *Journal of Cell Biology, 133,* 1205–1215.

Lindsley, D. L., & Zimm, G. G. (1992). *The genome of Drosophila melanogaster.* (pp. 492–497) San Diego, CA: Academic Press.

Lloyd, T. E., Atkinson, R., Wu, M. N., Zhou, Y., Pennetta, G., & Bellen, H. J. (2002). Hrs regulates endosome membrane invagination and tyrosine kinase receptor signaling in Drosophila. *Cell, 108,* 261–269.

Louvi, A., & Artavanis-Tsakonas, S. (2012). Notch and disease: A growing field. *Seminars in Cell & Developmental Biology, 23,* 473–480.

Mukherjee, A., Veraksa, A., Bauer, A., Rosse, C., Camonis, J., & Artavanis-Tsakonas, S. (2005). Regulation of Notch signalling by non-visual beta-arrestin. *Nature Cell Biology, 7,* 1191–1201.

Parks, A. L., Cook, K. R., Belvin, M., Dompe, N. A., Fawcett, R., Huppert, K., et al. (2004). Systematic generation of high-resolution deletion coverage of the Drosophila melanogaster genome. *Nature Genetics, 36,* 288–292.

Roman, G., He, J., & Davis, R. L. (2000). kurtz, a novel nonvisual arrestin, is an essential neural gene in Drosophila. *Genetics, 155,* 1281–1295.

Schneider, I. (1972). Cell lines derived from late embryonic stages of Drosophila melanogaster. *Journal of Embryology and Experimental Morphology, 27,* 353–365.

Schweisguth, F. (2004). Regulation of notch signaling activity. *Current Biology, 14,* R129–R138.

Thibault, S. T., Singer, M. A., Miyazaki, W. Y., Milash, B., Dompe, N. A., Singh, C. M., et al. (2004). A complementary transposon tool kit for Drosophila melanogaster using P and piggyBac. *Nature Genetics, 36,* 283–287.

Thompson, B. J., Mathieu, J., Sung, H. H., Loeser, E., Rørth, P., & Cohen, S. M. (2005). Tumor suppressor properties of the ESCRT-II complex component Vps25 in Drosophila. *Developmental Cell, 9,* 711–720.

Vaccari, T., & Bilder, D. (2005). The Drosophila tumor suppressor vps25 prevents non-autonomous overproliferation by regulating notch trafficking. *Developmental Cell, 9,* 687–698.

Vaccari, T., Lu, H., Kanwar, R., Fortini, M. E., & Bilder, D. (2008). Endosomal entry regulates Notch receptor activation in Drosophila melanogaster. *Journal of Cell Biology, 180,* 755–762.

Vaccari, T., Rusten, T. E., Menut, L., Nezis, I. P., Brech, A., Stenmark, H., et al. (2009). Comparative analysis of ESCRT-I, ESCRT-II and ESCRT-III function in

Drosophila by efficient isolation of ESCRT mutants. *Journal of Cell Science, 122,* 2413–2423.

Wilkin, M. B., Carbery, A. M., Fostier, M., Aslam, H., Mazaleyrat, S. L., Higgs, J., et al. (2004). Regulation of notch endosomal sorting and signaling by Drosophila Nedd4 family proteins. *Current Biology, 14,* 2237–2244.

Wilkin, M., Tongngok, P., Gensch, N., Clemence, S., Motoki, M., Yamada, K., et al. (2008). Drosophila HOPS and AP-3 complex genes are required for a Deltex-regulated activation of notch in the endosomal trafficking pathway. *Developmental Cell, 15,* 762–772.

Yamamoto, S., Charng, W. L., & Bellen, H. J. (2010). Endocytosis and intracellular trafficking of Notch and its ligands. *Current Topics in Developmental Biology, 92,* 165–200.

CHAPTER SEVENTEEN

Monitoring Notch/Delta Endosomal Trafficking and Signaling in Drosophila

Sylvain Loubéry[*,†], Marcos González-Gaitán[*,†,1]
[*]Department of Biochemistry, Faculty of Sciences, Geneva University, Geneva, Switzerland
[†]Department of Molecular Biology, Faculty of Sciences, Geneva University, Geneva, Switzerland
[1]Corresponding author: e-mail address: marcos.gonzalez@unige.ch

Contents

1. Introduction — 302
   1.1 Notch control by endocytosis and endosomal trafficking — 302
   1.2 The Drosophila notum: A powerful model system for Notch studies — 304
2. Antibody Uptake Assays to Monitor Notch and Delta Trafficking — 305
   2.1 The antibody uptake assay: Principle — 305
   2.2 The antibody uptake assay: Protocol — 308
   2.3 A refinement to the antibody uptake assay: The fibrin clot — 309
   2.4 Live imaging and quantification of Notch and Delta trafficking using the antibody uptake assay — 311
3. Correlative Imaging of Dividing SOP Cells and of Their Progeny — 313
   3.1 Correlative whole pupa imaging and scanning electron microscopy of adult flies — 313
   3.2 Correlative antibody uptake assays and immunostainings — 315
4. Concluding Remarks — 318
Acknowledgements — 319
References — 320

Abstract

The Notch signaling pathway plays important roles in many organisms and developmental contexts. The activities of the Notch receptor and of its ligand Delta are known to be regulated at several steps along the endocytic pathway. However, the precise molecular mechanism of Notch activation and the role played by endosomal sorting and trafficking remain elusive. We developed an antibody uptake assay to enable live imaging of endogenous internalized Notch and Delta in Drosophila tissues. In this chapter, we describe how to perform live antibody uptake assays in the Drosophila notum. In this tissue, Notch signaling plays a crucial role in the regulation of cell fate decisions in the lineage of sensory organ precursor (SOP) cells. We describe here how to do a correlative analysis of Notch/Delta live imaging in dividing SOPs and of the lineage of these particular SOPs. Combined with the wide range of genetic and chemical tools available in Drosophila research, these two methods will provide a better understanding of the role played by endocytic proteins and endosomal trafficking in Notch regulation, in terms of botch Notch trafficking and Notch signaling output.

1. INTRODUCTION

1.1. Notch control by endocytosis and endosomal trafficking

The Notch pathway is a highly conserved signaling pathway playing roles in a wide range of organisms, from flies and worms to mammals. It is involved in many decision-making events in the course of development, including control of proliferation, apoptosis, establishment of compartment boundaries, and cell fate decision (Artavanis-Tsakonas & Muskavitch, 2010). Accordingly, its deregulation has been linked to a variety of diseases, including cancer (Purow, 2012). In spite of these crucial and ubiquitous roles, the molecular mechanism of Notch activation remains elusive in many systems.

The Notch receptor undergoes a series of cleavages in order to be activated (Kopan & Ilagan, 2009). The first cleavage (S1) occurs in the *trans*-Golgi network. It is mediated by a furin-like protease that enables the heterodimerization of the receptor, which is presented with its extracellular domain noncovalently bound to its transmembrane domain. The Notch ligands belong to the so-called DSL family of proteins, named after some of its members: Delta and Serrate/Jagged in flies and LAG-2 in worms. Upon binding in *trans* to one of its ligands, the Notch receptor undergoes the second cleavage (S2), mediated by a protease of the ADAM family. This cleavage releases the Notch extracellular domain and prompts the receptor for the subsequent S3/S4 cleavage. The transmembrane gamma-secretase complex mediates this last event, which releases the Notch intracellular domain. This fragment translocates to the nucleus, where it binds CSL transcription factors (CBF1/RBPJk in mammals, Su(H) in flies, and LAG-1 in worms) and relieves the inhibition of the transcription of Notch target genes.

One peculiarity of the Notch signaling pathway is its lack of enzymatic amplification step. This may be the reason why the Notch receptor is haploinsufficient, and why *Notch* is one of the very rare Drosophila genes to be triplo-mutant: cells seem to be able to count *Notch* gene dosage (see Artavanis-Tsakonas & Muskavitch, 2010). Practically speaking, this means that one has to be particularly careful when using functional reporters for

Notch or one of its regulators, because having nonendogenous levels of Notch in a cell may not be buffered by the system and will be expected to induce changes in signaling levels.

Endocytosis and endosomal trafficking can play a role at different steps of Notch activation. It is widely admitted that the S2 cleavage depends on endocytosis of the Notch ligands, although the reason for this dependence is debated; this has been the subject of a recent review (Musse, Meloty-Kapella, & Weinmaster, 2012) and will not be discussed at length here. In short, according to the recycling model, endocytosis and recycling of the ligands are necessary either to prompt their activation by posttranslational modifications or to enable their targeting to plasma membrane domains favorable for interactions with the receptor. The alternative pulling force model relies on the fact that the ADAM cleavage site in the Notch receptor is embedded and protected in a fold of the protein; binding of a ligand followed by endocytosis of this ligand can generate a force sufficient to unfold the receptor and enable the S2 cleavage. Both models are supported by a number of studies, and it may well be that one, the other, or both can hold true depending on the system studied. For instance, it has been shown that in the Drosophila eye disc and ovary, Notch activation can occur in the absence of (or with reduced) ligand recycling (Banks et al., 2011; Windler & Bilder, 2010), which argues for a pulling force role for ligand endocytosis; however, endocytosis and recycling of the ligands are known to be necessary in epithelia such as the Drosophila notum, in which Delta has to be transcytosed and, from its initial baso-lateral localization, targeted to the apical membrane where Notch is positioned (Benhra, Vignaux, Dussert, Schweisguth, & Le Borgne, 2010; Giagtzoglou et al., 2012).

The S3 cleavage of Notch can also be controlled by endosomal trafficking. Indeed, gamma-secretase is present in endosomes and lysosomes (Pasternak et al., 2003; Vetrivel et al., 2004) and a number of studies indicate that the Notch receptor can undergo its S3/S4 cleavage in endosomal compartments (Hori, Sen, Kirchhausen, & Artavanis-Tsakonas, 2011; Tagami et al., 2008; Vaccari, Lu, Kanwar, Fortini, & Bilder, 2008). In parallel, endosomal sorting has been shown to regulate Notch signaling by controlling the differential targeting of the receptor either to the degradative pathway or to a compartment where it can be activated (Benhra et al., 2011; Hori et al., 2011; Vaccari et al., 2008; Windler & Bilder, 2010).

The role of endosomal trafficking for Notch activation is the subject of extensive studies in the sensory organ precursor (SOP) system of Drosophila (Furthauer & Gonzalez-Gaitan, 2009). Upon asymmetric division of each

SOP cell, correct cell fate determination of the pIIa and pIIb daughter cells is mediated by the exclusive activation of Notch in the pIIa cell only (see below). Four independent mechanisms control this asymmetric activation of Notch: (1) the protein Neuralized is inherited only by the pIIb cell and mediates internalization of Delta in this cell only (Le Borgne & Schweisguth, 2003); (2) a transient accumulation of recycling endosomes in the pIIb cell enhances recycling and activation of Delta in this cell (Emery et al., 2005); (3) endosomes marked by the protein Sara induce the asymmetric segregation of a signaling pool of Notch and Delta in the pIIa cell (Coumailleau, Furthauer, Knoblich, & Gonzalez-Gaitan, 2009); (4) the endocytic proteins α-adaptin and Numb are inherited by the pIIb cell and inhibit the Notch activator Sanpodo in this cell (Berdnik, Torok, Gonzalez-Gaitan, & Knoblich, 2002; Hutterer & Knoblich, 2005).

In the past years, it has thus become clear that Notch signaling can be regulated at different levels in the endosomal pathway. More and more endocytic proteins are found to play roles in Notch signaling control (Le Bras, Rondanino, Kriegel-Taki, Dussert, & Le Borgne, 2012). However, a precise understanding of the molecular mechanism of Notch sorting and activation in endosomes is still elusive. One reason for this is the difficulty to monitor in live samples the trafficking and interaction of Notch and its ligands.

1.2. The Drosophila notum: A powerful model system for Notch studies

The fruit fly is a very convenient and powerful model system to study Notch signaling. Since the description of the *Notch* locus by the group of Thomas Morgan in the late 1910s, and the later understanding that Notch is involved in signaling and in neurogenesis (Knust & Campos-Ortega, 1989), a vast number of proteins have been found to regulate Notch signaling in Drosophila (Guruharsha, Kankel, & Artavanis-Tsakonas, 2012); for most of these genes, a battery of genetics tools (mutants, RNAi, fluorescent reporters) is available.

Among the different tissues in which Notch plays a role in the fly, the pupal notum (the dorsal part of the thorax) is a system of choice for Notch imaging and signaling studies. Indeed, this tissue is easily accessible and can be used in high-quality microscopy studies with very little preparation time.

In the notum, Notch signaling plays a pivotal role in bristle development. The adult notum is covered with sensory organs made out of four cells: two outer cells (shaft and socket), and two inner cells (sheath and neuron) under the cuticle. Each of these organs arises from a single developmental unit: the SOP cell. Between approximately 16 h and 22 h after puparium formation (APF), the 200 SOPs of the pupal notum divide asymmetrically a number of times to form the sensory organs, in a stereotyped and well-characterized manner (Gho, Bellaiche, & Schweisguth, 1999; Gho, Lecourtois, Geraud, Posakony, & Schweisguth, 1996; Hartenstein & Posakony, 1989). Each SOP division gives rise to a pIIa and a pIIb daughter cells. The pIIb cell then divides to form a pIIIb cell and a glial cell that rapidly undergoes apoptosis. Finally, the pIIa and pIIIb cells divide to give rise, respectively, to the socket and shaft cells, and the sheath and neuron cells. Notch signaling plays a key role in this system by controlling cell fate decision between sibling daughter cells: a Notch signaling bias between the pIIa–pIIb, shaft-socket, and sheath-neuron pairs is essential so that each cell acquires its correct fate. As a result, any change in cell fate in these lineages can be attributed to perturbations of Notch signaling. And as the determination of organ composition in a given experimental condition enables one to trace back the signaling history of this organ (e.g., see Figure S15 in Coumailleau et al., 2009), any phenotype observed in the lineage can be directly accounted for and translated in a phenotype in Notch signaling.

In this chapter, we describe how to perform live antibody uptake assays in order to image the trafficking of endocytosed Notch and Delta proteins in the pupal notum. Our interest in this technique primarily lies in the understanding of signaling regulation during asymmetric cell division; however, it has to be emphasized that these uptake assays can be applied to the pupal notum at other stages than during sensory organ development, and also to the wing imaginal disc or to any other fly tissue accessible to antibodies. Furthermore, we describe how to do a correlative analysis of Notch/Delta live imaging in a dividing SOP and of the lineage of this SOP.

2. ANTIBODY UPTAKE ASSAYS TO MONITOR NOTCH AND DELTA TRAFFICKING

2.1. The antibody uptake assay: Principle

Notch and Delta trafficking can in principle be followed using GFP fusions of these proteins; such an approach has already been described for Notch (Kawahashi & Hayashi, 2010; Couturier, Vodovar, & Schweisguth, 2012). This method, however, suffers from three drawbacks: (1) as highlighted before, one has to be particularly careful regarding the putative effects of having non-endogenous levels of functional Notch; (2) Notch-GFP reporters highlight the entire population of Notch, which prevents from untangling the endosomal signaling population and the population in the secretory pathway; (3) GFP at low expression levels (which is a prerequisite for meaningful analysis in this context) bleaches rapidly and precludes any refined analysis of endosomal motility and distribution.

A technique has been developed in our lab to enable live imaging of endocytosed Notch and Delta: antibody uptake assays (Coumailleau et al., 2009). Briefly, the pupal notum is dissected and incubated in medium containing fluorescently tagged antibodies targeted toward the extracellular domain of Notch or Delta; after a short pulse, these reporters can be chased and imaged as they are internalized with their targets, enabling live visualization of Notch or Delta trafficking as they transit through the endocytic pathway. Antibody uptake assays in Drosophila tissues have initially been developed by the Cohen laboratory and by us to, respectively, monitor the morphogens Wingless and Decapentaplegic (Entchev, Schwabedissen, & Gonzalez-Gaitan, 2000; Strigini & Cohen, 2000); this technique has later been adapted by the Schweisguth laboratory for the study of Delta endocytosis in SOPs (Le Borgne & Schweisguth, 2003). In Coumailleau et al. (2009), the use of fluorescently labeled primary antibodies enabled for the first time live imaging of Notch and Delta endosomal trafficking. Importantly, this assay does not require any overexpression and allows monitoring the trafficking of endogenous Notch and Delta proteins (Fig. 17.1).

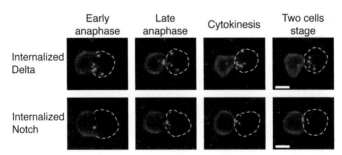

Figure 17.1 Anti-Delta and anti-Notch antibody uptake assays. Anti-Delta (top) and anti-Notch (bottom) antibody uptake assays have been performed as described in the text and imaged by spinning-disc confocal microscopy. Internalized Delta and Notch are marked by Alexa488 (green). The dotted white lines and mRFP-Pon (in red), respectively, indicate the parts of the SOP membrane inherited by the pIIa and the pIIb daughter cells. Bars: 5 μm. (See color plate.)

We use the Zenon technology from Life Technologies to tag our primary antibodies. This consists in mixing the anti-Notch or anti-Delta antibodies with Fabs targeted against the Fc portion of these IgGs and coupled to Alexa dyes. Alexa dyes are brighter and more photostable than GFP and have a better quantum efficiency, thus giving a better signal-to-noise ratio. The quantities of the different reagents in the antibody labeling mix have to be optimized each time a new antibody is used for uptake assays. Life Technologies recommends incubating approximately equal weights of primary antibody and of labeling Fab (resulting in a 3:1 molar excess of labeling Fabs versus primary IgG). Importantly, it has been shown that unbound labeling Fabs do not give any detectable signal in our imaging conditions (Coumailleau et al., 2009): this is probably due to the fact that receptor-mediated endocytosis of IgG–Fab complexes is more efficient than fluid-phase uptake of free Fabs.

Each time the antibody uptake assay is set up with a new primary antibody or on a new imaging system a "blank" control has to be performed, which consists in doing the assay using an antibody labeling mix prepared without any primary antibody. In this situation no detectable fluorescence signal should be observed, with two exceptions. First, the cuticle is very sticky and accumulates fluorescent Fabs, which can sometimes be observed at the apical-most side of the sample (depending on the final concentration used). Second, circulating cells of the immune system are retrieved with dissected nota and can sometimes be seen at the basal-most side of the sample; these cells nonspecifically accumulate both IgG-bound and -unbound Fabs, rendering them visible in endosomes. In a second step it is recommended to perform a complementary control, which consists in performing the antibody uptake assay in one of the following conditions: either in a mutant background in which the target protein is absent (in which case the observation should match the one of the "blank" control), or in a condition in which endocytosis of the target protein is inhibited (in which case the signal should be restricted to the plasma membrane of cells) (e.g., see Figures S1e' and S1g in Coumailleau et al., 2009).

This antibody uptake assay has been successfully and routinely performed with mouse monoclonal antibodies targeted against the proteins Delta, Notch, and Dally-like (Coumailleau et al., 2009). It could be extended to the use of any mouse monoclonal antibody targeting an antigen present at the cell surface. Life Technologies also provides Zenon kits to tag rabbit IgGs, opening the possibility to adapt antibody uptake assays with an even wider range of antibodies, provided these antibodies are pure enough (which is an issue with antisera and polyclonal mixes, which may need to be purified against their target antigen).

2.2. The antibody uptake assay: Protocol

2.2.1 More than 1 day before the experiment

Around 5 ml Clone 8 medium is needed for each experiment. To prepare 1 l of Clone 8 medium, mix 945 ml Shields and Sangs M3 medium (Sigma), 20 ml fetal bovine serum (Gibco), 25 ml fly extract, 10 ml penicillin/streptavidin (Gibco), and 500 µl insulin (Sigma). Filter the medium in sterile conditions, aliquot it and store it at 4 °C. Schneider's medium (Sigma) can be used instead of Clone 8 medium for the antibody uptake assay; this is, however, technically challenging and strongly not recommended, because Schneider's medium causes the nota to stick to glass and to plastic. Fly extract preparation has been described previously (Currie, Milner, & Evans, 1988, and see http://biology.st-and.ac.uk/sites/flycell/flyextract.html).

2.2.2 The day before the experiment

Five silanized hourglass dishes have to be prepared for each experiment; silanization renders the glass hydrophobic and prevents the tissues from sticking to it. Pour 1 ml of PlusOne Repel-Silane ES (GE Healthcare) into one hourglass dish, and transfer it from dish to dish during approximately 10 min, until it has evaporated. Then rinse the dishes with water (this step furthermore allows a quality control of the process), and let them dry in a dust-free place. Silanized hourglass dishes can be used until a few weeks after silanization.

Prior to the experiment, stage pupae so that sensory organ cells are at the stage of development desired for the experiment. Most of the SOPs accessible for imaging on the notum divide in a time window of approximately 2 h, and the first SOP divisions occur at the following times: 27 h APF at 18 °C, 19 h APF at 23 °C, 16 h APF at 25 °C, 14 h APF at 27 °C and 29 °C. A practical way to stage pupae is to collect animals as they enter pupariation, and isolate them in a new vial stored at the temperature of choice; collecting them hourly enables to discriminate easily prepupae from older specimens, while creating an imprecision on their age that is very acceptable, given the 2-h time span in which all SOPs in a given notum divide.

Prepare the antibody labeling mix and store it overnight at 4 °C protected from light. The conditions that have been optimized for the two antibodies that we use routinely in the lab are: Clone 8 medium to adjust the final volume to 25 μl + 1 μg (5 μl) Zenon Alexa Fluor anti-Mouse IgG1 labeling reagent + 1 μg (usually 5 μl, depending on the batch concentration) anti-Delta antibody (C594.9B from the Developmental Studies Hybridoma Bank)/or Clone 8 medium to adjust the final volume to 14 μl + 2 μg (10 μl) Zenon Alexa Fluor anti-Mouse IgG1 labeling reagent + 1 μg (usually 3.5 μl, depending on the batch concentration) anti-Notch antibody (C458.2H from the Developmental Studies Hybridoma Bank). This mix can be prepared the day of the experiment and incubated for 10 min at room temperature before following the rest of the procedure; however, the signal-to-noise ratio is better if it is incubated overnight at 4 °C prior to the experiment. The Zenon Alexa 488, 568, and 647 all give satisfying results and can be used as it suits best the experimental design.

2.2.3 The day of the experiment

Pupae dissection has been very precisely described by Jauffred and Bellaiche (2012). Proceed as such, with the only exception that 500 μl Clone 8 medium has to be used as dissection medium (instead of PBS). Once the notum is detached from the animal body, transfer it to 1 ml Clone 8 medium in a silanized hourglass dish to wash away fat bodies. Always use pipette tips that have been cut at the tip while transferring a notum from dish to dish, and pipette gently, so that the tissue sticks neither to the pipette tip nor to glass; also take care to aspirate as little medium as possible when doing this, so as not to dilute the receiving solutions.

In a second silanized hourglass dish, mix either: 295 μl Clone 8 medium and 5 μl of the anti-Delta-Zenon mix, or 87.5 μl Clone 8 medium and 12.5 μl of the anti-Notch-Zenon mix. Transfer the notum to this dish and homogenize the solution. Pulse for 5 min.

Wash three times in 1 ml Clone 8 medium, then transfer the notum to a drop of 500 μl Clone 8 medium in a glass-bottom Fluorodish culture dish (World Precision Instruments). The tissue is now ready to be imaged.

2.3. A refinement to the antibody uptake assay: The fibrin clot

The protocol above is as published in Coumailleau et al. (2009). One variation that we have developed consists in using a clot of fibrin to immobilize the tissue on the glass-bottom dish. This method was set up by Forer and Pickett-Heaps (1998) and also used for the imaging of other fly tissues (Januschke & Gonzalez, 2010). The principle is that adding thrombin to a solution of fibrinogen triggers the conversion of soluble fibrinogen into loose and sticky fibrin fibers; the clot thus formed can be used to spatially constrain tissues without exerting forces too intense that could have secondary effects on cell and tissue homeostasis.

Prior to the experiment, dissolve 500 U thrombin (Amersham) in 1 ml PBS, make 4 µl aliquots, and store at −80 °C. On the day of the experiment, dissolve 10 mg fibrinogen from bovine plasma (Calbiochem) in 1 ml Clone 8 medium; do not prepare this solution in advance, because the Clone 8 medium will slowly induce fibrin fiber formation.

Dissect the pupa and proceed with the pulse and the first two subsequent washes as described earlier. Do the third washing step in 1 ml Clone 8 medium + fibrinogen. Aspirate the notum along with approximately 100 µl medium and transfer it to a glass-bottom Fluorodish culture dish. Position the tissue with its dorsal side facing down and carefully aspirate the bulk of the medium using a P200 pipette: this way some medium remains with the notum, which does not get dehydrated. In this step, make sure that the tissue is positioned roughly in the middle of the remaining liquid, otherwise the fibrin clot will anchor it on the glass on one side only and not firmly enough. Add 1 µl thrombin at 0.5 U/µl and wait 2.5 min; while adding the thrombin, circle with the pipette tip around the tissue to spread the enzyme, so that the clot forms homogeneously and not only on one side of the tissue. Then gently add 12 µl Clone 8 medium + fibrinogen in a ring around the forming clot; thrombin from the clot will diffuse into this ring and trigger polymerization, which will anchor the initial clot on the dish and prevent it from detaching when medium is eventually added. Wait 2.5 min, then delicately add 300 µl Clone 8 medium. The tissue is now ready to be imaged.

One advantage of this fibrin clot is that it anchors the notum to the culture dish and prevents any movement of the tissue; in its absence, movements of the stage or any vibration in the imaging system can induce micrometer-size jumps of the floating tissue, which renders the imaging virtually impossible. Furthermore, up to three nota can be processed simultaneously for the antibody uptake assay and embedded in a single fibrin clot (in which case the volumes described in the clotting protocol should be doubled). This process can significantly increase the throughput of antibody uptake analysis. Moreover, it provides a decisive help for experiments in which there is a low probability to find a suitable cell of interest in a notum, for example, when studying the uptake of Notch or Delta in mutant SOPs in a mosaic background and with a mutation that induces a competitive disadvantage.

2.4. Live imaging and quantification of Notch and Delta trafficking using the antibody uptake assay

2.4.1 Live imaging of Notch and Delta trafficking

With or without the fibrin clot, the geometry of the system dictates the use of an inverted microscope for imaging. Besides, the thickness of the tissue as well as the strong autofluorescence of the cuticle poses the need for optical sectioning. The two simplest solutions are then to work with either a conventional confocal microscope or a spinning-disc confocal microscope. Although both are suitable for live imaging of the antibody uptake assay, a spinning-disc microscope is recommended, for two reasons: (1) imaging with a spinning-disc system induces considerably less photobleaching than with a conventional confocal microscope (this is especially important when using the Alexa647 dye, which bleaches faster than Alexa488 and Alexa568); (2) the much faster acquisition rate of the spinning-disc microscope is best suited to study the rapid dynamics of endocytosis and of endosomal compartments.

The acquisition conditions should be optimized according to the imaging system specifications and the experimental goals. If wanting to image the whole content of internalized Notch or Delta in a cell, z-stacks should be acquired with an interplane distance that is small enough so that consecutive planes overlap and all endosomes are imaged. Laser intensities and exposure times should be set to optimize the signal-to-noise ratio without inducing too much photobleaching, and while making sure that no pixel reaches saturation; the temporal sampling rate has to be determined depending on the goal of the experiment, and also to minimize photobleaching and associated phototoxicity.

Our preferred setup for antibody uptake imaging consists of a Marianas inverted spinning-disc confocal microscope (Intelligent Imaging Innovations). The microscope stand is an inverted AxioObserver Z1 (Zeiss) equipped with a piezo-driven stage. The spinning-disc head is a CSUX-M1 (Yokogawa), and we use an Evolve EM-CCD camera (Photometrics) set at 1×1 binning. Depending on the spatial resolution needed, we use a Plan Apochromat 1.4NA $63 \times$ or $100 \times$ objective. The $100 \times$ objective achieves a better resolution and is essential if wanting to process the images with deconvolution (in which case the Nyquist criterion, i.e., an oversampling of 2 in each spatial dimension, has to be satisfied); however, it collects less light than its $63 \times$ counterpart and necessitates longer exposure times. To monitor the movements of Notch- or Delta-positive endosomes, we typically use exposure times between 50 and 100 ms. We acquire z-stacks spanning 18 μm (and thus encompassing the whole height of a dividing SOP) with an interplane distance of 0.4 μm. If the tracking of individual compartments is not needed we acquire a stack every 20 s, which enables to image a cell during minutes without significant photobleaching. If wanting to follow the dynamics of individual endosomes, one should reduce this time interval; the easiest way to achieve this is to reduce the height of the z-stack and image only a portion of the cell.

2.4.2 Quantification of Notch and Delta trafficking

Using such antibody uptake assays, it has been shown in our lab that internalized Notch and Delta molecules chased between 10 and 40 min are found in a population of endosomes characterized by the adaptor protein Sara (Coumailleau et al., 2009). In the course of SOP mitosis, these Sara endosomes are asymmetrically segregated to the pIIa daughter cell, which contributes to the signaling bias between the pIIa and the pIIb daughter cells. This asymmetric distribution can be visualized using a Sara–GFP reporter, but also with the anti-Notch and anti-Delta antibody uptake assays (Coumailleau et al., 2009); the proportion of Sara–GFP/internalized Notch/internalized Delta endosomes targeted to the pIIa cell in cytokinesis can be measured and amounts to roughly 80% with each of these markers.

We use a custom-written ImageJ macro to semiautomatically measure this quantity in many cells (available upon request to S. L.); this macro proceeds as follows. For each cell, the z-stack corresponding to the first time point after cytokinesis is extracted from the movie and used for the analysis. The mean background value is measured and subtracted from the whole stack; importantly, this value has to be measured in the notum around the cells of interest and not in the extracellular space, so that both the camera noise and the autofluorescence of the tissue are taken into account. Because of noise fluctuations some background signal remains, which is removed by applying a high-pass threshold. A marker is needed to delineate the pIIa and pIIb daughter cells, in a channel that does not cross-talk with the one used by the labeling Fabs. Preferably, this can be a protein present at the plasma membrane of the two daughter cells; a polarity marker asymmetrically inherited by one of the two cells can also be used (such as mRFP-Pon; Coumailleau et al., 2009), but in this case, one has to take care during the acquisition that the signal in the cell that does not preferentially inherit the marker is strong enough to enable the visualization of its plasma membrane. This marker is used to manually delineate the two regions of interest in which the measurements in each z-plane are performed. The macro sums the integrated density in the pIIa and in the pIIb regions of interest to compute the total amount of internalized Notch or Delta in each daughter cell. The proportion of internalized Notch or Delta targeted to the pIIa daughter cell is then finally given by the ratio $I_{pIIa}/(I_{pIIa}+I_{pIIb})$, where I_{pIIa} and I_{pIIb}, respectively, represent the total amounts in each cell.

3. CORRELATIVE IMAGING OF DIVIDING SOP CELLS AND OF THEIR PROGENY

As outlined in Section 1, the Drosophila sensory organ system enables to establish direct correlations between Notch regulation on the one hand, and sensory organ lineage on the other hand. It is furthermore possible to image the division of a given SOP and to determine later on the lineage of that particular cell. This correlative analysis can be done in two different manners that we describe below.

3.1. Correlative whole pupa imaging and scanning electron microscopy of adult flies

The most straightforward way to image SOPs divisions and gain access to sensory organ lineage information is whole pupa imaging followed by scanning electron microscopy (SEM) of adults.

3.1.1 Whole pupa imaging of SOP division

Imaging of SOP division in a whole pupa has been described earlier (Gho et al., 1999). First, stick the pupa on its ventral side to a slide using double-sided scotch tape. Then carefully open the pupal case so that the whole notum and the anterior-most part of the abdomen are exposed. On each side of the pupa, place a spacer made out of four coverslips stuck together. Spread a droplet of Voltalef 10S oil on a glass coverslip and place it on the pupa and the spacers so that the apex of the notum touches the oil. Immobilize the coverslip using nail polish or Valap (a mix of equal masses of vaseline, lanolin, and paraffin). The pupa is ready for imaging.

When spreading the Voltalef 10S oil on the coverslip and applying it, it is crucial that the apex of the notum is immersed, otherwise most of the fluorescence will be diffracted at the glass/cuticle interface. At the same time, it is essential to keep the quantity of oil as small as possible, otherwise the pupa may die in a matter of minutes or hours (depending on the quantity of oil in excess).

Once the pupa is mounted, it can be imaged for several hours. Afterward, carefully remove the top coverslip and place the slide (with the pupa still stuck to it) in a protected and humidity-controlled environment: this can be a box kept at room temperature and stuffed with water-soaked paper, or an incubator at 25 °C. Sensory organ cells will all have been specified within 8 hours after SOP divisions: at that stage, the pupa can be dissected and the notum immunostained to assess for lineages. An alternative possibility is to let the animal reach the adult stage, which occurs around three days and a half after the SOPs divisions: the fly can then be processed for SEM, and the external cells of sensory organs examined.

3.1.2 SEM of adult flies

First, incubate the fly in diethyl ether for 20 min to sacrifice it. Then mount the fly on a SEM holder using double-sided carbon tape (Electron Microscopy Sciences). It is recommended to treat the animal with a gold sputter coater (such as JFC-1200 from JEOL) to avoid imaging artifacts due to charging of the specimen during the acquisition. We use a JSM-6510LV (JEOL) scanning electron microscope in the high vacuum mode with a spot size of 40, an acceleration of 15 kV, and a working distance of 12 mm. In these conditions the external parts of sensory organs (shaft and socket cells) can be readily identified. Perturbations of Notch signaling downstream the SOP division or the pIIa division induce either absent or supernumerary shafts or sockets: the number of shaft and socket cells in each organ thus tells the Notch signaling history of the external part of the organ (see Figure S15 in Coumailleau et al., 2009).

Two types of sensory organs are present on the notum: macrochaete and microchaete. Macrochaete arise from SOPs specified in the wing imaginal disc, and they have already appeared at 12 h APF (Hartenstein & Posakony, 1989). There are four macrochaete in the central part of the notum: the left and right anterior and posterior dorso-central bristles. They can be easily identified, either with an SOP-specific marker (e.g., using Neur-Gal4 > UAS-mRFP-Pon) or with a ubiquitous membrane marker (e.g., Ubi-DEcadherin-GFP). Consequently, these macrochaete can be used as landmarks to correlate dividing SOPs that are very close to them in the pupal notum and sensory organs on the adult notum (Fig. 17.2).

An alternative way of imaging SOP divisions and the external parts of sensory organs has been described (Zitserman & Roegiers, 2011); this method relies on a slightly different mounting of the pupa, and on the autofluorescence of the cuticle to image shaft and socket cells. This method

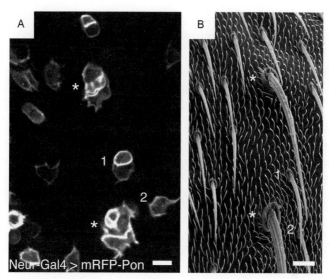

Figure 17.2 Correlative whole pupa imaging and scanning electron microscopy of adult flies. SOPs and pIIa/pIIb siblings have been imaged by spinning-disc confocal microscopy (A) and the corresponding fly has been imaged by scanning electron microscopy (B) as described in the text. Numbers indicate corresponding SOPs or pIIa/pIIb siblings (in A) and sensory organs (in B) that can be assigned without ambiguity. Asterisks indicate anterior and posterior dorso-central macrochaete. Bars: 10 μm (A) and 20 μm (B).

and ours, however, bear the caveats that the correlation between a dividing SOP and its progeny is: (1) limited to the cells in close proximity to dorso-central macrochaete, which is heavily restrictive; (2) prone to misattributions. In particular, SOPs and their daughter cells can undergo significant motion during pupal development (Renaud & Simpson, 2002), which adds uncertainty to the correlation. We have developed an alternative protocol for correlative imaging of SOP division and lineage determination that circumvents these two problems; we describe this method below.

3.2. Correlative antibody uptake assays and immunostainings

This correlative assay relies on the use of the fibrin clot described earlier that enables to cultivate dissected tissues; furthermore, the clot immobilizes the notum in a given position and orientation in which it can be reproductively imaged, first, while SOPs are dividing, and then, after sensory organ cells have been specified. Several folds running for tens of microns always appear in various places in dissected nota; together with the positions of macrochaete, these folds can be used as landmarks enabling to accurately determine which sensory organ was formed by a given SOP.

3.2.1 Reagents to prepare in advance

25 ml PEM is needed per sample. For 1 l PEM: dissolve 27.78 g Disodium PIPES (final concentration: 80 mM), 1.9 g EGTA (final concentration: 5 mM), and 0.2 g $MgCl_2 \times 6H_2O$ (final concentration: 1 mM) in 1 l water, and adjust the pH to 7.4. Store at room temperature.

2 ml *para*-formaldehyde (PFA) solution is needed per sample. For 200 ml PFA solution: dissolve 8 g PFA in 200 ml PEM (final PFA concentration: 4%) by gently stirring and heating. Once dissolved, adjust the pH to 7.4, make 2 ml aliquots and store at $-20\ °C$. PFA needs to be prepared and handled under a chemical hood.

A 20% Triton X-100 stock solution is needed. For a 40 ml stock solution, vigorously mix 8 ml Triton X-100 in 32 ml water and store at room temperature.

Prepare a 2.5 M NH_4Cl stock solution by dissolving 13.37 g NH_4Cl in 100 ml water. Store at room temperature.

3.2.2 Protocol

Process the pupa for the antibody uptake assay as described earlier. It is essential for the correlative assay to use the fibrin clot embedding described earlier, otherwise the notum will fold around itself in a matter of hours and become unusable for sensory organ analysis. Once the fibrin clot is polymerized, add 500 μl Clone 8 medium and proceed with the imaging of SOP divisions. While installing the mounted tissue under the microscope, it is strongly recommended to position it with an orientation that can be roughly reproduced in the second imaging session that will follow the immunostaining (see below), for example, with the antero-posterior axis of the notum in the axis of the microscope stand.

After having imaged the desired number of SOPs, add 1.5 ml Clone 8 medium in the culture dish and place it in the dark in a humidity-controlled incubator at 25 °C until sensory organ cells are fully specified (around 25 h APF).

Process the notum for fixation and immuno-histochemistry. First, remove the Clone 8 medium and add 1 ml PFA solution; incubate 30 min. This incubation and all the subsequent ones are done in the dark with gentle rocking and, unless stated otherwise, at room temperature. In the meantime, prepare 1 ml PFA-T by adding 20 μl of the 20% Triton

X-100 solution to 1 ml PFA solution (final Triton X-100 concentration: 0.4%, v/v).

Remove the PFA solution, and add 1 ml PFA-T; incubate 30 min. In the meantime, prepare:1 ml NH$_4$Cl solution: add 20 µl NH$_4$Cl 2.5 M to 1 ml PEM (final NH$_4$Cl concentration: 50 mM). 25 ml PEM-T: add 500 µl Triton 20% to 24.5 ml PEM and stir vigorously (final Triton X-100 concentration: 0.4%, v/v).

3 ml PEM-T-BSA: weigh 0.06 g BSA and dissolve in 3 ml PEM-T (final BSA concentration: 2%, m/v). For optimal results fresh solutions of PFA-T, PEM-T, and PEM-T-BSA should be used for each experiment.

Remove the PFA-T, rinse two times with 1 ml PEM-T, then add 1 ml NH$_4$Cl solution and incubate 10 min to quench any remaining PFA.

Remove the NH$_4$Cl solution and rinse two times with 1 ml PEM-T.

Remove the PEM-T and add 1 ml PEM-T-BSA. Incubate overnight at 4 °C. In the meantime, prepare the primary antibody; it should be diluted in PEM-T-BSA, and 500 µl is needed per culture dish.

Remove the PEM-T-BSA and add 500 µl primary antibody. Incubate 5 h.

Remove the primary antibody and rinse four times with 1 ml PEM-T. Then wash three times with PEM-T during 10, 20, and 30 min, respectively. In the meantime, prepare 2 ml PEM-T-NGS by adding 40 µl Normal Goat Serum (Vector Laboratories; final concentration: 2%, v/v) to 2 ml PEM-T.

Remove the PEM-T and add 1 ml PEM-T-NGS. Incubate overnight at 4 °C. In the meantime, prepare the secondary antibody; it should be diluted in PEM-T-NGS, and 500 µl is needed per culture dish.

Remove the PEM-T-NGS and add 500 µl secondary antibody. Incubate 5 h.

Rinse four times with 1 ml PEM-T, then wash three times with PEM-T during 10, 20, and 30 min, respectively. Remove the PEM-T, rinse once with PEM, then wash with PEM and incubate 5 min. The tissue is now ready for imaging.

Interestingly, the Alexa fluorophores used in the antibody uptake assay are degraded during the ∼8 h of culture of the notum, which allows using their channel for one of the lineage markers. The markers commonly used for lineage analysis are the following: mouse anti-Elav (9F8A9 from the Developmental Studies Hybridoma Bank) and rat anti-Elav (7E8A10 from the Developmental Studies Hybridoma Bank) to visualize neurons, mouse anti-Prospero (MR1A from the Developmental Studies Hybridoma Bank) to visualize sheath cells, and rat anti-Su(H) (from the laboratory of F. Schweisguth) for socket cells. A variation on the protocol is to stain for pIIa and pIIb cells; in this case, after SOP imaging, the notum should be kept in culture at 25 °C until 18 h APF (instead of 25 h APF as described earlier). The markers that can be used are rabbit anti-Pdm1 (from the laboratory of T. Préat) for pIIa cells and mouse anti-Prospero for pIIb cells. In all cases, rabbit anti-Sanpodo (from the laboratory of J. Skeath), guinea pig anti-Senseless (from the laboratory of H. Bellen), mouse anti-Achaete (Developmental Studies Hybridoma Bank), or mouse anti-Cut (2B10 from the Developmental Studies Hybridoma Bank) can be used to stain SOPs and their progeny.

Although the immunostaining can be observed using any confocal or spinning-disc microscope, it is strongly advised to use the same setup that was used for the imaging of SOPs divisions. Orient the tissue under the microscope in the same position as it was in the first imaging session (e.g., with the antero-posterior axis of the notum in the axis of the microscope stand). Under these conditions, it is very easy to correlate dividing SOPs observed in the first part of the experiment and immunostained sensory organs observed in the second part by using the folds of the tissue as landmarks: several folds running for tens of microns always appear in various places in dissected nota and can be used to accurately determine which sensory organ was formed by a given SOP (Fig. 17.3). Using different microscopes for the two imaging sessions can render this determination slightly more difficult, as the appearance of the folds may change depending on focal plane depth in particular.

4. CONCLUDING REMARKS

In this chapter, we describe how to perform live antibody uptake assays to image the trafficking of endocytosed Notch and Delta proteins. We routinely use this technique for the study of asymmetric cell division in the pupal notum; however, this simple procedure can be applied to any other fly tissue accessible to antibodies. Furthermore, we describe how to do a correlative analysis of Notch/Delta live imaging in a dividing SOP and of the lineage of this SOP. Importantly, this correlative assay enables to test the effect of chemical compounds on the trafficking of endogenous Notch or Delta and the putative effect of these compounds on Notch signaling and cell fate determination. The availability of a correlative assay

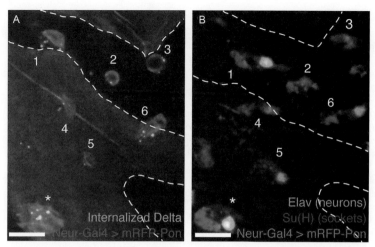

Figure 17.3 Correlative antibody uptake assay and immunostaining. The anti-Delta antibody uptake assay (A) and the correlative immunostaining (B) have been performed as described in the text and observed by spinning-disc confocal microscopy. Numbers indicate corresponding SOPs or pIIa/pIIb siblings (in A) and sensory organs (in B); the asterisks indicate a dorso-central macrochaete. Dotted lines indicate folds of the tissue. Bars: 15 μm. (See color plate.)

appears particularly useful in the case of compounds (or genetic treatments, such as RNA interference) that have a limited penetrance: this assay enables for example to correlate a change in cell fate decision with a certain threshold of disturbance of Notch or Delta dynamics.

Combined with the wide range of genetic and chemical tools available in Drosophila research, these two methods will provide a better understanding of the role played by endocytic proteins and endosomal trafficking in Notch regulation, in terms of botch Notch trafficking and Notch signaling output.

ACKNOWLEDGEMENTS

We are grateful to J. Januschke and C. Gonzalez (IRB, Barcelona) for teaching us the use of fibrin clots for primary cultures. We thank E. Derivery for critical reading of the manuscript and the bioimaging platform for technical help. This work was supported by the SNSF, the Swiss SystemsX.ch initiative, the SystemsX epiPhysX Grant, the ERC, the NCCR Frontiers in Genetics and Chemical Biology Programs, and the Polish-Swiss research program. S. L. was funded by a Marie Curie post-doctoral fellowship (FP7).

REFERENCES

Artavanis-Tsakonas, S., & Muskavitch, M. A. (2010). Notch: The past, the present, and the future. *Current Topics in Developmental Biology, 92,* 1–29.

Banks, S. M., Cho, B., Eun, S. H., Lee, J. H., Windler, S. L., Xie, X., et al. (2011). The functions of auxilin and Rab11 in Drosophila suggest that the fundamental role of ligand endocytosis in notch signaling cells is not recycling. *PLoS One, 6,* e18259.

Benhra, N., Lallet, S., Cotton, M., Le Bras, S., Dussert, A., & Le Borgne, R. (2011). AP-1 controls the trafficking of Notch and Sanpodo toward E-cadherin junctions in sensory organ precursors. *Current Biology, 21,* 87–95.

Benhra, N., Vignaux, F., Dussert, A., Schweisguth, F., & Le Borgne, R. (2010). Neuralized promotes basal to apical transcytosis of delta in epithelial cells. *Molecular Biology of the Cell, 21,* 2078–2086.

Berdnik, D., Torok, T., Gonzalez-Gaitan, M., & Knoblich, J. A. (2002). The endocytic protein alpha-Adaptin is required for numb-mediated asymmetric cell division in Drosophila. *Developmental Cell, 3,* 221–231.

Coumailleau, F., Furthauer, M., Knoblich, J. A., & Gonzalez-Gaitan, M. (2009). Directional Delta and Notch trafficking in Sara endosomes during asymmetric cell division. *Nature, 458,* 1051–1055.

Couturier, L., Vodovar, N., & Schweisguth, F. (2012). Endocytosis by Numb breaks Notch symmetry at cytokinesis. *Nature Cell Biology, 14,* 131–139.

Currie, D. A., Milner, M. J., & Evans, C. W. (1988). The growth and differentiation *in vitro* of leg and wing imaginal disc cells from *Drosophila melanogaster. Development, 102,* 805–814.

Emery, G., Hutterer, A., Berdnik, D., Mayer, B., Wirtz-Peitz, F., Gonzalez-Gaitan, M., et al. (2005). Asymmetric Rab 11 endosomes regulate delta recycling and specify cell fate in the Drosophila nervous system. *Cell, 122,* 763–773.

Entchev, E. V., Schwabedissen, A., & Gonzalez-Gaitan, M. (2000). Gradient formation of the TGF-beta homolog Dpp. *Cell, 103,* 981–991.

Forer, A., & Pickett-Heaps, J. D. (1998). Cytochalasin D and latrunculin affect chromosome behaviour during meiosis in crane-fly spermatocytes. *Chromosome Research, 6,* 533–549.

Furthauer, M., & Gonzalez-Gaitan, M. (2009). Endocytic regulation of notch signalling during development. *Traffic, 10,* 792–802.

Gho, M., Bellaiche, Y., & Schweisguth, F. (1999). Revisiting the Drosophila microchaete lineage: A novel intrinsically asymmetric cell division generates a glial cell. *Development, 126,* 3573–3584.

Gho, M., Lecourtois, M., Geraud, G., Posakony, J. W., & Schweisguth, F. (1996). Subcellular localization of Suppressor of Hairless in Drosophila sense organ cells during Notch signalling. *Development, 122,* 1673–1682.

Giagtzoglou, N., Yamamoto, S., Zitserman, D., Graves, H. K., Schulze, K. L., Wang, H., et al. (2012). dEHBP1 controls exocytosis and recycling of Delta during asymmetric divisions. *Journal of Cell Biology, 196,* 65–83.

Guruharsha, K. G., Kankel, M. W., & Artavanis-Tsakonas, S. (2012). The Notch signalling system: Recent insights into the complexity of a conserved pathway. *Nature Reviews Genetics, 13,* 654–666.

Hartenstein, V., & Posakony, J. W. (1989). Development of adult sensilla on the wing and notum of *Drosophila melanogaster. Development, 107,* 389–405.

Hori, K., Sen, A., Kirchhausen, T., & Artavanis-Tsakonas, S. (2011). Synergy between the ESCRT-III complex and Deltex defines a ligand-independent Notch signal. *Journal of Cell Biology, 195,* 1005–1015.

Hutterer, A., & Knoblich, J. A. (2005). Numb and alpha-Adaptin regulate Sanpodo endocytosis to specify cell fate in Drosophila external sensory organs. *EMBO Reports, 6,* 836–842.

Januschke, J., & Gonzalez, C. (2010). The interphase microtubule aster is a determinant of asymmetric division orientation in Drosophila neuroblasts. *Journal of Cell Biology, 188*, 693–706.

Jauffred, B., & Bellaiche, Y. (2012). Analyzing frizzled signaling using fixed and live imaging of the asymmetric cell division of the Drosophila sensory organ precursor cell. *Methods in Molecular Biology, 839*, 19–25.

Kawahashi, K., & Hayashi, S. (2010). Dynamic intracellular distribution of Notch during activation and asymmetric cell division revealed by functional fluorescent fusion proteins. *Genes to Cells, 15*, 749–759.

Knust, E., & Campos-Ortega, J. A. (1989). The molecular genetics of early neurogenesis in Drosophila melanogaster. *Bioessays, 11*, 95–100.

Kopan, R., & Ilagan, M. X. (2009). The canonical Notch signaling pathway: Unfolding the activation mechanism. *Cell, 137*, 216–233.

Le Borgne, R., & Schweisguth, F. (2003). Unequal segregation of Neuralized biases Notch activation during asymmetric cell division. *Developmental Cell, 5*, 139–148.

Le Bras, S., Rondanino, C., Kriegel-Taki, G., Dussert, A., & Le Borgne, R. (2012). Genetic identification of intracellular trafficking regulators involved in Notch-dependent binary cell fate acquisition following asymmetric cell division. *Journal of Cell Science, 125*, 4886–4901.

Musse, A. A., Meloty-Kapella, L., & Weinmaster, G. (2012). Notch ligand endocytosis: Mechanistic basis of signaling activity. *Seminars in Cell & Developmental Biology, 23*, 429–436.

Pasternak, S. H., Bagshaw, R. D., Guiral, M., Zhang, S., Ackerley, C. A., Pak, B. J., et al. (2003). Presenilin-1, nicastrin, amyloid precursor protein, and gamma-secretase activity are co-localized in the lysosomal membrane. *Journal of Biological Chemistry, 278*, 26687–26694.

Purow, B. (2012). Notch inhibition as a promising new approach to cancer therapy. *Advances in Experimental Medicine and Biology, 727*, 305–319.

Renaud, O., & Simpson, P. (2002). Movement of bristle precursors contributes to the spacing pattern in Drosophila. *Mechanisms of Development, 119*, 201–211.

Strigini, M., & Cohen, S. M. (2000). Wingless gradient formation in the Drosophila wing. *Current Biology, 10*, 293–300.

Tagami, S., Okochi, M., Yanagida, K., Ikuta, A., Fukumori, A., Matsumoto, N., et al. (2008). Regulation of Notch signaling by dynamic changes in the precision of S3 cleavage of Notch-1. *Molecular and Cellular Biology, 28*, 165–176.

Vaccari, T., Lu, H., Kanwar, R., Fortini, M. E., & Bilder, D. (2008). Endosomal entry regulates Notch receptor activation in Drosophila melanogaster. *Journal of Cell Biology, 180*, 755–762.

Vetrivel, K. S., Cheng, H., Lin, W., Sakurai, T., Li, T., Nukina, N., et al. (2004). Association of gamma-secretase with lipid rafts in post-Golgi and endosome membranes. *Journal of Biological Chemistry, 279*, 44945–44954.

Windler, S. L., & Bilder, D. (2010). Endocytic internalization routes required for delta/notch signaling. *Current Biology, 20*, 538–543.

Zitserman, D., & Roegiers, F. (2011). Live-cell imaging of sensory organ precursor cells in intact Drosophila pupae. *Journal of Visualized Experiments, 51*, e2706. http://dx.doi.org/10.3791/2706.

CHAPTER EIGHTEEN

Toll-Interacting Protein Pathway: Degradation of an Ubiquitin-Binding Protein

Miho Shimizu[*,†,1], **Asami Oguro-Ando**[‡], **Eri Ohoto-Fujita**[*,†,1], **Yoriko Atomi**[*,†,1,2]

[*]Graduate School of Information Science and Technology, The University of Tokyo, Tokyo, Japan
[†]Radioisotope Center, Cell to Body Dynamics Laboratory 1, The University of Tokyo, Tokyo, Japan
[‡]Department of Neuroscience and Pharmacology, Rudolf Magnus Institute of Neuroscience, University Medical Centre Utrecht, Utrecht, The Netherlands
[1]Present address: Material Health Science Laboratory, Department of Organic and Polymer Materials Chemistry, Tokyo University of Agriculture and Technology, Tokyo, Japan
[2]Corresponding author: e-mail address: atomi@bio.c.u-tokyo.ac.jp

Contents

1. Introduction 324
2. Expression of GFP-htt Protein and RFP-Tollip and How to Count Aggregation 325
   2.1 Tollip expression vector 325
   2.2 Htt expression constructs 325
   2.3 Knockdown of Tollip 325
   2.4 Detection of cell death 326
3. Cytoskeleton-Dependent Transport of Tollip to htt^{PQ}-Aggresome 327
   3.1 Biochemical analysis 327
   3.2 Microscopic analysis 327
4. Determinants of Early-/Late Endosomal Localization in the Cells 328
5. Summary 328
Acknowledgment 329
References 329

Abstract

The nine neurodegenerative disorders including Huntington disease (HD) are caused by the expansion of a trinucleotide CAG repeats (polyQ), which are located within the coding of the affected gene. Previous studies suggested that a gain of toxic function by polyQ repeats is widely thought to have a major role in pathogenesis. PolyQ-expanded htt induced ubiquitinated aggregates cause cell death in neuronal cells. Using a HD cellular model, we demonstrate that Tollip protects cells against the toxicity of polyQ-expanded htt and also protects cells from death (Oguro, Kubota, Shimizu, Ishiura, & Atomi, 2011). Tom1 which belongs to the VHS domain-containing protein family is also found to be directly binding to ubiquitin chains and Tollip (Katoh et al.,

2004; Yamakami, Yoshimori, & Yokosawa, 2003). Tollip recruits misfolded protein to aggresome via late endosome. The cell system can be used to determine if your protein of interest is controlled under a part of Tollip pathway or not among other cell homeostatic systems: molecular chaperones, autophagy, and endoplasmic reticulum (ER)-associated degradation (ERAD). Tollip can be used for polyQ cell toxicity sensor by detecting microtubule-dependent trafficking and aggresome colocalization of aggregated protein.

1. INTRODUCTION

The ubiquitin-binding protein Tollip (Toll-interacting protein) works as an endogenous "cellular drug," and overexpression of Tollip is proposed to protect neuronal cells from toxic misfolded protein aggregate through endosomal signaling (Oguro, Kubota, Shimizu, Ishiura, & Atomi, 2011). Tollip is a ubiquitously expressed conserved protein from vertebrate (human, mouse, rat, zebrafish, Xenopus) to worm (TLI-1 in *Caenorhabditis elegans*) and an inhibitory adaptor protein within Toll-like receptor (TLR) signaling (Bulut, Faure, Thomas, Equils, & Arditi, 2001). The TLR pathway is a part of the innate immune system that recognizes structurally conserved molecular patterns of microbial pathogens, leading to an inflammatory immune response. There is accumulating evidence that Toll/TLR signaling is linking to endocytic pathway (Lund & DeLotto, 2011).

In this endosomal trafficking system, various adapter proteins are involved. One of such protein is Tom1 which belongs to the VHS domain-containing family is also found to be directly binding to ubiquitin chains and Tollip (Katoh et al., 2004; Yamakami, Yoshimori, & Yokosawa, 2003). Huntington disease (HD) is a progressive autosomal dominant neurodegenerative disorder caused by expansion of polyglutamine (polyQ) in the huntingtin (htt) protein. Interestingly, Tollip is reported to be concentrated in polyQ aggregates (Doi et al., 2004), and the ubiquitin-binding protein p62 (also known as sequestosome 1) is known to mediate autophagy-dependent clearance of polyQ aggregates with accelerating httpQ aggregation (Komatsu et al., 2007). Recently, we showed that Tollip increases polyQ protein aggregates, but strangely protect cell death (Oguro et al., 2011). Tollip both binds ubiquitin protein via CUE domain and substrates via TDR facilitating unfolded/aggregate-proned protein to protein degradation through endocytic sorting on microtubule cytoskeleton. Neurodegenerative disease-related protein interaction of Tollip *in vivo* may include ATXN1 (Lim et al., 2006). Protein synthesis and degradation is essential for cell survival. Cytotoxic protein aggregates are involved in diverse neurodegenerative diseases such as polyQ diseases (HD, DRPLA, SCA), Alzheimer's disease, Parkinson's diseases, amyotrophic lateral sclerosis, and prion disease (Ross & Poirier, 2004). Although exact mechanisms are still unknown for the disease onset, efforts for developing a medical cure are intensive. Considering misfolded and aggregated polyQ as the excellent target for specific therapeutics, use of polyQ binding protein (QBP1) for enhancing toxic protein degradation through molecular chaperon has been successful at least in mouse model (Bauer et al., 2010; Nagai et al., 2000). This chapter will explain how you can study misfolded/aggregated protein sequestering machinery linked to endosomal signaling on microtubule cytoskeleton *in vitro* using expanded polyQ involving HD cellular model.

2. EXPRESSION OF GFP-htt PROTEIN AND RFP-TOLLIP AND HOW TO COUNT AGGREGATION

2.1. *Tollip* expression vector

Tollip overexpression is achieved by transiently transfecting with an RFP-Tollip expression vector [mouse Tollip cloned into the RFPc1 (Invitrogen)]. Untagged cDNA Tollip ORF clones in CMV expression vector are commercially available (e.g., OriGene Technologies, Inc., Rockville, MD). Anti-Tollip antibody is commercially available.

2.2. *Htt* expression constructs

The *htt* expression constructs, pIND-tNhtt-EGFP-60Q and pIND-tNhtt-EGFP-150Q, and the generation of the stable Neuro2a cell lines expressing http proteins are provided by Dr. Nukina (Jana & Nukina, 2005).

2.3. Knockdown of *Tollip*

1. To knockdown Tollip, cells were transiently transfected with Tollip stealth siRNA duplex oligoribonucleotides, 5-UCUCAAGGUAGA-ACGAGUCCACACC-3 and 5-GGUGUGGACUCGUUCUACC-UUGAGA-3.

2. To assess if levels of Tollip affected aggregate formation, cells were transiently cotransfected with either GFP-Tollip (or empty GFP cassette) and htt (20Q, 80Q, or 87Q) exon1 fused with a V5 tag.
3. All transfections are performed using the Lipofectamine 2000 reagent (Invitrogen).
4. Twelve hours after transfection, 1 mM of ponasterone A (Invitrogen) for induction of aggregation is added to the culture and then incubated for an additional 24 h.
5. Aggregate counting experiments were performed after cells were transiently transfected with Tollip stealth siRNA duplex or plasmid expression vector (transfection efficiency was almost 90% in Neuro2a cells). Aggregate containing cells are seeded into chamber slides, and aggregate containing cells are manually determined using a fluorescence microscope.

2.4. Detection of cell death

1. To test cell death, 5×10^5 cells were inoculated into each well of six-well plates, 48 h following transfection.
2. One day before transfection, place cells in 500 μl of growth medium without antibiotics such that they will be 30–50% confluent at the time of transfection.
3. Dilute each 1 μg DNA in 50 μl Opti-MEMR I Reduced Serum Medium without serum (final concentration of RNA when added to the cells is 33 nM). Mix Lipofectamine™ 2000 gently before use and then dilute 1 μl in 50 μl Opti-MEMR I Reduced Serum Medium. Mix gently and incubate for 5 min at room temperature. After the 5-min incubation, combine the diluted oligomer with the diluted Lipofectamine™ 2000. Mix gently and incubate for 20 min at room temperature. Add the oligomer–Lipofectamine™ 2000 complexes to each well containing cells and medium.
4. After transfection, cells were differentiated with 5 mM dibutyryl cyclic AMP in the medium and allowed to incubate for 3 days.
5. Aggregate containing cells were counted by Image J software and more than 200 cells were photographed using a fluorescence microscope. Dead cells with aggregations were counted by propidium iodide staining. Counting method's detail is provided in Image J instruction from NIH. (Do not fix the samples.) Also, you can use them for the cell titer blue cell viability assay (Promega).

3. CYTOSKELETON-DEPENDENT TRANSPORT OF TOLLIP TO htt^{PQ}-AGGRESOME

3.1. Biochemical analysis

1. After neuro2a cells are cultured over 24 h with MG132, proteasome inhibitor, which induces protein aggregates, at the final concentration of 0.5 mM, cells are harvested after keeping on ice, lysed in PBS, and supplemented with protease inhibitor cocktail (Sigma) on ice. Cell lysates are briefly sonicated, centrifuged for 10 min at 15,000 × g at 4 °C, and supernatants (soluble fraction) and pellet (insoluble fraction) are analyzed and fractionated with centrifugation.
2. Proteins in soluble and insoluble fractions are separated in SDS-PAGE and are analyzed with western blotting using anti-Tollip antibody and also anti-RFP antibody.
3. Maker of insoluble fraction is recognized with vimentin antibody and GAPDH is available for soluble fraction.
4. Relation of microtubule, which is used to transfer aggresome, is examined using nocodazole, which disassembles microtubule at the concentration of 10 mM for 6 h.
5. Similarly proteins in soluble and insoluble fractions are separated in SDS-PAGE and are analyzed with western blotting using anti-Tollip antibody and also anti-RFP antibody.

3.2. Microscopic analysis

1. After neuro2a cells are cultured over 24 h with MG132, proteasome inhibitor, which induces protein aggregates, at the final concentration of 0.5 mM, cells were stained with DAPI for cell nuclei, anti-Tollip antibody, antivimentin antibody.
2. Transport via microtubule, which is used to transfer aggresome, is examined using nocodazole, which disassembles microtubule at the concentration of 10 mM for 6 h.
3. Cells cultured under four conditions of control, no treatment, MG-132 treatment, nocodazole treatment, and both MG-132 and nocodazole treatments, are fixed with 4% formaldehyde and stained with various antibodies described earlier.
4. Colocalization between Tollip and vimentin is examined with fluorescent microscopy.

4. DETERMINANTS OF EARLY-/LATE ENDOSOMAL LOCALIZATION IN THE CELLS

1. For immunofluorescence experiments, cells were fixed with 4% *paraformaldehyde* in PBS for 20 min and blocked with 0.2% BSA in TBST (Tris-Buffered Saline Tween-20) for 1 h.
2. Fixed cells were incubated with antibodies against Tollip, vimentin (mouse monoclonal, Abcam), EEA1 (mouse monoclonal, BD transduction, early endosome marker), or syntaxin-7 (rabbit polyclonal, Abcam, late endosome marker) at 1:50 dilution (4 °C, overnight). After several washes with TBST, cells were incubated with Alexa488- or Alexa546-conjigated secondary antibodies (1:2000) for 1 h. After washes, cells were mounted in antifade solution.
3. Several other choices for endosomal markers either early and/or late are found in Mukhopadhyay et al. (2011).

5. SUMMARY

Protein misfolding/aggregate which appeared in various neurodegenerative diseases such as polyQ diseases, Altzheimer's diseases, and Parkinson's disease enhance endocytic pathway for denatured protein clearance in couple with protein quality control machinery (Fig. 18.1). Tollip is a bona fide endogenous cellular protective molecule ("natural drug") and

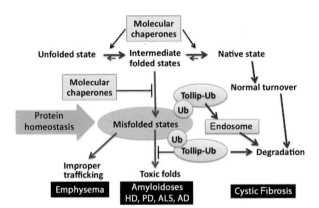

Figure 18.1 Interplay between protein quality control (transition of unfolded to intermediates to native states) clearance mechanism by the ubiquitin-binding protein Tollip as a new player in protein conformation disease. *Modified from Morimoto (2008).* (For color version of this figure, the reader is referred to the online version of this chapter.)

overexpression may be beneficial to human health. Inactivity along with high-fat diet stimulates IL-1RI-mediated inflammatory signaling in muscle and the level of Tollip expression is significantly reduced (Kim et al., 2010). On the other hand, autophagy, another protein clearance mechanism is found to be important for proper homeostatic response by endurance exercise (He et al., 2012). Thus, stimulating endosomal trafficking by exercise and nutrition (e.g., supplement) may contribute to not only preventing suffering from neurodegenerative diseases but also maintaining human body in a good health.

ACKNOWLEDGMENT
We thank the Radio Isotope Center, The University of Tokyo for providing wonderful research environment for past 3 years.

REFERENCES
Bauer, P. O., Goswami, A., Wong, H. K., Okuno, M., Kurosawa, M., Yamada, M., et al. (2010). Harnessing chaperone-mediated autophagy for the selective degradation of mutant huntingtin protein. *Nature Biotechnology*, *28*, 256–263.
Bulut, Y., Faure, E., Thomas, L., Equils, O., & Arditi, M. (2001). Cooperation of Toll-like receptor 2 and 6 for cellular activation by soluble tuberculosis factor and *Borrelia burgdorferi* outer surface protein A lipoprotein: Role of Toll-interacting protein and IL-1 receptor signaling molecules in Toll-like receptor 2 signaling. *Journal of Immunology*, *167*, 987–994.
Doi, H., Mitsui, K., Kurosawa, M., Machida, Y., Kuroiwa, Y., & Nukina, N. (2004). Identification of ubiquitin-interacting proteins in purified polyglutamine aggregates. *FEBS Letters*, *571*, 171–176.
He, C., Bassik, M. C., Moresi, V., Sun, K., Wei, Y., Zou, Z., et al. (2012). Exercise-induced BCL2-regulated autophagy is required for muscle glucose homeostasis. *Nature*, *18*, 511–515.
Jana, N. R., & Nukina, N. (2005). BAG-1 associates with the polyglutamine-expanded huntingtin aggregates. *Neuroscience Letters*, *378*, 171–175.
Katoh, Y., Shiba, Y., Mitsuhashi, H., Yanagida, Y., Takatsu, H., & Nakayama, K. (2004). Tollip and Tom1 form a complex and recruit ubiquitin-conjugated proteins onto early endosomes. *Journal of Biological Chemistry*, *279*, 24435–24443.
Kim, S. J., Choi, Y., Jun, H. S., Kim, B. M., Na, H. K., Surh, Y. J., et al. (2010). High-fat diet stimulates IL-1 type I receptor-mediated inflammatory signaling in the skeletal muscle of mice. *Molecular Nutrition & Food Research*, *54*, 1014–1020.
Komatsu, M., Waguri, S., Koike, M., Sou, Y. S., Ueno, T., Hara, T., et al. (2007). Homeostatic levels of p62 control cytoplasmic inclusion body formation in autophagy-deficient mice. *Cell*, *131*, 1149–1163.
Lim, J., Hao, T., Shaw, C., Patel, A. J., Szabó, G., Rual, J. F., et al. (2006). A protein–protein interaction network for human inherited ataxias and disorders of Purkinje cell degeneration. *Cell*, *19*, 801–814.
Lund, V. K., & DeLotto, R. (2011). Regulation of Toll and Toll-like receptor signaling by the endocytic pathway. *Small GTPases*, *2*, 95–98.
Morimoto, R. (2008). Proteotoxic stress and inducible chaperone networks in neurodegenerative disease and aging. *Genes & Development*, *22*, 1427–1438.

Mukhopadhyay, A., Nieves, E., Che, F. Y., Wang, J., Jin, L., Murray, J. W., et al. (2011). Proteomic analysis of endocytic vesicles: Rab1a regulates motility of early endocytic vesicles. *Journal of Cell Science, 124*(Pt. 5), 765–775.

Nagai, Y., Tucker, T., Ren, H., Kenan, D. J., Henderson, B. S., Keene, J. D., et al. (2000). Inhibition of polyglutamine protein aggregation and cell death by novel peptides identified by phage display screening. *Journal of Biological Chemistry, 275*, 10437–10442.

Oguro, A., Kubota, H., Shimizu, M., Ishiura, S., & Atomi, Y. (2011). Protective role of the ubiquitin binding protein Tollip against the toxicity of polyglutamine-expansion proteins. *Neuroscience Letters, 503*, 234–239.

Ross, C. A., & Poirier, M. A. (2004). Protein aggregation and neurodegenerative disease. *Nature Medicine, 10*, S10–S17.

Yamakami, M., Yoshimori, T., & Yokosawa, H. (2003). Tom1, a VHS domain-containing protein, interacts with Tollip, Ubiquitin, and Clathrin. *Journal of Biological Chemistry, 278*, 52865–52872.

CHAPTER NINETEEN

Measuring Interactions of FERM Domain-Containing Sorting Nexin Proteins with Endosomal Lipids and Cargo Molecules

Rajesh Ghai[*], Mehdi Mobli[†], Brett M. Collins[*,1]

[*]Institute for Molecular Bioscience, University of Queensland, St. Lucia, Queensland, Australia
[†]Centre for Advanced Imaging & School of Chemistry and Molecular Biosciences, The University of Queensland, St. Lucia, Queensland, Australia
[1]Corresponding author: e-mail address: b.collins@imb.uq.edu.au

Contents

1. Introduction 332
2. Secondary Structure-Based Domain Classification 333
   2.1 Identification of functional modules of PX-FERM proteins 334
3. Production of Recombinant PX-FERM Proteins 335
   3.1 Recombinant PX-FERM protein expression 335
   3.2 Purification of recombinant PX-FERM proteins 336
4. Measuring Interactions of PX-FERM Proteins with Phosphoinositide Lipids and Peptide Cargo Motifs by Isothermal Titration Calorimetry 338
5. Analysis of SNX17 PX Domain-PI3P Interactions by Nuclear Magnetic Resonance Spectroscopy 341
   5.1 ^{15}N and ^{13}C isotopic labeling of the SNX17 PX domain 342
   5.2 NMR data acquisition 343
   5.3 SNX17 PX domain-PI3P NMR titrations 344
   5.4 Identification of the PI3P-binding site using hybrid of structural methods 346
6. Summary 347
Acknowledgments 347
References 347

Abstract

Endosomal recycling pathways regulate cellular homeostasis via the transport of internalized material back to the plasma membrane. Phox homology (PX) and band 4.1/ezrin/radixin/moesin (FERM) domain-containing proteins are a recently identified subfamily of PX proteins that are critical for the recycling of numerous transmembrane cargo molecules. The PX-FERM subfamily includes three endosome-associated proteins called sorting nexin (SNX) 17, SNX27, and SNX31. These are modular peripheral membrane proteins that act as central scaffolds mediating protein–lipid interactions, cargo binding, and regulatory protein recruitment. This chapter outlines the methodology employed to classify the PX-FERM family using combined bioinformatics and structure prediction tools. It further details the application of isothermal titration calorimetry and nuclear magnetic resonance spectroscopy to understand the mechanisms that underpin their endosomal membrane recruitment and subsequent recognition of NPxY/NxxY peptide sorting motifs, present in many cargo receptors and required for their trafficking. It is now increasingly recognized that the formation of a stable trafficking complex is dictated by a multitude of coordinated protein–protein and protein–lipid interactions, and the approaches highlighted here will be useful for future studies aimed at understanding these biomolecular interactions in greater detail.

1. INTRODUCTION

The phox homology (PX) domain-containing family of proteins is characterized by the presence of the hallmark PX domain. These molecules are often referred to as sorting nexins (SNXs), a term that originally reflected the involvement of the founding member SNX1 in sorting of the epidermal growth factor receptor (Kurten, Cadena, & Gill, 1996). Subsequent studies have shown that many, if not most, PX proteins also regulate intracellular trafficking and signaling pathways (Cullen, 2008; Teasdale & Collins, 2012).

SNX17 is the founding member of the "PX-FERM" subfamily of the PX domain-containing molecules. SNX17 is a critical regulator of endosome to cell surface recycling of a variety of transmembrane receptor cargos. The PX-FERM subfamily has three members, SNX17, SNX27, and SNX31. These proteins possess the defining PX domain and a band 4.1/ezrin/radixin/moesin (FERM) domain with an atypical structure (Ghai, Falconer, & Collins, 2012; Ghai et al., 2011). In addition, SNX27 also contains an N-terminal class I PSD-95/disks large/zonula occludens (PDZ) domain, making it unique within the PX superfamily.

The PX-FERM proteins are restricted to early endosomal organelles and from there participate in the regulation of intracellular trafficking, and potentially signaling outcomes, in a coordinated manner. These proteins are highly modular scaffolds, possessing multiple domains with distinct functions in membrane and protein–protein interactions. The PX domain binds the phosphoinositide lipid phosphatidylinositol-3-phosphate (PI3P), an interaction that facilitates recruitment of these proteins to the endosome. The C-terminal FERM domain recognizes specific cargo sorting signals or peptide motifs (NPxY/NxxY) in the cytoplasmic tails of various receptors such as the amyloid precursor protein (APP), the adhesion receptor P-selectin, members of the lipoprotein receptor family such as low-density lipoprotein receptor, and a number of β-integrins to modulate their endosome to cell surface recycling (Bottcher et al., 2012; Czubayko, Knauth, Schluter, Florian, & Bohnensack, 2006; Donoso et al., 2009; Knauth et al., 2005; Lee et al., 2008; Steinberg, Heesom, Bass, & Cullen, 2012; Stockinger et al., 2002; van Kerkhof et al., 2005). The SNX27 PDZ domain, furthermore, binds class I PDZ-binding motifs (PDZbm) present in the intracellular region of numerous cargo receptors including G-protein coupled receptors and ion channels (Joubert et al., 2004; Lunn et al., 2007; Temkin et al., 2011). Here, we expand on methods employed to identify the PX–FERM family and how this identification assisted in the structural and functional characterization of each of the domains and subdomains using biochemical, biophysical, structural, and computational methods.

2. SECONDARY STRUCTURE-BASED DOMAIN CLASSIFICATION

Assignment of functional modules in a protein is obviously essential for predicting its overall functional properties, and allows for the accurate identification of related proteins with similar characteristics. The previous sequence-based domain annotations of the PX–FERM proteins, however, incorrectly assigned them with either C-terminal B41 or Ras association (RA) domains (Ghai et al., 2011; Teasdale & Collins, 2012). Described in this section is an approach combining both secondary structure predictions with sequence alignments, which allowed the identification of an atypical FERM domain present at the C-termini of SNX17 and the highly homologous SNX31 (40% identity), and the subsequent recognition that SNX27 possesses an analogous C-terminal structure. While the manual prediction and curation of secondary structures is outlined below, it is useful to note that several programs are available for performing similar secondary structure-based approaches for sequence alignment automatically, including the SBAL (Wang et al., 2012) and PRALINE (Simossis & Heringa, 2005) software packages. More generally, the combined use of both secondary structure and sequence homology allows greater accuracy in predicting domain boundaries of novel proteins for both bioinformatic analysis, and subsequent biochemical, and structural studies.

2.1. Identification of functional modules of PX-FERM proteins

1. Uniprot Reference ID numbers for human PX proteins are SNX17 (Q15036), SNX27b (Q96L92), and SNX31 (Q8N9S9).
2. Execute a multiple sequence alignment of SNX17, SNX27, and SNX31 using the Multalin server (http://multalin.toulouse.inra.fr/) (Corpet, 1988) and save the resultant alignment file in a fasta format.
3. Using Jpred (http://www.compbio.dundee.ac.uk/www-jpred/index.html) (Cole, Barber, & Barton, 2008), perform combined sequence alignment and secondary structure prediction analysis of SNX17, SNX27, and SNX31. Use the multiple sequence alignment fasta file generated by the program multalin as an input and the structure prediction calculations reveal that all the three proteins possess highly similar secondary structures irrespective of the low sequence identity. This suggests that these three proteins are related to each other and possess similar modular domains.
4. To identify domains other than the PX module in SNX17, SNX27, and SNX31, the C-terminal sequences following the PX domain are submitted for standard BLAST analysis via the NCBI server (http://www.ncbi.nlm.nih.gov/). The searches reveal the overall relationship of SNX17 with SNX31 and SNX27. However, these searches annotate the proteins with either B41 or RA domains. B41 and RA domains belong to the same fold class and resemble ubiquitin in structure. While the B41 domain is present in ERM proteins (ezrin/radixin/moesin) and forms the N-terminal submodule of the larger FERM domain (Yu, Zhang, Ye, & Jiang, 2011), sequence predictions do not reveal the presence of the remaining C-terminal FERM submodules.
5. Manually compare the secondary structure predictions of the regions of SNX17, SNX27, and SNX31 downstream of the PX domain with the FERM domain of the classical FERM protein radixin. Classical FERM domains possess three distinct structural submodules F1, F2, and F3. The comparison clearly demonstrates that the downstream regions of the SNX proteins have similar secondary structure elements, such that the N-terminus (predicted by sequence analysis to be a B41 or RA domain) resembles the FERM F1 submodule, whereas the C-terminal region contains each of the secondary structure elements of the phosphotyrosine-binding domain-like F3 module. By contrast, the F2 region of SNX proteins appears to be truncated, and it is this truncation that prevents standard sequence alignment searches from identifying the overall "atypical" FERM structure at the C-terminus of the proteins.

Sequence analysis in conjunction with structure predictions allowed the accurate identification of a novel subfamily of PX and FERM domain-containing molecules, including the three proteins SNX17, SNX27, and SNX31. Such a classification of functional domains has paved the way for studies of the functional characteristics of PX-FERM proteins, and the methods described could in general facilitate more detailed bioinformatics analyses of any given protein or protein family, for example, as has been done recently for the other PX molecules (Teasdale & Collins, 2012).

3. PRODUCTION OF RECOMBINANT PX-FERM PROTEINS

The structural, biophysical, and functional characterization of proteins demands highly pure and homogenous proteins in milligram quantities. Such large amounts can generally only be produced using recombinant prokaryotic or eukaryotic expression systems. *Escherichia coli*, however, has become the predominant expression system owing to the ease of handling, rapid and inexpensive growth, and well-established methods to genetically manipulate this expression system. Therefore, PX-FERM proteins were expressed in the *E. coli* expression system, and the detailed methodology employed to produce these proteins in milligram quantities is described below.

3.1. Recombinant PX-FERM protein expression

The cDNAs-encoding human SNX17 and mouse SNX27 and SNX31 PX-FERM proteins and their domain fragments were cloned in both pMCSG10 and pMCSG7 expression vectors using ligation-independent cloning (Eschenfeldt, Lucy, Millard, Joachimiak, & Mark, 2009; Ghai et al., 2011). These expression vectors drive protein production under the control of the T7 promoter. pMCSG10 and pMCSG7 contain, respectively, an N-terminal GST- and His-tag followed by a tobacco etch virus (TEV) cleavage sequence. Below is the detailed experimental protocol to express PX-FERM proteins.

1. Transform the plasmids encoding full-length PX-FERM constructs and domain truncations into BL21(DE3)/pLysS chemically competent *E. coli* cells and plate on a Luria Bertani (LB) (Amresco) agar plates supplemented with 50 μg/ml ampicillin (Sigma) and incubate at 37 °C overnight.

2. Pick a single colony and set up a starter culture by inoculating it into 50 ml of LB broth containing 50 μg/ml of ampicillin. Incubate the culture at 37 °C overnight in a shaking incubator at 200 rpm.
3. Inoculate 10 ml of the starter cultures into 2 l conical flasks containing 1 l LB broth (supplemented with 50 μg/ml ampicillin) each. In most cases typically 6–12 l yields sufficient material for subsequent biochemical and structural studies.
4. Incubate the flasks at 37 °C in a shaking incubator at 200 rpm until the OD_{600} reaches ~0.6. At this stage transfer the flasks to a cold room and bring down the temperature to 20 °C. This step is important to enhance soluble protein expression and reduce excessive formation of inclusion bodies.
5. Induce protein expression by adding 0.5 mM isopropyl-β-D-1-thiogalactopyranoside (IPTG) (Amresco) to each of the flasks and incubate the culture at 20 °C overnight in a shaking incubator at 200 rpm.
6. Harvest the cells the next morning (~16 h) by centrifugation at $9000 \times g$, 10 min, 4 °C using a precooled rotor.
7. Resuspend the pellet in 20 mM Tris (pH 8.0), 200 mM NaCl, 10% glycerol, 0.1% octylphenoxypolyethoxyethanol (IGEPAL CA-630), 50 μg/ml benzamidine (Sigma), two tablets of complete EDTA-free protease inhibitor (Roche), 100 units DNaseI (Roche), and 1 mM β-mercaptoethanol for GST-tagged proteins. Add 10 mM imidazole in the resuspension buffer for His-tagged proteins. Addition of glycerol and IGEPAL CA-630 are important for maintaining protein stability and solubility.
8. Snap freeze the cell suspensions using liquid nitrogen and store at −80 °C until required for purification. Typically this is done by gently pipetting the suspension directly into a flask of liquid nitrogen, retrieving the frozen cells via a strainer, and storing in zip-lock plastic bags until required. Pellets can be stored in this way for up to 6 months.

3.2. Purification of recombinant PX-FERM proteins

All purification procedures are performed at 4 °C. The buffers used in the procedures below are degassed and vacuum filtered using a 0.22 μm cut-off filter (Millipore).

1. Thaw the cell suspension and precool the cell disruptor (Constant Systems Limited) to 4 °C and equilibrate the cell with 20 mM Tris (pH 8.0), 200 mM NaCl, 10% glycerol, 0.1% IGEPAL, and 1 mM β-mercaptoethanol.

2. Lyse the cell suspension using the cell disruptor at 27,000 PSI, followed by lysate clarification by centrifugation at 50,000 × g for 30 min. Collect the supernatant.
3. Equilibrate 10 ml of glutathione sepharose resin (GE Healthcare) or nickel-nitrilotriacetic acid resin (Clonetech) with 10 column volumes of equilibration buffer (20 mM Tris (pH 8.0), 200 mM NaCl, 10% glycerol, and 1 mM β-mercaptoethanol) on a gravity column. For His-tagged proteins 10 mM imidazole is included.
4. Mix the supernatant with the equilibrated resin with gentle shaking for 2 h.
5. Wash the resin with 20 column volumes of wash buffer (20 mM Tris (pH 8.0), 200 mM NaCl, 10% glycerol, and 1 mM β-mercaptoethanol) to remove the unbound proteins. For His-tagged proteins 10 mM imidazole is included.
6. Elute the GST-tagged proteins using elution buffer (20 mM Tris (pH 8.0), 200 mM NaCl, 25 mM Glutathione, 10% glycerol, and 1 mM β-mercaptoethanol) and His-tagged proteins are eluted using 20 mM Tris (pH 8.0), 200 mM NaCl, 300 mM imidazole, 10% glycerol, and 1 mM β-mercaptoethanol.
7. When removal of affinity tags is required, cleave the GST or His-tags by adding 1 mg of TEV protease to 10 ml of resin and incubate overnight on a rocker.
8. Elute the cleaved proteins using wash buffer and analyze the eluted fractions by sodium dodecylsulfate-polyacrylamide gel electrophoresis (SDS-PAGE) for purity before pooling.
9. Combine the fractions containing the protein of interest and concentrate using a 15 ml 10 kDa cut-off centrifugal filter (Millipore).
10. Equilibrate a Superdex 200 16/60 gel filtration column (GE Healthcare) with 20 mM Tris (pH 8.0), 200 mM NaCl, 5% glycerol, and 2 mM dithiothreitol (DTT).
11. Load the concentrated protein onto the Superdex 200 column and perform gel filtration chromatography at a flow rate of 1 ml/min. Collect the purified fractions and analyze by SDS-PAGE. A representative figure of the chromatogram and SDS-PAGE analyzed fractions is shown in Fig. 19.1 for SNX17.
12. Pool all pure fractions and measure the concentration using absorbance at 280 nm (A_{280}). Extinction coefficients at 280 nm were calculated from protein sequences using the ExPASy Protparam portal (http://web.expasy.org/protparam/).

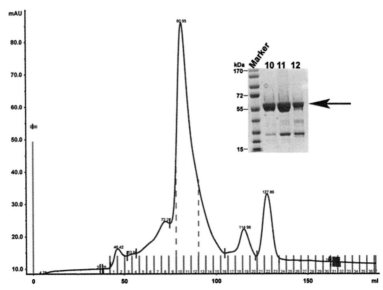

Figure 19.1 Representative figure of a gel filtration chromatogram of the full-length SNX17 PX-FERM protein and Coomassie-stained SDS-PAGE analysis of the fractions. (For color version of this figure, the reader is referred to the online version of this chapter.)

13. Proteins can be used immediately or stored at −80 °C in 100 μl aliquots after snap freezing sample tubes in liquid nitrogen.

4. MEASURING INTERACTIONS OF PX-FERM PROTEINS WITH PHOSPHOINOSITIDE LIPIDS AND PEPTIDE CARGO MOTIFS BY ISOTHERMAL TITRATION CALORIMETRY

PX-FERM proteins reside at the PI3P-enriched early endosomes, and from there regulate cargo trafficking by sensing the specific sorting motifs (NPxY/NxxY) in the cytoplasmic domains of cargo receptors (Adachi and Tsujimoto, 2010 #24; Bottcher et al., 2012; Donoso et al., 2009; Ghai et al., 2012, 2011; Knauth et al., 2005; Lee et al., 2008; Steinberg et al., 2012; Stockinger et al., 2002; van Kerkhof et al., 2005). How PX-FERM proteins are recruited to early endosomes via interaction with PI3P and bind to the short peptide sequences for cargo sorting can be quantified using isothermal titration calorimetry (ITC). ITC is a gold standard method used to measure molecular associations, and is the only technique that provides direct quantitative thermodynamic energetic information by measuring the heat absorbed and released during a binding event (Ghai et al., 2012). A detailed methodology and practical considerations is described below:

1. Equilibrate a Superdex 200 gel filtration column with 100 mM Tris (pH 8.0) and 200 mM NaCl (ITC buffer) and load the protein onto the column to exchange the protein buffer. Alternatively dialysis can be used, although gel filtration provides superior buffer exchange and also gives qualitative information regarding protein homogeneity and the presence of protein aggregation that may interfere with experiments. Avoid addition of high concentrations of reducing agents in the buffer, including DTT and β-mercaptoethanol. The most crucial step toward obtaining reliable ITC data is buffer matching between the sample in the cell and the ligand to be injected from the syringe.

2. Concentrate the proteins using centrifugal ultrafiltration and measure the concentration via A_{280} absorbance measurements as described above. As it is important to know the protein concentration accurately, measuring concentration by colorimetric protein assay (BioRad) can be used to independently verify the A_{280} estimate.

3. Dissolve the short acyl chain (diC8) phosphatidylinositol phospholipids (PIPs) (Echelon) or the sorting motif-containing short peptide sequences (Sigma) in the ITC buffer such that the final concentration is 2 mM. Make sure that the pH of the ligand and protein match, by spotting 1 μl onto a pH strip. The presence of trace amounts of trifluoroacetic acid used during reversed phase chromatographic purification of the peptides in particular can dramatically alter the pH if the buffering capacity is not high enough.

4. Before performing the experiments make sure the ITC sample cell and syringe are thoroughly clean, by sequentially washing the cell and the syringe with 5% Decon 90 detergent, water, and ITC buffer three times each.

5. Evaluate the concentration of protein required to carry out the experiment keeping in mind the C value where $C = K_a \times [M]$ (K_a refers to the predicted association constant). Typical C values range from 10 to 1000. Of equal importance is the determination of optimal ligand concentration. The ligand concentration must be at least 10- to 50-fold of the protein in order to obtain complete saturation of the binding reaction. Typical concentrations are 20 and 50 μM of protein for 500 μM of PIPs and 1–2 mM of peptides, respectively.

6. Centrifuge at least 300 μl of proteins and 70 μl ligands at 17,000 × g for 3 min to remove particulate matter as it can give rise to nonspecific heat signals.
7. Carefully load the protein and the ligand in the sample cell and the syringe, respectively. It is imperative to ensure that there are no bubbles in the cell and the syringe.
8. Fill the reference cell with double distilled filtered and degassed water. ITC buffer can also be used.
9. Set up the experimental parameters to run the ITC experiment including the number and the volume of ligand injections, interval between injections reference cell, stirring speed, and temperature. In this case the number and volume of injections should be set as 13 injections of 3.1 μl spaced at 180 s intervals. Conduct the protein–PIP-binding experiments at 25 °C, whereas the protein–peptide interactions are performed at 10 °C using a Microcal iTC200 instrument (GE Healthcare).
10. Repeat the experiment at least three times to obtain statistically validated-binding parameters.
11. Perform a blank ITC run by titrating the ligand into the buffer to confirm that the binding heat signals are specific to the protein–ligand interaction. Often it is recommended to use this blank data to subtract the dilution heats from the experimental-binding data; however, our preference is to instead subtract the heats measured for the final experimental data points after binding reaches saturation (Ghai et al., 2012).
12. Analyze the data using the ORIGIN7™ software supplied by the instrument manufacturer. Examine the raw data carefully to confirm that the automatic baseline estimations in the software have worked correctly. Manual adjustments of the baseline may be required in some instances.
13. Integrate the area under the peaks and fit the curve by selecting a one site-binding model to generate a nonlinear regression fit (Fig. 19.2). Iterative fitting will provide the best-fit values for the equilibrium association constant (K_a), enthalpy of binding (ΔH), and stoichiometry of binding (N). The Gibbs free energy of binding (ΔG) is obtained from the relation $\Delta G = -RT\ln K_a$. The entropy of binding ΔS is then calculated from the equation $\Delta G = \Delta H - T\Delta S$.

The PX domain mediates PI3P binding and the FERM domain recognizes sorting motifs required for cargo recycling. Therefore, titration of both wild type and mutant PX-FERM proteins using ITC can be used to examine the role of specific residues in regulating lipid or cargo binding. Such

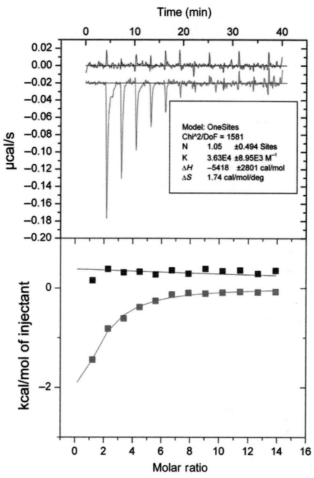

Figure 19.2 Representative ITC experiment showing interaction of the SNX17 PX-FERM protein with a peptide harboring the NPxY/NxxY motif from the APP transmembrane cargo protein (red). The blue curve depicts a blank run (titration of peptide into the buffer). (See color plate.)

experiments allow direct comparison of changes in affinity for direct protein–peptide and protein–PIP interactions (Ghai et al., 2012, 2011).

5. ANALYSIS OF SNX17 PX DOMAIN-PI3*P* INTERACTIONS BY NUCLEAR MAGNETIC RESONANCE SPECTROSCOPY

Nuclear magnetic resonance (NMR) serves as a highly sensitive tool to probe molecular interactions at atomic resolution. While providing confirmation of other biophysical methods such as ITC, it also provides direct spatial information concerning biomolecular interactions. The experimental information generated from this method assisted in identifying the molecular determinants involved in the SNX17 PX-PI3P binding. To study protein–ligand interactions by NMR, isotopically labeled (^{15}N/^{13}C) protein samples are required. Protein samples were prepared using a method modified from Marley, Lu, and Bracken (2001).

5.1. ^{15}N and ^{13}C isotopic labeling of the SNX17 PX domain

1. A comprehensive list of materials required to prepare minimal media for ^{15}N and ^{13}C isotopic labeling is listed as follows:

M9-A salts (1 l) (5 × concentration)	
KH$_2$PO$_4$	15 g
Na$_2$HPO$_4$·7H$_2$O	64 g
NaCl	2.5 g

Make up to 1 l with double distilled filtered water.

M9-B minimal growth media (250 ml)	
M9-A salts	241 ml
MEM Vitamin solution (Invitrogen)	2.5 ml
1 M MgSO$_4$	0.4 ml
1 M CaCl$_2$	20 µl
0.25 g/ml NH$_4$Cl[a] (filter sterilized)	1 ml
0.2 g/ml D-glucose[b] (filter sterilized)	5 ml

[a]Use ^{15}NH$_4$Cl for ^{15}N labeled samples.
[b]Use ^{15}NH$_4$Cl and ^{13}C-D-glucose for ^{15}N/^{13}C labeled samples.

2. Briefly, grow cells in 1 l LB broth as normal at 37 °C until reaching an OD$_{600}$ of 0.7–0.8 (as described in Section 3.1).
3. Using sterilized 1 l centrifuge bottles spin down cells at 5000 × g for 15 min, 4 °C, and gently resuspend in 500 ml M9-A salts to wash cells.
4. Pellet the cells a second time and resuspend in 250 ml M9-B media.
5. Grow for 1.0–1.5 h at 37 °C, reduce the temperature to 20 °C, and induce expression by adding 1 mM IPTG. Incubate the culture overnight at 20 °C, shaking at 200 rpm.

6. Harvest and lyse the cells to purify the ^{15}N single or ^{15}N/^{13}C double-labeled SNX17 PX domain as described in Sections 3.1 and 3.2. For NMR studies the His-tag is removed by TEV cleavage.
7. Gel-filter the SNX17 PX domain using Superdex 200 column into NMR buffer (10 mM HEPES (pH 7.0), 100 mM NaCl, 2 mM DTT) with and without 10 mM Li$_2$SO$_4$. Concentrate the protein to 0.8 mM. Li$_2$SO$_4$ has been found to improve the NMR spectra dramatically as outlined below.

5.2. NMR data acquisition

1. Collect the ^{15}N-HSQC spectrum at 298 K on a high-field (900 MHz) spectrometer (Bruker) equipped with a cryoprobe and Z-axis gradients.
2. Collect triple resonance spectral data in 90%/10% H$_2$O/D$_2$O using the ^{15}N/^{13}C double-labeled protein.
3. Obtain sequential backbone assignments by recording three-dimensional HNCACB, CBCA(CO)NH, HNCO, and HN(CA)CO spectra. All of the aforementioned experiments were collected using the nonuniform sampling (NUS) approach subsequently processed using the maximum entropy method (Mobli, Maciejewski, Gryk, & Hoch, 2007a, 2007b). Process all the spectra using the Rowland NMR toolkit (Mobli et al., 2007a, 2007b) and analyze with the program XEASY (Bartels, Xia, Billeter, Güntert, & Wüthrich, 1995).

In initial experiments, several expected resonances were absent and the spectral quality was generally poor, with most signals dominated by line broadening under standard buffer conditions (10 mM HEPES (pH 7.0), 100 mM NaCl, 2 mM DTT). This is consistent with dynamic conformational changes in certain regions of the protein and therefore, complete sequence-specific resonance assignment could not be achieved. To overcome such problems we performed buffer optimization experiments in order to stabilize the protein and thus improve spectral quality.

We have crystallized the SNX17 PX domain and the crystallization condition contained a high concentration of Li$_2$SO$_4$. Moreover, the crystal structure reveals a bound sulfate molecule in the PI3P-binding site, a region that undergoes dynamic structural changes during PI3P binding (Ghai et al., 2011). Therefore, the protein was gel filtered into 10 mM HEPES (pH 7.0), 100 mM NaCl, 2 mM DTT, and 10 mM Li$_2$SO$_4$ with the view that addition of Li$_2$SO$_4$ might reduce the dynamic exchange in the protein. Remarkably, the spectra obtained after adding 10 mM Li$_2$SO$_4$ was of high quality and

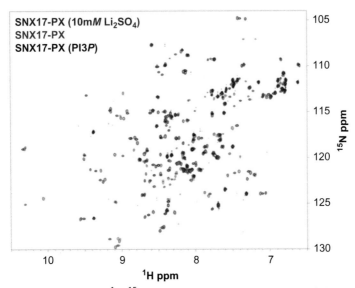

Figure 19.3 Representative ^{1}H–^{15}N two-dimensional HSQC spectra of the SNX17 PX domain under different sample conditions. The spectrum of the unliganded protein (red) is overlaid with spectra of SNX17 in the presence of 10 mM LiSO$_4$ or a molar excess of diC8-PI3P. This figure is reproduced from Ghai et al. (2011) with permission from the Proceedings of the National Academy of Sciences USA. (See color plate.)

facilitated sequential backbone assignments, suggesting stabilization of protein dynamics (Fig. 19.3).

Importantly, following the near complete assignment of the backbone atoms in the protein, the assignments of the SO_4^{2-} bound domain could be readily transferred to the apo-SNX17 PX data, allowing lipid-binding experiments to be performed as described in the following section. The resonance assignments for the SNX17 PX domain are available from the Biological Magnetic Resonance Bank under the ID 17497.

5.3. SNX17 PX domain-PI3P NMR titrations

The SNX17 PX domain binding to the water soluble diC8-PI3P (Echelon Biosciences) was investigated at 298 K by NMR. As the SO_4^{2-} binds to and occludes the PI3P-binding site, it was necessary to perform the PI3P titrations in the absence of Li$_2$SO$_4$.

1. Record the ^{15}N HSQC spectra of ^{15}N-labeled SNX17 PX domain at a concentration of 0.05 mM in 90%/10% H$_2$O/D$_2$O NMR buffer lacking Li$_2$SO$_4$.
2. Subsequently measure the ^{15}N HSQC spectra of the protein after adding increasing amounts of diC8-PI3P up to a final molar ratio of 8:1. See Fig. 19.3 for an example of the spectra obtained from the unbound and bound SNX17 PX domain.
3. Compute the difference in the ^{1}H and ^{15}N chemical shifts between the free and the ligand-saturated state using the formula:

$$\Delta \delta_{ave} = \left[\left(\Delta \delta_{HN}^2 + (\Delta \delta_N / 5)^2 \right) / 2 \right]^{1/2}$$

$\Delta \delta_{HN}$ and $\Delta \delta_N$ refers to the chemical shift differences in proton and nitrogen dimension, respectively, and the nitrogen dimension is scaled to better agree with the chemical shift dispersion of the proton dimension.

4. Evaluate the change in resonance intensity between apo and sequential PI3P-bound forms.
5. Plot the binding isotherms for amino acids that exhibit maximum and saturable shifts. Titrations were fitted to the explicit solution of the quadratic equation describing the $P + L \Leftrightarrow PL$-binding equilibrium, where K_d is defined as $K_d = [P][L]/[PL]$. In this case the concentration of SNX17 PX was fixed $[P_0]$, and increasing concentrations of PI3P $[L_0]$ were added. Total concentrations $[P_0]$ and $[L_0]$ are the sum of free and bound species. Under these conditions, the chemical shift is described by following equation (Tnimov et al., 2012):

$$F = F_{min} + \frac{(F_{max} - F_{min})\left(([P_0] + [L_0] + K_d) - \sqrt{([P_0] + [L_0] + K_d) - 4[P_0][L_0]}\right)}{2[P_0]}$$

where F is the measured chemical shift and F_{max} and F_{min} refer to its maximal and minimal values. The parameters $[P_0]$, F_{max}, F_{min}, and K_d were allowed to vary during the fit by nonlinear least square regression, which can be performed using Grafit 5.0 (Erithacus software).

5.4. Identification of the PI3P-binding site using hybrid of structural methods

The experimental information from the SNX17 PX domain crystal structure, ITC mutagenesis data, NMR titrations, and bioinformatic analysis were used to model the precise SNX17-PI3P-binding site and mechanism. Using the computational modeling program HADDOCK, the mode of SNX17-PX binding to PI3P was investigated in more detail by molecular docking (Dominguez, Boelens, & Bonvin, 2003). Such an analysis can provide important insights into the structural basis of ligand binding.

1. The SNX17 crystal structure (3LUI) (Ghai et al., 2011) and PI3P molecule from the SNX9 PX-PI3P (2RAK) (Pylypenko, Lundmark, Rasmuson, Carlsson, & Rak, 2007) crystal structure were used as starting structures.
2. Topology and parameter files for the PI3P molecule were generated using the PRODRG server (Schuttelkopf & van Aalten, 2004).
3. Identify "active" (i.e., lipid-interacting) residues based on their NMR chemical shift perturbation, ITC mutagenesis data, solvent accessibility, and sequence conservation. In this case SNX17 residues selected as active were Arg36 and Lys62.
4. Recognize passive restraints (i.e., residues not interacting directly) based on the NMR titration data. The passive residues were also selected based on the solvent accessibility (>50%) which was calculated by NACCESS (Lee & Richards, 1971).
5. Using these active and passive residues generate ambiguous interaction restraints (AIRs) with a 2 Å upper distance limit (the maximum distance between any atom of an active residue of one molecule to any atom of an active or passive residue of the second molecule).
6. Run the HADDOCK program by employing the AIR files, parameter, and topology files in order to force proximity between SNX17 active residues and P1 and P3 phosphate atoms of the PI3P molecule.
7. HADDOCK utilizes a rigid body energy minimization protocol and generates 1000 solutions initially and selects 200 structures with the lowest total energy. Run simulated semiflexible annealing refinement of these structures.
8. Refine these 200 solutions by an explicit water refinement step.
9. Analyze the solutions by running a series of CNS (Brunger et al., 1998; Dominguez et al., 2003) analysis scripts that also generate clusters of structure solutions.
10. Visualize and analyze the structure of the complex using molecular graphics programs such as PYMOL or CCP4MG.

6. SUMMARY

Sequence and structure prediction analyses have allowed the definition of a new family of multidomain PX-FERM proteins. Such methods combining both secondary structure and sequence homology considerations may also be more generally applicable to the bioinformatic study of other proteins. The ongoing dissection and characterization of the individual functional modules of PX-FERM proteins are beginning to provide mechanistic explanations for the varied functions carried out by the PX-FERM proteins in endosomal trafficking and recycling, and their potential roles in endosomal signal transduction. In particular, the application of structural methods in conjugation with biophysical techniques including ITC and NMR provide important details needed to understand the mechanisms underpinning their endosomal localization and recognition of NPxY/NxxY sorting motif-containing cargo. More generally, the question of how peripheral membrane proteins coordinate interactions with membrane-spanning receptors, membrane-tethered regulatory proteins, and membrane lipids is one of the central questions in the field of endosome transport and signaling, and the methods described here may be useful for future studies of these processes.

ACKNOWLEDGMENTS

This work was supported by funds from the Australian Research Council (ARC) (DP0985029), and National Health and Medical Research Council (NHMRC) of Australia Dementia Seed Funding (APP1025538). B. M. C. is supported by an ARC Future Fellowship Award (FT100100027). M. M. is supported by an ARC Future Fellowship Award (FTl10100925).

REFERENCES

Adachi, H., & Tsujimoto, M. (2010). Adaptor protein sorting nexin 17 interacts with the scavenger receptor FEEL-1/stabilin-1 and modulates its expression on the cell surface. *Biochimica et Biophysica Acta, 1803*, 553–563.

Bartels, C., Xia, T. H., Billeter, M., Güntert, P., & Wüthrich, K. (1995). The program XEASY for computer-supported NMR spectral analysis of biological macromolecules. *Journal of Biomolecular NMR, 5*, 1–10.

Bottcher, R. T., Stremmel, C., Meves, A., Meyer, H., Widmaier, M., Tseng, H. Y., et al. (2012). Sorting nexin 17 prevents lysosomal degradation of beta1 integrins by binding to the beta1-integrin tail. *Nature Cell Biology, 14*(6), 584–592.

Brunger, A. T., Adams, P. D., Clore, G. M., DeLano, W. L., Gros, P., Grosse-Kunstleve, R. W., et al. (1998). Crystallography & NMR system: A new software suite for macromolecular structure determination. *Acta Crystallographica Section D: Biological Crystallography, 54*(Pt. 5), 905–921.

Cole, C., Barber, J. D., & Barton, G. J. (2008). The Jpred 3 secondary structure prediction server. *Nucleic Acids Research*, *36*(Web Server issue), W197–W201.

Corpet, F. (1988). Multiple sequence alignment with hierarchical clustering. *Nucleic Acids Research*, *16*(22), 10881–10890.

Cullen, P. J. (2008). Endosomal sorting and signalling: An emerging role for sorting nexins. *Nature Reviews Molecular Cell Biology*, *9*(7), 574–582.

Czubayko, M., Knauth, P., Schluter, T., Florian, V., & Bohnensack, R. (2006). Sorting nexin 17, a non-self-assembling and a PtdIns(3)P high class affinity protein, interacts with the cerebral cavernous malformation related protein KRIT1. *Biochemical and Biophysical Research Communications*, *345*(3), 1264–1272.

Dominguez, C., Boelens, R., & Bonvin, A. M. (2003). HADDOCK: A protein-protein docking approach based on biochemical or biophysical information. *Journal of the American Chemical Society*, *125*(7), 1731–1737.

Donoso, M., Cancino, J., Lee, J., van Kerkhof, P., Retamal, C., Bu, G., et al. (2009). Polarized traffic of LRP1 involves AP1B and SNX17 operating on Y-dependent sorting motifs in different pathways. *Molecular Biology of the Cell*, *20*(1), 481–497.

Eschenfeldt, W. H., Lucy, S., Millard, C. S., Joachimiak, A., & Mark, I. D. (2009). A family of LIC vectors for high-throughput cloning and purification of proteins. *Methods in Molecular Biology*, *498*, 105–115.

Ghai, R., Falconer, R. J., & Collins, B. M. (2012). Applications of isothermal titration calorimetry in pure and applied research–survey of the literature from 2010. *Journal of Molecular Recognition*, *25*(1), 32–52.

Ghai, R., Mobli, M., Norwood, S. J., Bugarcic, A., Teasdale, R. D., King, G. F., et al. (2011). Phox homology band 4.1/ezrin/radixin/moesin-like proteins function as molecular scaffolds that interact with cargo receptors and Ras GTPases. *Proceedings of the National Academy of Sciences of the United States of America*, *108*(19), 7763–7768.

Joubert, L., Hanson, B., Barthet, G., Sebben, M., Claeysen, S., Hong, W., et al. (2004). New sorting nexin (SNX27) and NHERF specifically interact with the 5-HT4a receptor splice variant: Roles in receptor targeting. *Journal of Cell Science*, *117*(Pt. 22), 5367–5379.

Knauth, P., Schluter, T., Czubayko, M., Kirsch, C., Florian, V., Schreckenberger, S., et al. (2005). Functions of sorting nexin 17 domains and recognition motif for P-selectin trafficking. *Journal of Molecular Biology*, *347*(4), 813–825.

Kurten, R. C., Cadena, D. L., & Gill, G. N. (1996). Enhanced degradation of EGF receptors by a sorting nexin, SNX1. *Science*, *272*(5264), 1008–1010.

Lee, J., Retamal, C., Cuitino, L., Caruano-Yzermans, A., Shin, J. E., van Kerkhof, P., et al. (2008). Adaptor protein sorting nexin 17 regulates amyloid precursor protein trafficking and processing in the early endosomes. *Journal of Biological Chemistry*, *283*(17), 11501–11508.

Lee, B., & Richards, F. M. (1971). The interpretation of protein structures: Estimation of static accessibility. *Journal of Molecular Biology*, *55*(3), 379–400.

Lunn, M. L., Nassirpour, R., Arrabit, C., Tan, J., McLeod, I., Arias, C. M., et al. (2007). A unique sorting nexin regulates trafficking of potassium channels via a PDZ domain interaction. *Nature Neuroscience*, *10*(10), 1249–1259.

Marley, J., Lu, M., & Bracken, C. (2001). A method for efficient isotopic labeling of recombinant proteins. *Journal of Biomolecular NMR*, *20*(1), 71–75.

Mobli, M., Maciejewski, M. W., Gryk, M. R., & Hoch, J. C. (2007a). An automated tool for maximum entropy reconstruction of biomolecular NMR spectra. *Nature Methods*, *4*(6), 467–468.

Mobli, M., Maciejewski, M. W., Gryk, M. R., & Hoch, J. C. (2007b). Automatic maximum entropy spectral reconstruction in NMR. *Journal of Biomolecular NMR*, *39*(2), 133–139.

Pylypenko, O., Lundmark, R., Rasmuson, E., Carlsson, S. R., & Rak, A. (2007). The PX-BAR membrane-remodeling unit of sorting nexin 9. *The EMBO Journal, 26*(22), 4788–4800.

Schuttelkopf, A. W., & van Aalten, D. M. (2004). PRODRG: A tool for high-throughput crystallography of protein-ligand complexes. *Acta Crystallographica Section D: Biological Crystallography, 60*(Pt. 8), 1355–1363.

Simossis, V. A., & Heringa, J. (2005). PRALINE: A multiple sequence alignment toolbox that integrates homology-extended and secondary structure information. *Nucleic Acids Research, 33*(Web Server issue), W289–W294.

Steinberg, F., Heesom, K. J., Bass, M. D., & Cullen, P. J. (2012). SNX17 protects integrins from degradation by sorting between lysosomal and recycling pathways. *The Journal of cell biology, 197*(2), 219–230.

Stockinger, W., Sailler, B., Strasser, V., Recheis, B., Fasching, D., Kahr, L., et al. (2002). The PX-domain protein SNX17 interacts with members of the LDL receptor family and modulates endocytosis of the LDL receptor. *The EMBO Journal, 21*(16), 4259–4267.

Teasdale, R. D., & Collins, B. M. (2012). Insights into the PX (phox-homology) domain and SNX (sorting nexin) protein families: Structures, functions and roles in disease. *Biochemical Journal, 441*(1), 39–59.

Temkin, P., Lauffer, B., Jager, S., Cimermancic, P., Krogan, N. J., & von Zastrow, M. (2011). SNX27 mediates retromer tubule entry and endosome-to-plasma membrane trafficking of signalling receptors. *Nature Cell Biology, 13*(6), 715–721.

Tnimov, Z., Guo, Z., Gambin, Y., Nguyen, U. T., Wu, Y. W., Abankwa, D., et al. (2012). Quantitative analysis of prenylated RhoA interaction with its chaperone, RhoGDI. *Journal of Biological Chemistry, 287*(32), 26549–26562.

van Kerkhof, P., Lee, J., McCormick, L., Tetrault, E., Lu, W., Schoenfish, M., et al. (2005). Sorting nexin 17 facilitates LRP recycling in the early endosome. *The EMBO Journal, 24*(16), 2851–2861.

Wang, C. K., Broder, U., Weeratunga, S. K., Gasser, R. B., Loukas, A., & Hofmann, A. (2012). SBAL: A practical tool to generate and edit structure-based amino acid sequence alignments. *Bioinformatics, 28*(7), 1026–1027.

Yu, H., Zhang, Y., Ye, L., & Jiang, W. G. (2011). The FERM family proteins in cancer invasion and metastasis. *Frontiers in Bioscience, 16*, 1536–1550.

AUTHOR INDEX

Note: Page numbers followed by "*f*" indicate figures and "*t*" indicate tables and "*np*" indicate footnotes.

A

Aasland, R., 4–5
Abankwa, D., 345
Abdelmoti, L., 54–55
Abraham, T., 66, 72–73, 73*f*, 75
Acar, M., 284
Ackerley, C. A., 303
Ackland, M. L., 81
Adachi, 338–340
Adams, P. D., 346
Adams, S. E., 134–135, 140–141
Adamska, M., 246–247, 247*t*, 249–251
Adesnik, M., 196
Agarwal, K., 212–213
Ahmad, A., 35–38
Ahmad, R., 35–38
Ahmed, K. A., 208–210, 211
Ahringer, J., 96, 101, 103
Aizenman, E., 78
Akam, E. C., 149
Al Rawi, S., 97
Alberti, A., 95, 97, 101, 103, 110, 111, 112–113, 114–116
Albinson, B., 123–124, 127
Alexa, A., 54–55
Al-Nedawi, K., 208
Alvarez, F., 120–123
Ambros, V., 98–99
Ambudkar, S. V., 208–209
Amherdt, M., 134–135, 140–141
Amit, I., 48
Ammoury, A., 120–123
Anastassiadis, K., 245–260
Andreini, C., 78
Andrews, G. K., 78, 79–81, 87
Andrews, N. W., 27
Anitei, M., 224–225, 238–239
Annunziata, F., 27
Antrobus, R., 172–174

Aoki, K., 144–147
Aras, M. A., 78
Arditi, M., 324
Arenzana, N., 87
Arias, C. M., 332–333
Arjonen, A., 276
Arnold, P. Y., 208–209
Arrabit, C., 332–333
Artavanis-Tsakonas, S., 284, 285–286, 289, 295, 297, 302–303, 304
Ashwell, G., 196
Askwith, C. C., 263–264, 268, 271
Aslam, H., 284
Atkinson, R., 289
Atomi, Y., 324–325
Attar, N., 157, 158
Au, C. E., 151
Aureli, M., 27–28, 35–38
Ayadi, A., 248–249
Aydemir, T. B., 79–81

B

Bacia, K., 238–239
Backer, J. M., 4
Bagshaw, R. D., 303
Bai, Y., 250
Balch, W. E., 138–139, 151, 224–225
Baldassarre, M., 134–135, 136–137, 151
Balkin, D. M., 216
Ballabio, A., 27
Banci, L., 78
Bandilla, P., 225–226
Banks, S. M., 303
Bannykh, S., 151
Barber, J. D., 334
Barlow, L. D., 224–225
Barnett, P., 263–264
Barois, N., 95, 103, 111, 112–113, 114–116

Baron, M., 286
Bar-Sagi, D., 182–183
Bartel, B., 81
Bartels, C., 343
Barthet, G., 332–333
Barton, G. J., 334
Bashashati, A., 75
Bass, M. D., 332–333, 338–340
Bassereau, P., 238–239
Bassik, M. C., 328–329
Basso, L., 27, 31, 33
Batte, K., 212–213
Battey, J. F., 245–260
Bauer, A., 285, 286
Bauer, P. O., 324–325
Baumann, K., 182–183
Baust, T., 226–227, 230np, 231–233, 238–239
Bawolak, M.-T., 120–125, 127, 128, 129
Beaudet, A. L., 27
Beckers, C. J., 138–139
Bell, C., 245–260
Bellaiche, Y., 304–305, 309, 313
Bellen, H. J., 284, 289
Belvin, M., 285–286
Benet, L. Z., 123–124
Benhra, N., 303
Benkemoun, L., 97, 101, 110
Bennett, H., 276
Bennetzen, M. V., 112–113
Berbari, N. F., 263–264, 268, 271
Berdnik, D., 303–304
Bergelson, J. M., 75
Bergeron, J. J., 151
Bergmann, J. E., 138–139
Berns, A., 83–84
Berson, J. F., 5
Bertini, I., 78
Beuchat, M. H., 81
Beznoussenko, G. V., 19, 134–135, 136–137
Bhattacharyya, R. P., 48–49, 54–55
Biebermann, H., 262
Bilder, D., 284, 303
Billeter, M., 343
Bilsland, E., 129
Bin, B. H., 85–86
Bird, A. W., 225–226

Birling, M. C., 248–249
Bishop, G. A., 263–264
Bivona, T. G., 151
Blanpain, A., 230np
Blouin, C. M., 182–183
Bode, F., 182–183
Boelens, R., 346–347
Bohnensack, R., 332–333
Bolino, A., 247t, 254
Bolis, A., 247t, 254
Bolte, S., 214–215
Bonangelino, C. J., 246, 249–250
Bond, L. M., 134–135, 136, 140–141
Bonfanti, L., 134–135, 136–137
Bonifacino, J. S., 224–225
Bonvin, A. M., 346–347
Booth, A. M., 211
Boroomand, S., 75
Bortot, B., 28, 35–40
Bosomworth, H. J., 81
Bossi, G., 27
Bossios, A., 208–209
Botelho, 246
Bottcher, R. T., 332–333, 338–340
Bottomley, J., 248–249
Bouche, V., 27
Boulianne, G. L., 286
Bourel-Bonnet, L., 226–227, 230np, 231–233
Bouthillier, J., 120–125, 127, 128, 129
Boxer, S. G., 238–239
Boyd, J. H., 66, 68, 75
Boyd, L., 109–110
Bozzi, Y., 151
Braakman, I., 151
Bracken, C., 341–342
Bradford, M. M., 31
Bradley, A., 245–260
Bray, S. J., 284
Brearley, C. A., 246
Brech, A., 5, 6, 284
Bremnes, B., 4
Breusegem, S. Y. A., 171–174
Bridges, D., 246–247, 247t, 254
Bright, N. A., 159, 172–174
Brinkman, M., 246, 249–250
Broder, U., 333
Brodersen, D. E., 208

Brodsky, L., 123–124
Bronson, R., 246–247, 247t, 252–253
Brown, M. D., 54–55
Brugger, B., 134–135, 140–141
Brunger, A. T., 346
Bruni, C. B., 276
Bruyneel, E., 151
Bryant, N. J., 4, 6
Bu, G., 332–333, 338–340
Bucci, C., 276
Buccione, R., 151
Buchkovich, N. J., 5, 224–225
Buchner, 255
Buerstedde, J. M., 82, 85, 86, 87
Bugarcic, A., 332, 333, 335–336, 340–341, 343–344, 344f, 346
Bujny, M. V., 157–158
Bulleid, N. J., 151
Bulut, Y., 324
Bunnett, N. W., 209, 263–264
Burd, C. G., 4
Burkart, C. A., 123–124
Burton, D. Y., 81
Buss, F., 134–135
Bussell, J., 248–249
Bussini, S., 247t, 254

C

Cacace, A. M., 48–49
Cadena, D. L., 332
Cai, Y., 77–92
Caler, E. V., 27
Calero, M., 158
Camonis, J., 285, 286
Campeau, 4–5, 9, 256–257
Campelo, F., 224–225
Campos-Ortega, J. A., 304
Cancino, J., 134–136, 137–138, 147, 148, 149f, 332–333, 338–340
Capestrano, M., 134–135, 136, 137–138, 138f, 145f, 147
Capitani, M., 134–136, 137–138, 138f, 145f, 147, 148, 149f
Carbery, A. M., 284
Carlsson, S. R., 346
Carter, R., 48
Caruano-Yzermans, A., 332–333, 338–340
Cassonnet, P., 81

Cavalieri, C., 27, 31, 33
Ceron, J., 101
Cerri, F., 246–247, 254
Chaba, T., 120–123
Challiss, R. A., 149
Chan, Y. H., 238–239
Chaput, N., 212
Charng, W. L., 284
Chawla, A., 4
Che, F. Y., 328
Chen, D., 216–217
Chen, J. W., 120–123, 129
Chen, P. M., 120–123, 129
Chen, Y., 216
Cheng, C. Y., 183–184, 188, 190–191, 191f
Cheng, H., 303
Cheng, X., 249–250, 256
Chevallet, M., 35
Chi, Z. H., 77–92
Chiantia, S., 5
Chiariello, M., 276
Chicanne, G., 246–247, 254
Chigorno, V., 27–28
Childress, J. L., 284
Chimienti, F., 79
Chinchore, Y., 289
Cho, B., 303
Choi, Y., 328–329
Chou, M. M., 72–73
Choudhury, A., 271, 276
Chow, C. Y., 246–247, 247t, 249–251, 252–253, 254–256
Christoforidis, S., 209
Ciccarone, V., 27, 28
Ciechanover, A., 196
Cimermancic, P., 332–333
Civic, N., 79–81
Clackson, T., 134–135, 140–141
Claeysen, S., 332–333
Clark, S., 81
Clarke, L., 120–123
Clemence, S., 284
Clore, G. M., 346
Cobb, M. H., 49
Cocucci, E., 208
Cohen, S. M., 284, 306
Cole, C., 334
Cole, N. B., 139

Colland, F., 216–217
Collinet, C., 209
Collins, B. M., 332, 333, 335, 338–341
Collins, L. L., 182–183
Collinson, L. M., 4
Colvin, R. A., 78
Coneyworth, L. J., 81
Connolly, C. N., 135, 147
Cook, K. R., 285–286
Cooke, F. T., 246
Copeland, T., 48–49
Cordelieres, F. P., 214–215
Corpet, F., 334
Corvera, S., 4, 6
Cotton, M., 303
Cottrell, G. S., 209
Coumailleau, F., 216–217, 284, 289, 303–305, 306, 307–308, 309–310, 312–313, 314
Courage, N. L., 134–135, 140–141
Cousins, R. J., 79–81
Couturier, L., 305–306
Cox, J., 225–226, 236
Coyne, C. B., 75
Crowder, C. M., 112–113
Cuajungco, M. P., 81
Cuitino, L., 332–333, 338–340
Culetto, E., 95, 97, 103, 111, 112–113, 114–116
Cullen, P. J., 157–158, 332–333, 338–340
Currie, D. A., 308
Cutler, D. F., 135, 147
Czubayko, M., 332–333, 338–340
Czupalla, C., 226–227

D

Dacks, J. B., 224–225
Dahl, R. H., 149
Dakhlallah, D., 212–213
Dansen, T. B., 135, 147
Daro, E., 196, 271, 276
D'Arrigo, A., 4–5
Das, S. R., 208–209
Dati, G., 247t, 254
Daukas, G., 182–183
Dautry-Varsat, A., 196, 205
Davidson, M. W., 209–210
Davis, D. M., 208, 211

Davis, R. J., 54–55
Davis, R. L., 286
Davisson, M., 246, 247t, 252–253
Dawson, T., 249–250
Day, R. N., 209–210
de Barsy, T., 120–123
De Camilli, P., 216, 224–225, 246
De Corte, V., 151
De Duve, C., 120–123
De Lemos-Chiarandini, C., 196
De Matteis, M. A., 4–5, 151, 225
De Renzis, S., 263
Dean, A., 196
Deeb, R., 246
DeLano, W. L., 346
Dell'Angelica, E. C., 79
DeLotto, R., 324
Delpire, E., 266–267
Delvecchio, K., 247t, 252, 253–254
Deng, W., 77–92
Déry, O., 263–264
Dhanasekaran, D. N., 54–55
Di Fiorce, P. P., 182–183
Di Giandomenico, D., 134–135, 136
Di Paolo, G., 225
Di Pentima, A., 134–135, 136
Di Tullio, G., 151
Dikic, I., 182–183
Dina, G., 246–247, 247t, 254
Dinger, M. E., 208–209
Djeddi, A., 95, 97, 103, 111, 112–113, 114–116
Dobson, P. D., 129
Doglia, S. M., 123–124, 127
Dohgu, S., 127
Doi, H., 324–325
Dolph, P. J., 289
Dominguez, C., 346–347
Dominguez, M., 271, 276
Dompe, N. A., 285–286
Doms, R. W., 138–139
Donahue, L. R., 248
Donaldson, J. G., 196
Dondapati, R., 246
Dong, X. P., 249–250, 256
Donoso, M., 332–333, 338–340
Douglas, M. R., 246
Dove, S. K., 246

Dowling, J. J., 246–247, 247t, 249–252, 256–257
Draper, R. K., 198–199
Driver, S. E., 101
D'Souza, B., 208
D'Souza-Schorey, C., 72–73
Duden, R., 224–225
Duex, J. E., 246, 249–250
Dunbar, C. E., 209
Dussert, A., 303, 304

E

Ehsanian, R., 123–124, 127, 129
Eichelsdoerfer, J. L., 81
Eide, D. J., 79, 81, 87
Eisenberg, T., 112–113
Ekroos, K., 29–30
Ekstrom, K., 208–209
El Andaloussi, S., 211
Elisei, F., 28
Ellenberg, J., 138–139, 167
Ellis, C. D., 81, 87
Emery, G., 303–304
Emiliani, C., 27–28, 31, 35
Emin, M. T., 208–209
Emr, S. D., 4, 5, 224–225, 246, 249–250
English, J. M., 49
Entchev, E. V., 306
Eppig, J. T., 248–249
Equils, O., 324
Erdtmann-Vourliotis, M., 263–264, 268, 271
Ericsson, M., 135
Eschenfeldt, W. H., 335–336
Esfandiarei, M., 75
Espinosa, E. J., 158
Estacion, M., 266–267
Eun, S. H., 303
Evans, C. W., 308
Evans, J. A., 81

F

Fabbri, M., 151
Facchiano, F., 151
Falconer, R. J., 332, 338–341
Falcon-Perez, J. M., 79
Falkenberg, C. V., 182–183
Fallon, J. K., 211
Falus, A., 208
Fang, Y., 81, 211
Farrer, L. A., 171–172
Fasching, D., 332–333, 338–340
Faure, E., 324
Fava, E., 65–76
Favre, M., 81
Fawcett, R., 285–286
Feller, S. M., 123–124, 127, 129
Ferguson, C. J., 246–247, 247t, 249–253, 254–257
Fernandes, M. J. G., 125, 126f, 127, 129
Fey, G. H., 196
Finney, L., 79
Finney, M., 106
Fire, A., 98–99, 101
Fjorback, A. W., 159
Fleming, A., 112–113
Fligger, J., 252
Florian, V., 332–333, 338–340
Fogarty, K., 6
Fontaine, C. P., 78
Ford, D., 81
Forer, A., 309–310
Fortini, M. E., 284, 303
Foster, S. A., 159
Fostier, M., 284, 286
Fraldi, A., 27
Frame, M. C., 182–183
Francisco, G. C., 224–225
Franklin, C., 248
Fransen, J. A., 276
Fray, M., 248–249
Freeman, C., 172–174
Frias, S., 81
Fu, L., 264–265, 268f, 273f
Fuchs, A., 35
Fuchs, H., 248–249
Fujiwara, N., 85–86
Fukada, T., 78, 79–81, 85–86, 89
Fukuda, M., 196, 198–199, 200, 202, 276
Fukumori, A., 303
Fukunaka, A., 79–81, 85–86
Furthauer, M., 216–217, 303–305, 306, 307–308, 309–310, 312–313, 314
Fürthauer, M., 284, 289
Furuichi, T., 79–81
Fusella, A., 134–135, 136–137

Futter, C. E., 4, 135, 147
Fuwa, T. J., 284, 286

G

Gagné, V., 125, 126*f*, 127, 129
Gagné-Henley, A., 120–125, 129
Gaither, L. A., 79
Gallai, V., 27
Galli, T., 196, 271
Gallop, J. L., 224–225
Galperin, E., 48, 49–51, 52, 54–55, 58–59
Gambin, Y., 345
Garmaroudi, F. S., 75
Gary, J. D., 246, 249–250
Gasparia, A. V., 123–124
Gasser, R. B., 333
Gaudreault, R. C., 120–125, 129
Gaullier, J. M., 4, 6
Gensch, N., 284
Geraud, G., 304–305
Geuze, H. J., 12
Ghai, R., 332, 333, 335–336, 338–341, 343–344, 344*f*, 346
Gho, M., 304–305, 313
Giagtzoglou, N., 303
Giannotta, M., 134–136, 137–138, 138*f*, 145*f*, 147, 148, 149*f*
Gibson, A., 135, 147
Gigengack, S., 12
Giger-Mateeva, V. I., 250–251
Gill, G. N., 182–183, 332
Gill, R., 264–265, 268*f*, 273*f*
Gillette, J. M., 209, 211, 213, 214–215
Gillooly, D. J., 4, 6, 209
Giner, A., 209
Giunta, C., 79–81, 87
Gleeson, P. A., 151
Glick, B. S., 224
Go, M. J., 284, 286
Godi, A., 4–5, 225
Goel, M., 266–267
Goh, L. K., 48
Gokool, S., 159
Goldberg, J., 151
Gombart, Z. J., 120–123, 129
Gomez, M., 276
Gonnord, P., 182–183

Gonzalez, C., 309–310
Gonzalez, M. A., 208–209
Gonzalez, S., 208–209
Gonzalez-Gaitan, M., 216–217, 303–305, 306, 307–308, 309–310, 312–313, 314
González-Gaitán, M., 284, 289
Gordon, P. B., 95, 114–116
Goswami, A., 324–325
Goto-Silva, L., 209
Gottesman, M. M., 208–209
Goud, B., 135, 276
Gould, G. W., 182–183
Gould, R., 4, 6
Gould, S. J., 211
Grabner, H., 65–76
Grandjean, C., 230*np*
Grant, A. E., 246–247, 247*t*, 249–250, 251–252, 254–257
Grässlund, A., 123–124, 127
Graves, H. K., 303
Griffiths, G., 135, 276
Griffiths, G. M., 27
Groizeleau, C., 216–217
Gros, P., 346
Grosse-Kunstleve, R. W., 346
Grossi, M., 134–136, 137–138, 147, 148, 149*f*
Gruenberg, J., 4, 5, 6, 17
Gryk, M. R., 343
Guan, K. L., 54–55
Guiral, M., 303
Gullans, S. R., 266–267
Güntert, P., 343
Guo, B., 77–92
Guo, J., 250
Guo, Z., 345
Gurevich, E. V., 262
Gurevich, V. V., 262
Guruharsha, K. G., 304
Gustafsen, C., 159
Gustafson, K., 286
Gutierrez-Vazquez, C., 208–209, 211

H

Haas, A., 151
Haase, H., 78, 89
Habermann, B., 209

Hackett, T. L., 66, 68, 75
Hagihara, M., 79–81
Haglund, K., 182–183
Hagn, M., 248
Hajjar, C., 97, 109–110
Halder, G., 284
Haltiwanger, R. S., 151
Hamburger, A., 216–217
Han, M., 54–55
Handel, M., 263–264, 268, 271
Hannus, M., 65–76
Hanson, B., 332–333
Hanyaloglu, A. C., 262–263
Hao, T., 101, 324–325
Hara, T., 324–325
Harbour, M. E., 171–174, 177
Harper, D. C., 5
Harrison, C., 135–136
Hartenstein, V., 304–305, 314
Hartwieg, E., 109–110
Hasdemir, B., 209
Hasegawa, A., 81
Hasilik, A., 27–28, 31, 35
Hastings, K. L., 120–123
Hatada, M., 134–135, 140–141
Hauber, J., 196
Haucke, V., 225
Hayakawa, A., 6
Hayashi, S., 305–306
Hayes, S. J., 6
He, C., 328–329
He, J., 286
Heesom, K. J., 332–333, 338–340
Helenius, A., 66, 94, 138–139, 271
Helfer, E., 177
Henderson, B. S., 324–325
Henne, W. M., 5, 224–225
Hennigar, S. R., 79
Henning, R., 182–183
Henriot, V., 177
Hepler, J. R., 262
Herbst, J. J., 182–183
Heringa, J., 333
Hermsdorf, T., 262
Hersh, B. M., 109–110
Hershfinkel, M., 78
Higashiyama, H., 79–81
Higgs, J., 284, 286

Higuchi, S., 127
Hildreth, J. E., 211
Hiort, C., 123–124, 127
Hirano, T., 78, 79, 89
Hirozane-Kishikawa, T., 101
Hirschberg, K., 138–139
Hirst, J., 224–225
Hiscox, S., 79–81
Hobert, O., 98
Hoch, J. C., 343
Hocke, G. M., 196
Hoekstra, D., 211
Hoflack, B., 224–225, 226–227, 276
Hofmann, A., 333
Hogstrand, C., 79–81
Hojyo, S., 81, 85–86
Höllt, V., 263–264
Holmes, W. R., 78
Hong, W., 332–333
Honsho, M., 29–30
Hopkins, C. R., 4, 135, 147
Hori, K., 284, 285–286, 289, 295, 297, 303
Horvitz, H. R., 109–110
Hosaka, T., 85–86
Howell, K. E., 149
Hrabe de Angelis, M., 248
Hsu, C., 5
Hsu, V. W., 147
Hu, W., 96
Huang, F., 48
Huang, Y., 123–124, 265–266
Huang, Z., 86–87
Hubbard, K. B., 262
Huber, L. A., 49, 182–183
Hubner, N. C., 225–226, 236
Hudrisier, D., 208
Hull, M., 276
Huotari, J., 94
Huppert, K., 285–286
Hurbain, I., 5
Hurley, J. H., 224–225
Hutagalung, A. H., 276
Hutterer, A., 303–304

I

Iguchi, H., 212
Ikonomov, O. C., 246, 247t, 252, 253–254

Ikuta, A., 303
Ilagan, M. X., 302
Illuzzi, G., 27, 35–38
Irvine, K. D., 151
Ishida, Y., 87
Ishihara, K., 85–86, 87, 88*f*
Ishikawa, H. O., 151
Ishiura, S., 324–325
Islam, M. N., 208–209
Ismail, N., 212–213
Ito, M., 284, 286
Itoh, N., 87
Itoh, T., 196, 224–225
Ivaska, J., 276

J

Jacob, Y., 81
Jacobs, S., 263–264
Jacq, X., 216–217
Jaenisch, R., 83–84
Jager, S., 332–333
James, D. E., 12
Jana, N. R., 325
Janes, K. A., 75
Januschke, J., 309–310
Jauffred, B., 309
Jean-Alphonse, F., 262–263
Jensen, K. J., 75
Jensen, T. H., 208
Jeong, J., 79–81, 87
Jiang, W. G., 334
Jiang, X., 48
Jin, J. P., 247*t*, 253–254
Jin, L., 328
Jin, N., 246–247, 247*t*, 249–251, 252–253, 254–256
Joachimiak, A., 335–336
Johannes, L., 157–158, 238–239
Johnson, A. D., 263–264, 268, 271
Joly, E., 208
Jones, A. T., 276
Jones, J. M., 246–247, 247*t*, 249–250, 251–252, 254–257
Jones, S. M., 149
Jorge, 255
Joubert, L., 332–333
Jun, H. S., 328–329

K

Kahn, R. A., 151
Kahr, L., 332–333, 338–340
Kahya, N., 238–239
Kallio, K., 276
Kamath, R. S., 96, 101, 103
Kambe, T., 78, 79–81, 84–86, 89
Kamino, S., 79–81
Kang, G., 208–209
Kankel, M. W., 304
Kanwar, R., 284, 303
Kanzaki, M., 246
Kaplan, J., 198–199
Kashef, K., 54–55
Katoh, Y., 324–325
Katona, I., 250
Katz, M., 48
Kauffman, E. J., 246
Kaufman, E., 246–247, 254
Kaufman, R. J., 87
Kaufmann, A. M., 128
Kaulen, H. D., 182–183
Kawahashi, K., 305–306
Keenan, T., 134–135, 140–141
Keene, J. D., 324–325
Kell, D. B., 129
Kelleher, S. L., 79, 81
Keller, D. S., 138–139, 235–236
Kempf, C., 196
Kenan, D. J., 324–325
Kendrick-Jones, J., 134–135
Khalili, A., 75
Kieffer, S., 35
Kille, P., 79–81
Kim, B. M., 328–329
Kim, S. J., 328–329
Kimura, T., 87
King, G. F., 332, 333, 335–336, 340–341, 343–344, 344*f*, 346
Kinne, R., 182–183
Kirchhausen, T., 284, 285–286, 289, 295, 297, 303
Kirsch, C., 332–333, 338–340
Kiyokawa, E., 144–147
Klausner, R. D., 147, 196
Klein, D. E., 6
Klionsky, D. J., 95
Klychko, E., 213

Knaust, A., 234
Knauth, P., 332–333, 338–340
Knoblich, J. A., 216–217, 284, 289, 303–305, 306, 307–308, 309–310, 312–313, 314
Knust, E., 226–227, 304
Kobayashi, T., 81
Kohda, A., 87, 88f
Koike, M., 324–325
Kolotuev, I., 108–109
Komatsu, M., 324–325
Kopajtich, R., 151
Kopan, R., 302
Korcsmaros, T., 182–183, 209
Koreth, J., 101
Kosaka, N., 212
Kostas, S. A., 101
Krämer, H., 208, 289
Kramer, J. M., 98–99
Krause, E., 226–227
Krauss, M., 225
Krausz, E., 65–76
Kreienkamp, H. J., 263–264
Kreis, T. E., 138–139, 276
Kreuzer, O. J., 263–264
Kriegel-Taki, G., 304
Krijnse-Locker, J., 135
Krisch, B., 263–264
Krise, J. P., 128
Krishnakumar, S., 81
Krogan, N. J., 332–333
Kubota, H., 324–325
Kuno, T., 81
Kuroda, M., 212–213
Kuroda, T. S., 202
Kuroiwa, Y., 324–325
Kurokawa, Y., 81, 85–86
Kurosawa, M., 324–325
Kurten, R. C., 332
Kushner, J. P., 198–199
Kweon, H. S., 134–135, 136–137

L

Labouesse, M., 108–109
Laird, P. W., 83–84
Lake, R. J., 284
Lakisic, G., 177
Lallet, S., 303
Lamant, L., 120–123
Lamaze, C., 182–183
Lamers, W. H., 267
Landry, K., 35–38
LaPointe, P., 224–225
Lappano, R., 262
Larochelle, A., 209
Lasswell, J., 81
Latterini, L., 28
Lauffenburger, D. A., 182–183
Lauffer, B., 332–333
Lawe, D. C., 4, 6
Lazarczyk, M., 81
Le Borgne, R., 303–304, 306
Le Bras, S., 303, 304
Le Roy, C., 182–183
Le, T. L., 182–183
Lecourtois, M., 304–305
Lee, B., 346
Lee, C. M., 54–55
Lee, J., 332–333, 338–340
Lee, J. H., 303
Lee, J. J., 208–209
Lee, M. C., 151
Lee, W. M., 183–184, 191f
Lee, W. W., 246–247, 247t, 254
Lee, Y., 211
Lefebvre, C., 95, 97, 101, 108, 110, 114–116
Legname, G., 123–124
Legouis, R., 95, 97, 101, 103, 108, 110, 111, 112–113, 114–116
Lemaire, K., 79
Lemansky, P., 27, 31, 33
Lemmon, M. A., 6
Lenk, G. M., 246–247, 247t, 249–253, 254–257
Lesort, M., 264–265, 268f, 273f
Leto, D., 182–183
Levchenko, A., 208–210
Lewis, J. S., 263–264, 268, 271
Lewis, M. J., 135
Lewis, R. E., 72–73
Li, J., 196, 250
Li, M. W. M., 183–184
Li, T., 303
Li, T. F., 216–217
Li, W., 54–55
Li, X., 249–250, 256

Li, Y., 87
Li, Z., 96
Liang, D., 77–92
Liang, F., 35–38
Lichten, L. A., 79
Lie, P. P. Y., 183–184, 190–191
Lienhard, G. E., 12
Lim, J., 324–325
Lim, J. W., 208–209
Lim, W. A., 48–49, 54–55
Lin, J., 75
Lin, W., 303
Linders, K., 83–84
Lindsay, M., 4, 6, 81
Lindsley, D. L., 286
Lippard, S. J., 86–87
Lippincott-Schwartz, J., 139, 144–147, 209
Liu, L., 246–247, 247t, 252–253
Liuzzi, J. P., 79–81
Lloyd, K. C., 248
Lloyd, T. E., 289
Loberto, N., 27–28, 35–38
Lodge, R., 120–125, 129
Lodish, H. F., 138–139, 196
Loeser, E., 284
Loew, L. M., 182–183
Loose, L. D., 182–183
Lotvall, J. O., 208–209
Lou, H., 75
Loukas, A., 333
Louvet-Vallée, S., 97
Louvi, A., 284
Lu, H., 284, 303
Lu, M., 341–342
Lu, W., 332–333, 338–340
Lucy, S., 335–336
Luini, A., 134, 151
Lukacs, M., 54–55
Lund, V. K., 324
Lundmark, R., 346
Lunn, M. L., 332–333
Luo, H., 75
Luo, Z., 66, 68, 72–73, 73f, 75
Lütcke, A., 17, 276
Luton, F., 151
Lykke-Andersen, S., 208
Lyons, T., 81

M

Ma, Y., 81
MacDiarmid, C. W., 81
Machida, Y., 324–325
Maciejewski, M. W., 343
Maggiolini, M., 262
Magini, A., 27–28, 31, 33, 35–40
Magnuson, T., 248
Mahuran, D. J., 27, 28
Mâle, P., 276
Malerod, L., 6
Malhotra, V., 224–225
Manabe, S., 182–183
Mangeat, P., 211
Manil-Segalen, M., 95, 114–116
Mann, 236
Mannie, M. D., 208–209
Mansueto, G., 27
Marceau, F., 120–125, 126f, 127, 128, 129
Marchant, D., 66, 68, 72–73, 73f, 75
Marchese, A., 27, 182–183
Maret, W., 78
Mariño, G., 112–113
Mark, I. D., 335–336
Marks, M. S., 72–73
Marley, J., 341–342
Martella, O., 134–135, 136–137
Martinez, C., 271, 276
Martinez-Menarguez, J. A., 134–135, 136–137
Martino, S., 27
Masilamani, A. P., 27, 35–38
Masuda, S., 79–81, 87
Mather, I. H., 276
Mathes, K., 48–49
Mathieu, J., 284
Mathivanan, S., 208–209
Matsuda, M., 144–147
Matsui, T., 196, 198–199, 200
Matsuki, Y., 212
Matsumoto, N., 303
Matsuno, K., 284, 286
Matsuura, W., 79–81, 87, 88f
Mattick, J. S., 208–209
Maxfield, F. R., 205
May, B. C., 123–124
Mayer, B., 303–304

Mazaleyrat, S. L., 284
McCartney, 246
McClellan, S., 79–81
McClure, M., 48
McCormick, L., 332–333, 338–340
McCormick, N. H., 79, 81
McGraw, T. E., 205
McIvor, W. E., 81
McLeod, I., 332–333
McMahon, H. T., 224–225
Medina, D. L., 27
Meehan, B., 208
Megalou, E., 112–113
Megirian, R., 182–183
Mehmedbasic, A., 159
Mehta, B. M., 208–209
Meisler, M. H., 246, 250–251, 252–253
Meldolesi, J., 208
Mellman, I., 196, 271, 276
Mello, C. C., 98–99, 101
Melnyk, O., 230np
Meloty-Kapella, L., 303
Mencarelli, S., 27, 31, 33
Mendoza, J. A., 81
Menut, L., 284
Mercer, J., 66
Mercer, T. R., 208–209
Merino-Trigo, A., 151
Meves, A., 332–333, 338–340
Meyer, H., 332–333, 338–340
Meyerhof, W., 263–264
Miaczynska, M., 182–183, 209
Michaud, N. R., 48–49
Michaud, S., 120–123
Michelet, X., 95, 97, 101, 103, 110, 111, 112–113, 114–116
Michell, R. H., 246
Migaki, H., 85–86, 87, 88f
Mikoryak, C. A., 198–199
Milash, B., 285–286
Millard, C. S., 335–336
Millay, M., 213
Miller, C. M., 138–139
Miller, E. A., 151
Milligan, G., 262
Milner, M. J., 308
Minta, J. O., 125
Mironov, A. A. Jr., 19, 134–135, 136–137

Misener, S., 213
Mishima, K., 79–81
Mitra, A., 289
Mitsuhashi, H., 324–325
Mitsui, K., 324–325
Mittelbrunn, M., 208–209, 211, 214
Miyamoto, A., 208
Miyawaki, A., 144–147
Miyazaki, W. Y., 285–286
Mlak, K., 246
Mobli, M., 332, 333, 335–336, 340–341, 343–344, 344f, 346
Moldovan, L., 212–213
Montgomery, M. K., 101
Moorman, A. F., 267
Moresi, V., 328–329
Morgan, A., 75
Morimoto, R., 328f
Morimoto, S., 182–183
Morissette, G., 123–125, 127, 128
Moritz, R. L., 208–209
Morrison, D. K., 48–49, 54–55
Morrow, I. C., 4, 6
Morselli, E., 112–113
Mostov, K. E., 151
Mothes, W., 208
Motley, A., 159
Motoki, M., 284
Mougin, C., 216–217
Mruk, D. D., 183–184, 188, 191f
Mukherjee, A., 285, 286
Mukhopadhyay, A., 328
Munegowda, M. A., 208
Murakami, M., 78, 79, 89
Murphy, C., 209
Murphy, J. E., 209
Murray, J. W., 328
Murray, S. A., 248–249
Muskavitch, M. A., 302–303
Musse, A. A., 303
Mykytyn, K., 263–264, 268, 271

N

Na, H. K., 328–329
Nagai, Y., 324–325
Nagao, M., 79–81, 85–86, 87
Nahorski, S. R., 149
Naito, M., 127

Nakae, S., 77–92
Nakamura, T., 144–147
Nakanishi, T., 87
Nakayama, K., 324–325
Nardella, J. P., 208–209
Narita, K., 271, 276
Nassirpour, R., 332–333
Nau, J. J., 246, 249–250
Neefjes, J. J., 12
Nelson, D. C., 81
Neto, H., 182–183
Nezis, I. P., 284
Nguyen, U. T., 345
Nichols, J. T., 208, 209–210
Nicholson, G., 246–247, 257
Nicholson, R. I., 79–81
Nicot, A.-S., 101
Nicoziani, P., 134–135, 136–137
Niehage, C., 226–227
Nielsen, E., 263
Nieves, E., 328
Nigam, S. K., 266–267
Nilsson, T., 135, 151
Nishida, K., 78, 79, 81, 89
Nishimura, N., 182–183
Niu, X., 208–209
Noda, T., 112–113
Nord, 248
Nordeen, M. H., 149
Nordén, B., 123–124, 127
Norwood, S. J., 332, 333, 335–336, 340–341, 343–344, 344*f*, 346
Novick, P. J., 276
Nukina, N., 303, 324–325

O

Oakley, J., 158
Ober, R. J., 271, 276
Oboki, K., 77–92
O'Brien, J. S., 27
Ochiya, T., 212
O'Connell, K., 109–110
Oda, K., 81, 87
O'Donnell, J. M., 81
Oguro, A., 324–325
Ohashi, W., 81, 85–86
Ohyashiki, J. H., 212–213
Ohyashiki, K., 212–213
Okada, S., 27
Okada, T., 214–215
Okano, H., 284, 286
Okochi, H., 123–124
Okochi, M., 303
Okumura, F., 87
Okumura, K., 87, 88*f*
Okuno, M., 324–325
Oliver, S. G., 129
Olsen, S. L., 208
Olson, S. D., 208–209
Oorschot, V., 12
Orci, L., 151
Orlacchio, A., 27
Ostrowski, M., 208

P

Padilla, B. E., 209
Padilla-Parra, S., 48
Paigen, 247–248
Paing, M. M., 27
Pak, B. J., 303
Pal, M., 123–124
Palfy, M., 182–183, 209
Pallinger, E., 208
Palmer, A. E., 79–81, 87
Palmiter, R. D., 81
Pambrun, L., 125
Panopoulou, E., 209
Pap, E. H., 135, 147, 208
Park, J. G., 79–81, 87
Parker, P. J., 246
Parks, A. L., 285–286
Parshina, I., 238–239
Parton, R. G., 17, 196, 276
Pasternak, S. H., 303
Pasztoi, M., 208
Patel, A. J., 324–325
Patel, D. M., 208–209
Patki, V., 4
Patrushev, N., 81
Paul, C., 120–123
Pecheur, E. I., 230*np*
Peden, A. A., 134–135
Pelham, H. R., 135
Pelicci, G., 27
Pelkmans, L., 66, 75, 209
Pellinen, T., 276

Pennetta, G., 289
Pepperkok, R., 167
Perera, R. M., 216
Pessin, J., 246
Peterkofsky, B., 136
Peters, P. J., 12
Pfeffer, S. R., 158
Pfeiffer, M., 263–264
Phair, R. D., 138–139
Philips, M. R., 151
Phistry, M., 289
Pickett-Heaps, J. D., 309–310
Pilecka, I., 209
Piper, R. C., 246–247, 247t, 253
Pirruccello, M., 216
Pisanu, A., 134–135, 136, 137–138, 138f, 145f, 147
Ploegh, H. L., 12
Pocha, S. M., 226–227
Pockrandt-Hemstedt, H., 182–183
Poinas, A., 35
Poirier, M. A., 324–325
Polchi, A., 27–28, 31, 35–40
Polishchuk, R. S., 134–135, 136, 137–138, 138f, 145f, 147
Polo, S., 182–183
Pons, C., 81
Poole, B., 120–123
Popoff, V., 157–158
Porto, C. S., 184
Posakony, J. W., 304–305, 314
Poser, I., 225–226
Pozzan, T., 144
Preisinger, C., 151
Presley, J. F., 138–139
Previtali, S. C., 247t, 254
Prichard, M., 264–265, 268f, 273f
Prinetti, A., 27–28, 35–38
Prockop, D. J., 208–209
Prost-Squarcioni, C., 120–123
Prusiner, S. B., 123–124
Prywes, R., 87
Pulvirenti, T., 134–136, 137–138, 138f, 145f, 147, 148, 149f, 151
Puri, V., 271, 276
Purmal, A. A., 123–124
Purow, B., 302
Pylypenko, O., 346

Q

Qiao, W., 87
Quie, P. G., 182–183

R

Racanicchi, L., 27
Racchetti, G., 208
Raiborg, C., 4, 94
Rajendran, L., 5
Rak, A., 346
Rak, J., 208
Rand, M. D., 284
Raposo, G., 238–239
Rappolee, D., 247t, 253–254
Rasmuson, E., 346
Rauchman, M. I., 266–267
Ravazzola, M., 134–135, 140–141
Read, E., 172–174
Reasor, M. J., 120–123
Recheis, B., 332–333, 338–340
Reddel, S. W., 246–247, 257
Reddy, A., 27
Reddy, E. P., 54–55
Reggiori, F., 95
Reid, E., 172–174
Reinsch, S., 196
Remenyi, A., 48–49, 54–55, 182–183, 209
Rempe, 249
Ren, H., 96, 324–325
Ren, M., 196, 205
Renaud, O., 314–315
Retamal, C., 332–333, 338–340
Rhinn, M., 209
Richards, F. M., 346
Richardson, S. C., 246–247, 247t, 253
Riehle, E., 246–247, 247t, 254
Rietdorf, J., 263
Rink, L., 78, 89
Rivera, V. M., 134–135, 140–141
Robertson, S. E., 72–73
Robinson, M. S., 159, 224–225
Rodgers, W., 28–29
Roegiers, F., 314–315
Rogg, L. E., 81
Rollins, C. T., 134–135, 140–141
Roman, G., 286
Römpler, H., 262

Rondanino, C., 304
Rongey, C., 81
Rørth, P., 284
Rosati, E., 27
Rosato, A., 78
Rose, J. K., 28–29
Rosenfeld, M. G., 182–183
Rosenthal, N., 248–249
Rosorius, O., 196
Ross, C. A., 324–325
Rosse, C., 285, 286
Rothman, J. E., 224
Roudier, N., 95, 97, 101, 108, 110
Rouquié, D., 35
Roy, C., 125, 126f, 127, 129
Rual, J.-F., 101, 324–325
Rubin, G. M., 48–49
Rubinsztein, D. C., 112–113
Rudnicki, M. A., 83–84
Ruggiero, C., 134–136, 137–138, 147, 148, 149f
Rusten, T. E., 284
Rutledge, E. A., 198–199
Ruvkun, G., 106

S

Sabatini, D. D., 196
Sabatini, R., 27
Sachdeva, M. M., 6
Sachse, M., 97
Sacks, D. B., 54–55
Sadowski, L., 209
Sahenk, Z., 250
Sahlender, D. A., 159, 224–225
Sahoo, S., 213
Sailler, B., 332–333, 338–340
Saito, H., 77–92
Sakuraba, H., 81
Sakurai, T., 303
Salazar, G., 81
Sall, A., 66, 72–73, 73f, 75
Sallese, M., 134, 135, 151
Saltiel, A. R., 182–183
Samie, M., 249–250, 256
Samokhvalov, V., 112–113
Sanchez-Cabo, F., 208–209
Sanchez-Madrid, F., 211
Sandilands, E., 182–183

Sangkuhl, K., 262
Sankaran, V. G., 6
Santini, G., 151
Santone, I., 151
Santoni, V., 35
Saranadasa, M., 123–124
Sarchielli, P., 27
Sasamura, T., 286
Sato, T. K., 246, 249–250
Sauna, Z. E., 208–209
Sawada, Y., 127
Sayers, L. G., 246
Sbrissa, D., 246, 247t, 252, 253–254
Scandroglio, F., 27, 35–38
Schechter, A. N., 196
Schekman, R., 151
Schelhaas, M., 66
Schenck, A., 209, 216, 238–239
Scherfeld, D., 238–239
Schiefermeier, N., 182–183
Schilling, W. P., 266–267
Schink, K. O., 4, 6
Schlierf, B., 196
Schluter, T., 332–333, 338–340
Schmidt, H., 263–264, 268, 271
Schneider, I., 289
Schoenfish, M., 332–333, 338–340
Schöneberg, T., 262
Schreckenberger, S., 332–333, 338–340
Schreff, M., 263–264, 268, 271
Schroeder, S., 112–113
Schroer, T. A., 139
Schuck, S., 29–30
Schuit, F., 79
Schultz, C., 6–7
Schultz, K. M., 213
Schulz, A., 262
Schulz, S., 263–264, 268, 271
Schulze, K. L., 303
Schuttelkopf, A. W., 346
Schwab, Y., 108–109
Schwabedissen, A., 306
Schweisguth, F., 284, 303–306, 313
Schwille, P., 238–239
Scott, B. A., 112–113
Scriver, C. R., 27
Seaman, M. N. J., 156, 157, 159, 171–174, 175–176, 224–225

Sebben, M., 332–333
Seglen, P. O., 95, 114–116
Segura, E., 208
Seidel-Rogol, B., 81
Seki, T., 284
Sekler, I., 78, 81
Sellers, J. R., 134–135
Sem Wegner, C., 5
Sen, A., 284, 285–286, 289, 295, 297, 303
Sens, P., 238–241
Settembre, C., 27
Setty, S. R., 72–73
Seyrantepe, V., 35–38
Shah, N., 147
Shaham, S., 103
Sharma, D. K., 271, 276
Shaw, C., 324–325
Sheff, D., 276
Shen, D., 249–250
Shen, J., 87
Shen, L., 75
Sherer, N. M., 208
Shevchenko, A., 29–30, 209, 234
Shiba, Y., 324–325
Shiga, K., 246–247, 247t, 249–251
Shimizu, M., 324–325
Shimoda, S., 79–81
Shin, J. E., 332–333, 338–340
Shinahara, W., 182–183
Shisheva, A., 246, 252
Short, B., 151
Shrager, P., 250–251
Shy, M. E., 250
Si, X., 66, 72–73, 73f, 75
Sieburth, D. S., 54–55
Siggia, E. D., 138–139
Simons, K., 27–28, 29–30, 276
Simonsen, A., 4
Simossis, V. A., 333
Simpson, E. M., 248–249
Simpson, J. C., 276
Simpson, P., 314–315
Simpson, R. J., 208–209
Singer, M. A., 285–286
Singh, C. M., 285–286
Singhera, G. K., 66, 68, 75
Sinkins, W. G., 266–267
Sitaram, A., 72–73

Siu, E. R., 184
Sjostrand, M., 208–209
Slaugenhaupt, S. A., 81
Slot, J. W., 12
Slutsky, S. G., 246–247, 247t, 253
Sly, W. S., 27
Smedley, D., 248–249
Soling, H. D., 135
Sone, T., 87
Sönnichsen, B., 263
Sonnino, S., 27–28, 35–40
Sorbi, S., 27
Sorkin, A., 48, 49–51, 52, 54–55, 58–59, 151, 262–263
Sou, Y. S., 324–325
Sousa-Blin, C., 177
Sowinski, S., 208
Spampanato, C., 27
Spang, A., 224–225
Spees, J. L., 208–209
Splettstoesser, B., 225–226
Srivedi, K., 158
St. George-Hyslop, P., 171–172
Stagg, S. M., 224–225
Stamatos, N. M., 35–38
Stanarius, A., 263–264, 268, 271
Stange, C., 238–239
Stathakis, D. G., 81
Steele-Mortimer, O., 17
Stefan, C. J., 246, 249–250
Steffen, J., 87
Steinberg, F., 332–333, 338–340
Stenmark, H., 4–5, 6, 17, 94, 209, 284
Stinchcomb, D., 98–99
Stirling, J. L., 27
Stockinger, W., 332–333, 338–340
Stoffel, W., 182–183
Stow, J. L., 182–183
Strakova, J., 246
Strasser, V., 332–333, 338–340
Stremmel, C., 332–333, 338–340
Strigini, M., 306
Stuffers, S., 5, 6
Stumm, R., 263–264
Stunnenberg, H., 276
Su, L., 183–184
Su, W. H., 183–184
Sudharshan, E., 6

Sudol, M., 151
Sugiura, R., 81
Sun, K., 328–329
Sun, L., 208–209
Sun, Q., 54–55
Sun, S., 183–184
Sung, H. H., 284
Surh, Y. J., 328–329
Suzuki, T., 78, 79, 85–86, 87, 88*f*
Szabó, G., 324–325
Szakacs, G., 208–209
Sze, K. L., 184
Sztul, E., 264–265, 268*f*, 273*f*

T

Tagami, S., 303
Takata, F., 127
Takatsu, H., 324–325
Takeda, S., 82, 85, 86, 87
Takeshita, F., 212
Takeuchi, H., 151
Tan, J., 332–333
Tan, X., 216–217
Tancini, B., 27–28, 31, 33, 35–40
Tang, Q., 271, 276
Taniguchi, M., 79–81
Tao, C., 284
Tassi, C., 27
Tattersall, D., 172–174
Taylor, G., 120–123
Taylor, K. M., 79–81
Taylor, R. S., 149
Teasdale, R. D., 151, 332, 333, 335–336, 340–341, 343–344, 344*f*, 346
Teis, D., 49, 182–183
Temkin, P., 332–333
Temple, B. R., 27
Tenza, D., 5
Terai, T., 182–183
Teranishi, F., 81
Tetrault, E., 332–333, 338–340
Theos, A. C., 5
Therrien, M., 48–49
Thery, C., 208, 212
Thibault, S. T., 285–286
Thomas, E. C., 151
Thomas, L., 324
Thompson, B. J., 284
Thorne, T., 213

Thornton, J. K., 81
Tian, E., 96
Tian, Y., 96
Timmons, L., 101
Tirasophon, W., 87
Tnimov, Z., 345
Toh, B. H., 4–5
Tongngok, P., 284
Toomre, D., 27–28, 216
Torok, T., 303–304
Tower, C., 264–265, 268*f*, 273*f*
Towne, C. F., 246–247, 257
Townsley, F. M., 135
Traer, C. J., 158
Trajkovic, K., 5
Tramier, M., 48
Traub, L. M., 224–225
Traynor, J. R., 135–136
Trejo, J., 27, 182–183
Tronchère, H., 246–247, 254
Trouet, A., 120–123
Trouplin, V., 216–217
Trucco, A., 134–135, 136–137
Truschel, S. T., 5
Tseng, H. Y., 332–333, 338–340
Tsuboi, T., 200
Tucker, T., 324–325
Tuft, R., 6
Tulipano, G., 263–264
Tulkens, P., 120–123
Turinsky, J., 182–183
Turner, J. R., 75
Tushir, J. S., 72–73

U

Udenfriend, S., 136
Ueba, O., 68
Ueno, T., 324–325
Ullrich, O., 196
Ulrich, R. G., 120–123
Ulug, E. T., 246
Umeno, H., 81
Umezu, T., 212–213
Urbanelli, L., 27–28, 31, 35–40
Urbé, S., 196
Utokaparch, S., 66, 68, 75
Uttenweiler-Joseph, S., 209

V

Vaccari, I., 246–247, 254
Vaccari, T., 284, 303
Vaccaro, C., 271, 276
Vaerman, J. P., 151
Valadi, H., 208–209
Valentine, R. A., 81
Vallance, H., 120–123
Valle, D., 27
van Aalten, D. M., 346
van den Hoff, M. J., 267
van der Sluijs, P., 196, 271, 276
Van Hoof, F., 120–123
van Kerkhof, P., 332–333, 338–340
van Renswoude, J., 196
van Summeren, R., 135, 147
van Vliet, C., 151
Van Waes, C., 123–124, 127, 129
Vardarajan, B. N., 171–172
Varga, J., 54–55
Velasquez, V., 79
Veraksa, A., 285, 286
Verges, M., 151
Verkhusha, V. V., 48
Verma, R., 246–247, 247t, 254
Vetrivel, K. S., 303
Vidal, M., 211
Vignaux, F., 303
Villafania, L., 208–209
Villarroya-Beltri, C., 208–209, 211
Virbasius, J. V., 4
Vodovar, N., 305–306
Vogt, S., 79
Volceanov, L., 177
Volchuk, A., 134–135, 140–141
Volpicelli-Daley, 246
von Zastrow, M., 48, 54–55, 151, 262–263, 332–333
Vonderheit, A., 271
Vuoriluoto, K., 276

W

Waguri, S., 324–325
Walker, J. M., 79–81, 87
Walsh, B. J., 182–183
Wandinger-Ness, A., 276
Wang, C. K., 333
Wang, F., 79–81, 87
Wang, H., 303

Wang, J., 328
Wang, M., 216–217
Wang, X., 134–135, 140–141, 249–250, 256
Wang, Y., 87, 212–213, 216
Ward, E. S., 271, 276
Ward, J. H., 198–199
Wardwell, S., 134–135, 140–141
Wassmer, T., 158, 226–227
Watanabe, K., 79–81
Way, D., 208–209
Weaver, B. P., 79
Webster, P., 276
Weeratunga, S. K., 333
Wei, M., 208–209
Wei, Y., 328–329
Weigert, R., 196
Weinmaster, G., 208, 303
Weisman, L. S., 246, 249–250
Wen, L., 216
Wenzel, D., 5
Wessling-Resnick, M., 276
Westphalen, K., 208–209
Wheatley, C. L., 271, 276
White, G. A., 208–209
Whiteford, C. C., 246
Whitney, M. J., 208–209
Widmaier, M., 332–333, 338–340
Wieland, F., 5
Wieland, F. T., 224
Wiley, H. S., 182–183
Wilkin, M. B., 284
Willars, G. B., 265–266
Wilson, D. W., 135
Windler, S. L., 303
Winters, J. J., 246–247, 247t, 250–252, 256–257
Wiradjaja, F., 246, 249–250
Wirtz, K. W., 135, 147
Wirtz-Peitz, F., 303–304
Wong, B. W., 75
Wong, E. W. P., 183–184
Wong, H. K., 324–325
Wong, J., 75
Wood, M. J., 211
Woolfson, D. N., 134–135, 140–141
Wrana, J. L., 182–183, 209
Wright, T. R., 81
Wu, C. Y., 87, 159

Wu, D., 216
Wu, M. N., 289
Wu, W., 66, 72–73, 73f, 75
Wu, Y. W., 345
Wunderlich, W., 49
Wurmser, A. E., 246, 249–250
Wüthrich, K., 343

X
Xia, T. H., 343
Xiang, J., 208
Xiao, X., 191f
Xie, X., 303
Xie, Y., 208, 247t, 253–254
Xu, A., 216
Xu, G., 196
Xu, H., 54–55
Xu, S., 101

Y
Yada-Matsushima, T., 81
Yaffe, M. B., 75
Yamada, K., 284
Yamada, M., 324–325
Yamaguchi-Iwai, Y., 79–81, 85–86, 87
Yamakami, M., 324–325
Yamamoto, S., 284, 303
Yamamoto, Y., 182–183
Yamasaki, S., 78, 79, 81, 85–86, 89
Yamauchi, A., 127
Yamazaki, T., 79–81, 85–86, 87
Yan, H. H. N., 183–184
Yan, Q., 250
Yanagawa, B., 75
Yanagida, K., 303
Yanagida, Y., 324–325
Yang, J. M., 211
Yang, Y., 216
Yao, C., 208

Yap, A. S., 182–183
Yarden, Y., 48
Ye, L., 334
Yeh, B. J., 48–49, 54–55
Yeung, A. C., 196
Yokosawa, H., 324–325
Yoshimori, T., 112–113, 324–325
Yoshioka, Y., 212
Yu, H., 334
Yuan, J., 75

Z
Zaal, K. J., 139
Zaccolo, M., 144
Zeke, A., 54–55
Zeng, J., 196
Zerial, M., 17, 196, 209, 263, 276
Zhang, J., 75
Zhang, M., 75, 216–217
Zhang, Q., 249–250, 253
Zhang, S., 303
Zhang, X., 249–250, 256
Zhang, Y., 246–247, 247t, 249–251, 253, 254, 334
Zhang, Y. E., 216–217
Zhao, Y., 5, 96
Zhou, J., 271, 276
Zhou, Y., 289
Zigmond, S. H., 182–183
Zijderveld, A., 83–84
Zimm, G. G., 286
Zinchuk, O., 214–215
Zinchuk, V., 214–215
Zitserman, D., 303, 314–315
Zolov, S. N., 246–247, 247t, 252–253, 254
Zoncu, R., 216
Zou, Z., 328–329
Zuo, C. D., 266–267

SUBJECT INDEX

Note: Page numbers followed by "*f*" indicate figures and "*t*" indicate tables.

A

Alkaline phosphatase (ALP)
 description, 87
 measurement of, 87–88
ALP. *See* Alkaline phosphatase (ALP)
Amphisomes visualization
 endosomal and autophagic
  compartments, 114–116
 hypotheses, 114–116
 LGG-1 and VPS-27, 114–116,
  115*f*
Antibodies
 primary, 96–97
 secondary, 97
Antibody uptake assays
 anti-CD8 antibody-uptake, 159,
  161*f*
 anti-CIMPR antibodies, 159
 anti-Delta and anti-Notch, 306, 306*f*
 CDMPR, 157–158
 CD8-reporter proteins, 159, 160*f*
 CIMPR, 157–158
 day before experiment, 308–309
 1 day before experiment, 308
 day of experiment, 309
 description, 305–306
 endosome-to-golgi recycling pathway,
  157
 fibrin clot, 309–310
 high-throughput scale
  antibody-uptake and staining, 166
  automated imaging, 167–169
  cell seeding and siRNA transfection,
   165
  endosome-to-golgi retrieval assays,
   163–164, 164*f*
  library/mini-library siRNAs, 163–164
  mini-library screen, 163
  mouse hybridoma cells, 164
  robotic liquid handling system, 163
 whole-genome screen, 164–165
 IgG-bound and -unbound Fabs, 307
 medium scale, 169
 Notch and Delta trafficking
  live imaging, 311–312
  quantification, 312–313
 pupal notum, 306
 rabbit IgGs, 307–308
 small scale, 157–169
 trans-golgi network (TGN), 158
 Zenon technology, 307
APPL endosomes, 216
Autophagosomes
 amphisome visualization, 114–116
 and endo-lysosomal system, 95
 in vivo fluorescence/
  immunofluorescence, 110–111
 mammalian cells, 95
Autophagy analysis
 Atg8, recruitment, 110–111
 autophagosomal maturation, 111
 flux, 111–112
 LGG-1 and LGG-2 dots, 111
 pharmacological modulation
  3-MA and Wortmannin treatments, 114
 spermidine plates setup and *C. elegans*,
  113–114

B

Band 4.1/ezrin/radixin/moesin (FERM)
 domain
 description, 332
 phosphoinositide lipids and peptide cargo
  motifs, 338–341
 recombinant PX-FERM proteins,
  332–333, 335–338
 SNX27 PDZ, 332–333
 SNX17 PX domain-PI3*P* interactions,
  341–346

369

Biolistic transformation
 bombardment, 100
 gold particles preparation and DNA
  coating, 99–100
 plating worms, 100
 screening, 100
 worms preparation, 99
Blood–testis barrier (BTB)
 Cdc42-dominant negative mutant,
  183–184
 integral membrane proteins, 184–187
 mammalian testis, 183–184
 permeability function, 183–184
 Sertoli cell, 183–184
 spermatogenesis, 183–184
BTB. See Blood–testis barrier (BTB)

C

CAR. See Coxsackievirus and adenovirus
  receptor (CAR)
Cargo proteins, 140–141
Carrier biogenesis, synthetic membranes
 binding, cytosolic proteins, 233
 coupling, peptides (see Peptides to
  liposomes, coupling)
 flotation density centrifugation, 226–227
 hydrazino-peptides, 226–227
 liposomes preparation, 229–230
 mouse brain cytosol preparation, 227
 purification, coated liposomes, 233–234
 wild-type/mutated cytoplasmic domains,
  226–227
 6xHis-MBP-Tev-Cys-tagged proteins,
  227–229
Cation-dependent mannose-6-phosphate
  receptor (CDMPR), 157–158
Cation-independent mannose-6-phosphate
  receptor (CIMPR), 157–158
Cation trapping
 antineoplasic applications, 129
 diffusion, drugs, 129
 endosome-lysosome continuum,
  120–123
 hydrophilic local anesthetics, 120–123
 macroautophagic accumulation, cells,
  128–129
 organic amines, 120–123
 pharmacologic actions, 129

quinacrine uptake (see Quinacrine uptake,
  cells)
 trans-Golgi network and derived
  organelles, 120–123
 triethylamine series, 120–123, 121f
 V-ATPase- mediated concentration,
  120–123, 122f
CDMPR. See Cation-dependent mannose-
  6-phosphate receptor (CDMPR)
Cell-cell communication. See Intercellular
  transfer (ICT)
Cellular acidic compartments. See Cation
  trapping
Charcot-Marie-Tooth disease, 246–247
CIMPR. See Cation-independent mannose-
  6-phosphate receptor (CIMPR)
CLEM. See Correlative light and electron
  microscopy (CLEM)
Core machineries
 carrier biogenesis (see Carrier biogenesis,
  synthetic membranes)
 label-free quantitative proteomics, 234–237
Correlative light and electron microscopy
  (CLEM)
 cell lines, 17
 cell seeding, gridded coverslips, 17
 chemical fixation, 16–17
 correlation and image processing, 21
 electron microscopy, 21
 Epon embedding, 19–20
 fixation, staining and mounting, 18
 high-pressure freezing, 16–17
 light microscopy and registration, cells
  confocal microscopy, 18
  super-resolution light microscopy, 19
 live cell microscopy, 17
Correlative pupa imaging
 antibody uptake assays and
  immunostainings
  protocol, 316–318, 319f
  reagents, 316
 SEM, adult flies, 314–315
 SOP division, 313–314
Coxsackievirus and adenovirus receptor
  (CAR)
 anti-JAM-A, 190
 endocytosis, 186f
Cultured adherent cells, 124–125

D

Degradation. See Lysosomes, TfR degradation
Delta trafficking. See Notch signaling pathway
Drosophila notum
  asymmetric cell division, 305
  neurogenesis, 304
  Notch signaling, 304–305
  pIIIb cells, 304–305
DT40 cells deficient
  functional analysis, zinc mobilization, 86–88
  ZnTs (see Zinc transporters (ZnTs))

E

EGFR. See Epidermal growth factor receptor (EGFR)
Electron microscopy (EM)
  description, 108
  low-melting agarose pads, 108
  VPS-27::GFP and GFP::LGG-1, 108–109
EM. See Electron microscopy (EM)
Endocytosis
  assay
    biotinylation, cell surface proteins, 184–187, 185f
    buffers, 187–188
    CAR and JAM-A, 184–187, 186f
    cell lysates, 184–187
    cell staining, 190
    immunofluorescence microscopy, 190–191, 191f
    materials, 187, 291–292
    methods, 188–190, 292
    sulfo-NHS-biotin, 184–187
  asymmetric activation, 303–304
  description, 302
  Drosophila genes, 302–303
  furin-like protease, 302
  Notch activation, 303
  S3 cleavage, 303
  SOP system, 303
Endomembrane-initiated signaling
  biomolecular sensors, 135

carboxy-terminal sequence, 135
ER-to-golgi traffic-synchronization protocols, 136–141
FRET, 135
GPCRs, 135–136
KDELR (see KDELR signaling)
membrane trafficking, 134–135
regulatory circuits, 134
traffic-generated signaling, 141–146
transport synchronization assays, 134–135
Endosomal–lysosomal system
  cell surface glycosphingolipids, 43–44
  cell surface receptor regulation, 27
  confocal microscope analysis, 28
  fluorescence microscopy analysis, 38–42
  hydrolysis, terminal N-linked galactosyl moiety, 27
  immunoaffinity capture, 28
  intracellular trafficking, 43–44
  lipid microdomain-associated glycohydrolases, 31–34
  lysosomal enzymes, 27
  neurodegenerative disorders, 27
  plasma membrane-associated isoforms, 28
  purification, lipid microdomains, 28–31
  signal transduction pathways, 27–28
Endosomal maturation
  amphisome visualization, 114–116
  antibodies, 96–97
  and autophagy analysis, 110–114
  Caenorhabditis elegans, 95
  description, 94
  endocytosis and autophagy, 95
  ESCRT mutants, 100–103
  fluorescent-tagged protein construction and transgenesis, 98–100
  macroautophagy, 95
  MVBs, 94
  RNAi clones, 96
  solutions, 97–98
  strains, 95
  vesicular compartments, 104–110
Endosomal protein sorting
  antibody-uptake assays (see Antibody uptake assays)
  automated imaging, 170–171
  cargo molecules, 157

Endosomal protein sorting (*Continued*)
  cell biological imaging-based methods, 177
  classification, 169–170
  generation, immunity and autophagy, 156
  membrane proteins, 156
  morphological changes, 169–170
  native IP, 172–176
  pellet *vs.* supernatant, 171–172
  retromer-mediated trafficking, 157
  Spotdetector bio-application, 170–171
  transient transfection, native immunoprecipitations, 176–177
  whole cell stain image, 170–171
Endosomal signaling
  biochemical-based model, 183–184
  BTB, 183–184
  Cdc42, 183–184
  description, 182–183
  dual-labeled immunofluorescence, 182–183
  endocytosis assay (*see* Endocytosis)
  IL-1α, 184
  microdomain, 182–183
  polarity proteins, 183–184
  protein ubiquitination, 182–183
  radiolabeled proteins, 182–183
  Sertoli cells, 183–184
Endosomal sorting complex required for transport (ESCRT). *See* ESCRT mutants
Endosomal trafficking, GPCRs.
  See G-protein-coupled receptors (GPCRs)
Endosome-to-golgi recycling pathway, 157
Epidermal growth factor receptor (EGFR)
  degradation assay
    knockdown cells, 204
    materials, 204
  endocytic pathway, 203–205
  and PtdIns3P localization
    confocal microscopy, 7–8, 8*f*
    immuno-EM, 10–12
    pre-embedding immuno-EM, 13–16
    super-resolution light microscopy, 9–10
  recycling pathway, 196
Epifluorescence microscopy, 127

Epitope tag
  SSTR3 and Rabs, 264
  wild-type/mutant Rab, 269
ER-to-golgi traffic-synchronization protocols
  PC-I traffic synchronization, 136–137
  PC-IV traffic synchronization, 137–138
  polymerization/depolymerization, cargo proteins, 140–141
  VSVG traffic synchronization, 138–139
ESCRT mutants
  developmental phenotypes, 103, 104*f*
  genetics, 100–101
  RNAi experiments, 101–103
Exosomes
  microvesicles
    isolation, 211–212
    labeling after purification, 212
    labeling, de novo, 211
  signaling
    APPL endosomes, 216
    SARA endosomes, 216–217

F

Fibrin clot
  advantages, 310
  experiment, 310
  P200 pipette, 310
  tissue, glass-bottom dish, 309–310
Fig4
  conditional knockout, 251–252
  FIG4-I41T, 252
  genetic interactions, 254–255
  neurons *vs.* astrocytes, 251
  pale tremor mouse (*see* Pale tremor mouse)
  strain background, genetic effects, 255, 256*f*
FIG4-I41T, human disease mutation, 252
Fluorescence microscopy analysis
  cholesterol depletion and isolation, lipid microdomains, 40
  colocalization studies, 42, 43*f*
  DEAE-chromatography, 38–40
  GFP construct, transfection and cell selection, 38
  HEXA cDNA preparation, 38
  intracellular distribution, recombinant Hex A-GFP, 42, 42*f*

Fluorescence techniques measurement, ICT
  confocal microscopy, 214–215, 215f
  flow cytometry, 214
Fluorescent resonance energy transfer
  (FRET), 135
Fluorescent-tagged protein
  construction, 98
  transformation procedures, 98–100
Freeze-cracking permeabilization, 104–105
FRET. *See* Fluorescent resonance energy
  transfer (FRET)

G

Genetic modifier screens
  β-arrestin homolog, 285
  description, 284
  endocytic trafficking, 284
  exelixis collection, 285–286
  localization, Notch, 289–292
  materials, 286
  modifiers, Dx–Krz action
    primary screen, 288
    screening strategy validation, 288, 288f
    secondary screens, 289
    validation, screening phenotype, 287,
      288f
  mutations, 284
  Notch pathway, 284
  optical approaches, 292–295
  ubiquitination (*see* Ubiquitination, Notch
    signaling)
GFP. *See* Green fluorescent protein (GFP)
GFP-virus-infected cells
  antibody staining, influenza method, 70
  cellular infectivity, 69–70
  high-throughput segmentation
    algorithms, 70
  human respiratory viruses, 68
  influenza infectivity, flow cytometry, 69–70
  pharmacological kinase inhibitors, 68, 69t
Giant unilamellar vesicles (GUVs)
  cytosol preparation, 239
  description, 238–239
  imaging, protein dynamics, 241, 242f
  liposomes, 239–241
  micromanipulation, 238–239
Glycohydrolase Hex and Gal assay, 30–31,
  30f, 32f

GPCR kinases (GRKs), 262–263
GPCRs. *See* G-protein-coupled receptors
  (GPCRs)
G-protein-coupled receptors (GPCRs)
  description, 262
  and GFP, 264
  and GRKs, 262–263
  heterotrimeric Gαβγ proteins, 262
  KDELR folding, 135–136
  and Rab GTPases, 263
  Rabs and SSTR3, 269–272
  and SSTR3 (*see* Somatostatin receptor 3
    (SSTR3))
Green fluorescent protein (GFP)
  polyclonal antibody, 268f
  tagged GPCR, 264
GRKs. *See* GPCR kinases (GRKs)
GUVs. *See* Giant unilamellar vesicles
  (GUVs)

H

HD. *See* Huntington disease (HD)
HFs. *See* Human skin fibroblasts (HFs)
Htt. *See* Huntingtin (Htt)
Human skin fibroblasts (HFs), 136–137
Huntingtin (Htt)
  polyQ aggregates, 324–325
  and RFP-Tollip, 325–326
Huntington disease (HD), 324–325

I

ICT. *See* Intercellular transfer (ICT)
IMCD-3 cells
  stably expressing SSTR3, 267–269, 268f
  transfection, electroporation, 266–267
Immuno-electron microscopy (EM)
  cell preparation, 10–12
  Epon embedding, 14–16
  gelatin, 10, 11f
  labeling, anti-GFP, 12
  localization, GFP2xFYVE and EGFR,
    12, 12f
  pre-embedding labeling, 13–14
  primary antibodies, 12
Immunofluorescence confocal microscopy
  cell staining, 74
  nuclear staining, 74
  virus staining, 74

Immunofluorescence on embryos
  antibody staining, 105
  methanol-acetone fixation, 105
  paraformaldehyde fixation, 105
  permeabilization, freeze-cracking, 104–105
Immunolocalization protocol
  antibody staining, 107–108
  mounting worms, 108
  permeabilization and fixation, 106–107
Immunostaining
  materials, 289–290
  methods, 290–291
Intercellular transfer (ICT)
  cell communication, 208
  description, 208
  fluorescence techniques measurement, 214–215
  labeling
    cellular proteins and lipids, 209–211, 210f
    exosomes/microvesicles, 211–212
    nucleic acids, 212–213
  regulation, signaling, 209
  signaling endosomes, 216–217
Intracellular ZnTs
  B cell receptor signaling pathway, 79–81
  biological functions, 81
  cytosolic signaling, 79–81
  endosomal signaling, 81
  expression, zinc transporters, 79–81
  L-type calcium channel, 81
  regulation, zinc, 79–81
  secretory pathway homeostasis, 81
  zinc mobilization, 79
  zinc-requiring enzyme activities, 81
In vivo analysis, acidification, 112

K

KDAR. See Knockdown and reconstitution (KDAR)
KDELR signaling
  BODIPY, 147
  genetically encoded artificial ligand, 148
  Golgi-SFKs activation, 147, 149f
  KDELR-induced [^{35}S]GTPγS loading, G$\alpha_{q/11}$, 149–151
  membrane-permeant peptides, 148
  receptor overexpression, 147–148
  signal-transduction mechanisms, 147
  stimulation, 135
  transduction machinery, 149–151
Knockdown and reconstitution (KDAR)
  MEK2 analysis, 49–54
  SHOC2 protein, 54–61

L

Label-free quantitative proteomics
  LC-MS acquisition, 235–236
  postacquisition workflow
    crumbs 2- and PI3P-dependent manner, 236, 238t
    cytosolic proteins, 236, 237f
    MaxQuant (V1.2.2.5), 236
    proteinGroups.txt-file, 236
    volcano plot, 236, 237f
  SDS-PAGE and in-gel digestion, 234–235
Lipid microdomain-associated glycohydrolases
  affinity chromatography, 34
  biotinylation, plasma membrane proteins, 33–34
  characterization, Hex and Gal assay, 30–31, 30f, 32f
  in vivo assay, 35–37, 37f
  immunology capture, 35
  lipid microdomains concentration, 34
  natural and synthetic substrates, 31
  protein biotinylation, 31
  purification, 31, 33f
  sucrose-density gradient centrifugation
    biochemical method, 28–29
    lipid microdomain extraction methods, 29–30, 30f
    lipid–protein interactions, 28–29
    types, detergents, 28–29
Liposome-based assays
  carrier biogenesis, synthetic membranes, 226–234
  coat protein complexes, 224–225
  description, 224
  GTPases, 224–225
  protein dynamics (see Giant unilamellar vesicles (GUVs))
  subcellular proteomics, 225–226

transport carrier biogenesis, 225
tubular-vesicular element, 224
Lysosomes, TfR degradation
  constitutive degradation, western
      blotting, 197–199
  immunofluorescence method, 199–200
  materials, 197
  Rab (*see* Rab proteins)
  western blotting, 199

M

Macroautophagic accumulation, cells
  anti-human LC3B antibodies, 129
  giant vacuoles, 128
MaxQuant (V1.2.2.5), 236
MEK2 analysis
  expression constructs, 49–51
  fluorescent microscopy, 52
  MEK2-GFP-tagged protein, 52–54
  RNA interference, 51–52
  western blot analysis, 52
Membrane-permeant peptides, 148
Membrane trafficking pathway.
      See Recycling endosomes
Microinjection transformation, 98–99
MicroRNAs (miRs)
  labeling, 213
  transcriptional communication, 208–209
Microvesicles. *See* Exosomes
miRs. *See* MicroRNAs (miRs)
Mouse brain cytosol preparation, 227
Mouse models
  altered metabolism, PI, 246–247, 247*t*
  C57BL/6 J, 249
  Charcot-Marie-Tooth disease, 246–247
  CRE recombinase, 248–249
  Fig4 and *Vac14*, 247–248
  gene-trap mice, 248
  PI(3,5)P$_2$ deficiency, 255–257
  targeted knockout lines, 248–249
  tissue-specific promoter fragment, 248
MTMR2. *See* Myotubularin-related 2
      (MTMR2)
Multivesicular bodies (MVBs)
  depletion, ESCRT genes, 103
  endosomal sorting, 94
  lysosomes, 94
Multivesicular endosomes (MVEs)

biogenesis, 5
CLEM, 16–21
early endosomal antigen1 (EEA1), 4–5
EGFR, 4
endocytic pathway, 4
FYVE domain-containing proteins, 4
microscopical techniques, 21
PtdIns3P and EGFR, 5–16
UIM, 5
MVBs. *See* Multivesicular bodies (MVBs)
MVEs. *See* Multivesicular endosomes
      (MVEs)
Myotubularin-related 2 (MTMR2)
      phosphatase, 254–255

N

Native IP, endosomal proteins, 172–177
Notch signaling pathway
  antibody uptake assays, 305–313
  Drosophila notum, 304–305
  Dx and Krz, 289
  endocytosis and endosomal trafficking,
      302–304
  endocytosis assay, 291–292
  endosomal markers, 289
  immunostaining, 289–291
  and ligand molecules, 293–294
  ubiquitination, 295–296
Nucleic acids
  description, 212–213
  labeling, miRs, 213

O

Optical approaches
  description, 292–293
  fluorescently tagged Notch and ligand
      molecules, 293–294
  stable cell lines
    materials, 294
    methods, 294
    rescue experiments, 295

P

Pale tremor mouse
  autophagic pathway, 250
  cellular phenotype, 249–250, 250*f*
  description, 249–250
  pathological characteristics, 250–251

Pale tremor mouse (*Continued*)
  vacuole membranes, 249–250
Paraformaldehyde fixation, 105
PC-I traffic synchronization, 136–137
PC-IV traffic synchronization, 137–138
Pellet *vs.* supernatant, endosomes, 171–172
Peptides to liposomes, coupling
  alpha-hydrazino acetyl group, 231–233
  cysteine, 230–231, 232*f*
Phosphatidylinositol 3-phosphate
    (PtdIns3P)
  and EGFR localization (*see* Epidermal
    growth factor receptor (EGFR))
  endocytosis, 5
  fluorescently labeled, 6–7
  GFP, 6
  protein–lipid interaction, 6
Phosphoinositide lipids, 338–341
Phox homology (PX) domain.
    See PX-FERM proteins
Pikfyve mutant
  conditional knockout, 254
  hypomorphic gene-trap allele, 253–254
Primary antibodies, 96–97
Protein endocytosis
  Cdc42, 183–184
  Par3/Par6, 183–184
  TGF-β3-mediated disruptive effects,
    183–184
PtdIns3P. *See* Phosphatidylinositol
    3-phosphate (PtdIns3P)
PX-FERM proteins
  functional modules identification,
    334–335
  phosphoinositide lipids and peptide cargo
    motifs
    description, 338–340
    ITC experiment, 338–340, 341*f*
    methodology and practical
      considerations, 338–340
    recombinant, 335–338

Q

Quinacrine uptake, cells
  antiprotozoal drug, 123–124
  cell fractionation, 128
  cell uptake kinetics, 123–124
  cultured adherent cells, 124–125

DNA-binding antineoplasic agents,
  123–124
epifluorescence microscopy, 127
fluorescein, optical instruments, 123–124
microscopic, cytofluorometric correlates,
  127–128
pigmented/fluorescent amines, 123–124
pinocytosis, 128
suspension, cells, 125–127
undifferentiated, nonadherent U937
  monoblastoid cells, 125, 126*f*
in vivo metabolites, 123–124

R

Rab12 knockdown
  description, 203
  EGFR degradation assay, 204
  Tf recycling assay, 204, 205
Rab proteins
  colocalization rates determination
    EGFP-Rabs and Lamp-1, 201*f*, 202
    TfR and EGFP-Rabs, 200–201, 201*f*
  mammalian kidney cells
    confocal microscopy colocalization,
      SSTR3, 271–272
    microscope setup, 270–271
    transfection, IMCD-3 cells, 269–270
  SSTR3 transit
    half-time residency analysis, 274–276,
      275*f*
    time-lapse dual color imaging,
      272–273, 274*f*
  TfR expression level, 202–203, 203*f*
  trafficking, SSTR3
    dynamics, cells expression, 277–278
    mutant forms, endosome, 276–277
Recombinant PX-FERM proteins
  expression, 335–336
  purification, 336–338
Recycling endosomes
  degradation, TfR (*see* Transferrin receptor
    (TfR))
  description, 196
  genome-wide screening, 196
  Rab12 knockdown, 203–205
  screening methods, Rab proteins,
    200–203
Respiratory syncytial virus (RSV), 68
RNA interference (RNAi)

Subject Index

clones, 96
dsRNA transcription, 101–103, 102f
feeding, 103
injection, 103
RSV. *See* Respiratory syncytial virus (RSV)

S

Scanning electron microscopy (SEM)
  adult flies, 314–315
  and SOPs (*see* Sensory organ precursor (SOP) cells)
Secondary antibodies, 97
Secretory and endosomal-lysosomal compartments
  analysis, proteome, 78
  cellular integrity and homeostasis, 78
  functional analysis, zinc mobilization
    cytosolic zinc levels and zinc-related homeostasis, 87
    immunofluorescence analysis, 86–87
    measurement, ALP activity (*see* Alkaline phosphatase (ALP))
    zinc-dependent enzymes, 87
  intracellular ZnTs, 79–81
  molecular functions, 78–79
  types, immune cells, 78
  zinc-related phenotypes, 78–79
SEM. *See* Scanning electron microscopy (SEM)
Sensory organ precursor (SOP) cells
  correlative antibody uptake assay, 319f
  daughter cells, 314–315
  Delta endocytosis, 306
  pupa imaging, 313–314
Sertoli cell endosome. *See* Endosomal signaling
SHOC2 protein
  cell sorting, 60
  DNA transfections and lentivirus production, 58
  ERK1/2 pathway, 54–55
  experimental strategy, 59
  expression constructs
    cell lines, lentiviruses, 55, 56f
    enzymatic digest and dideoxynucleotide sequencing, 57
    fluorescently labeled version, 55
    silent mutations, 57
    T4 polynucleotide kinase (PNK), 55–57

expression system, 58–59
fluorescent protein tag, 60
lentivirus-mediated KDAR cells, 59–60
visualization, endosomal Shoc2-tRFP, 60–61
western blotting, 59
Signaling proteins, endosomes
  adaptors/scaffolds, 48–49
  cell surface biotinylation, 61–62
  MEK2 (*see* MEK2 analysis)
  optical approaches, 48
  SHOC2 (*see* SHOC2 protein)
  signal transduction cascades, 48
  spatial distribution, 48
SMAD anchor for receptor activation (SARA)
  endosomes, 216–217
  ICT to SARA-positive signaling endosomes, 217
  positive signaling endosomes, 217
SNX. *See* Sorting nexin (SNX)
Somatostatin receptor 3 (SSTR3)
  description, 264–265
  generation, tagged construct, 265–266
  IMCD-3 cells transfection, 266–267
  and Rabs (*see* Rab proteins)
SOP cells. *See* Sensory organ precursor (SOP) cells
Sorting nexin (SNX)
  ^{15}N and ^{13}C isotopic labeling, 342–343
  NMR data acquisition, 343–344, 344f
  PI3P binding site, 346
  SNX17 PX domain-PI3P NMR titrations, 344–345
SSTR3. *See* Somatostatin receptor 3 (SSTR3)

T

TfR. *See* Transferrin receptor (TfR)
Time-lapse dual color imaging, 272–273, 274f
Toll-interacting protein (Tollip)
  cytoskeleton-dependent transport
    biochemical analysis, 327
    microspical analysis, 327
  description, 324
  detection, cell death, 326
  early/late endosomal localization, 328
  endosomal trafficking system, 324–325

Toll-interacting protein (Tollip) (*Continued*)
  expression vector, 325
  HD, 324–325
  and Htt expression constructs, 325
  knockdown, 325–326
  protein quality control, 328, 328*f*
  and QBP1, 324–325
Tollip. *See* Toll-interacting protein (Tollip)
Traffic-induced tyrosine phosphorylation
  FRET analysis, 144–146
  IF microscopy, 141–143
  western blotting, 143–144
Transferrin receptor (TfR)
  assay
    knockdown cells, 205
    materials, 204
  degradation, lysosomes, 197–200
  endocytosis, 196
  recycling, 196
Transgenesis. *See* Fluorescent-tagged protein

U

Ubiquitination, Notch signaling
  assay, 296
  Dx and Shrub, 295
  ESCRT-III complex, 295
  materials, 295–296

V

Vac14, missense mutation
  genetic interactions, 254–255
  ingls, 252–253
  null gene-trap allele, 253
Vesicular stomatitis virus G glycoprotein (VSVG), 138–139
Virus–host cell receptor interactions
  inhibition, toll-like receptor 4 signaling, 71–72
  MAPK phosphorylation, 71
  pattern recognition receptors, 70–71
  receptor neutralization, ERK signaling, 72
Virus-induced signaling
  endocytosis pathways, 66
  ERK MAP kinase activation, 66
  histocompatibility complex type 1, 66
  immunofluorescence confocal microscopy, 73–74
  p38 MAP kinase signaling, 66

primary pulmonary fibroblasts, 67
purification
  protocol optimization, 67
  sucrose purification, RSV, 68
  three-dimensional modeling and rendering, 74–75
Virus replication
  Arf6-mediated endocytosis pathway, 66
  GFP-virus-infected cells (*see* GFP-virus-infected cells)
Virus staining
  immunofluorescence confocal microscopy, 74
Vital dyes and fluorescent construct, 109–110
VSVG. *See* Vesicular stomatitis virus G glycoprotein (VSVG)
VSVG traffic synchronization, 138–139

W

Western blot analysis
  assay, 71
  autophagic flux, 111–112
Wortmannin treatments, 114

X

6xHis-MBP-Tev-Cys-tagged proteins, 227–229

Z

Zinc transporters (ZnTs)
  characteristics, 82
  construction and preparation, knockout vectors, 82
  electroporation, 82–83
  estrogen receptor chimeric Cre recombinase, CreER, 84
  genomic DNA preparation, clones, 83–84
  human orthologs, ZIP, 85
  immunoprecipitation, 85–86
  intracellular (*see* Intracellular ZnTs)
  phenotypes, 84–85
  transient transfection, 86
ZIPs. *See* Zrt/Irt-like proteins (ZIPs)
Zrt/Irt-like proteins (ZIPs)
  immunofluorescence analysis, zinc mobilization, 86–87
  and ZnT transporters, DT40 mutants, 85

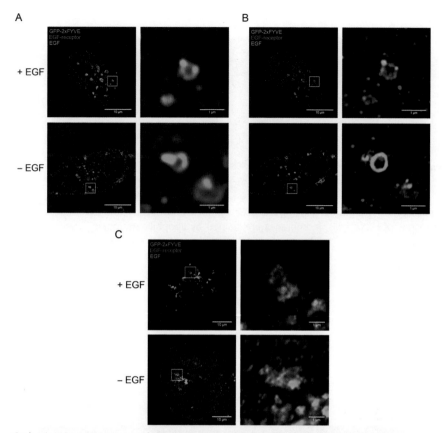

Catherine Sem Wegner et al., Figure 1.1 Localization of GFP2xFYVE, EGF, and its receptor studied using super-resolution and confocal microscopy. U2OS-GFP2xFYVE cells seeded on coverslips. After about 20 h the cells were either fixed directly (A, B, C, lower) or following a 15 min pulse with 50 ng EGF-Alexa647 per ml medium (A, B, C, upper). Cells were fixed in 4% FA/0.1% GA (A and B) or 3% FA (C) and labeled with Sheep anti-EGFR (1/1000) followed by a secondary Alexa Fluor 555 antibody. (A) Deconvolved widefield SIM images, (B) reconstructed SIM images, and (C) deconvolved confocal images. Bars: 10 μm (overviews, A, B, C, left), 1 μm (enlarged areas, A, B, C, right).

Alessandro Magini et al., Figure 2.10 Intracellular distribution of recombinant Hex A-GFP. Fluorescent microscopy images of pEGFP-N1-HEXA cells. Nucleus is evidenced by DAPI staining. Magnification, 100×.

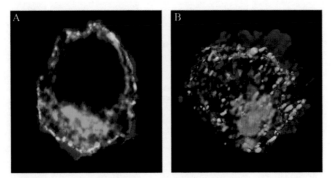

Alessandro Magini et al., Figure 2.12 Confocal microscopy and 3D-reconstruction images of HEXA cells. Membranes are stained by lipophilic membrane dye DiQ (red). (A) Confocal microscopy image of pEGFP-N1-HEXA cells. Colocalization is reported in yellow. (B) Three-dimensional reconstruction of confocal microscopy image reported in (A). Colocalization is reported in blue.

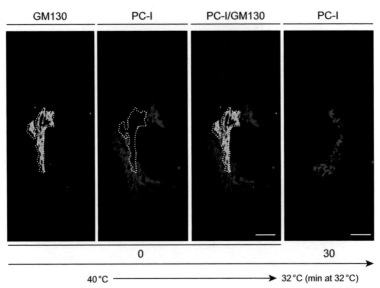

Carmen Ruggiero et al., Figure 8.1 Synchronized transport of PC-I. HFs were incubated for 3 h at 40 °C, shifted to 32 °C for 30 min, and double stained for GM130 (green) and PC-I (red). Scale bars, 10 μm. *Modified from Pulvirenti et al. (2008).*

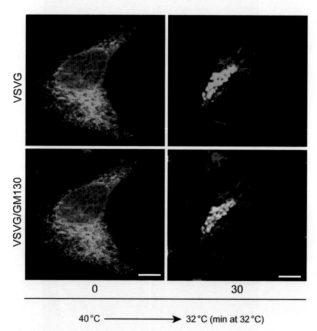

Carmen Ruggiero et al., Figure 8.2 Synchronized transport of VSVG. HeLa cells were transfected with VSVG–GFP, incubated overnight at 40 °C, shifted to 32 °C for 30 min, and stained for GM130 (red). Scale bars, 10 μm.

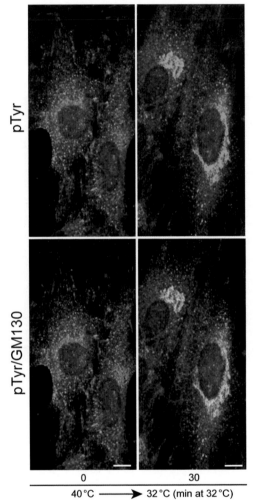

Carmen Ruggiero et al., Figure 8.3 Transport-dependent tyrosine phosphorylation by immunofluorescence microscopy. HFs were incubated for 3 h at 40 °C, shifted to 32 °C for 30 min, and double stained for pTyr (green) and GM130 (red). Scale bars, 10 μm.

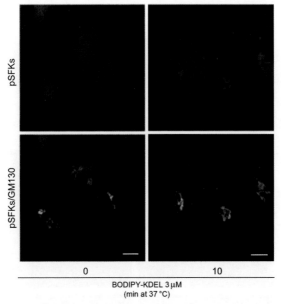

Carmen Ruggiero et al., Figure 8.5 BODIPY-KDEL triggers Golgi-SFKs activation. HeLa cells were treated with the membrane-permeant KDELR agonist BODIPY-KDEL (3 μM) at 37 °C, for 10 min. The cells were then stained for active SFKs (pSFKs, red) and GM130 (green). Scale bars, 10 μm. Modified from Giannotta et al. (2012).

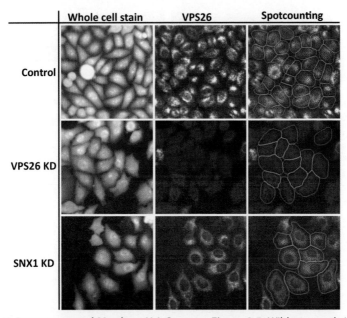

Sophia Y. Breusegem and Matthew N.J. Seaman, Figure 9.5 Wild-type and siRNA KD HeLa cells were stained with anti-VPS26 rabbit antibodies and Alexa 555 anti-rabbit antibodies as well as a blue whole cell stain. Two hundred and fifty cells were imaged on a Cellomics Arrayscan. Overlays were generated using the *Spotdetector* bio-application, where the whole cell stain defines the cell boundaries (green) and a fixed intensity threshold defines the VPS26-positive endosomes (red).

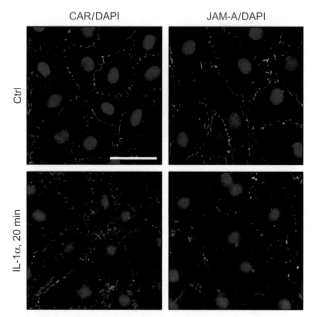

Xiang Xiao et al., Figure 10.3 A study by immunofluorescence microscopy to assess changes in protein distribution at the Sertoli cell cell–cell interface following treatment with IL-1α. Sertoli cells were cultured at 0.05×10^6 cells/cm^2 for 4.5-day on Matrigel-coated coverslips which were placed in 12-well dishes with 2-ml DME/F-12 medium per well (supplemented with growth factors) to allow the establishment of a functional TJ-permeability that mimicked the Sertoli cell BTB *in vivo* as described (Xiao, Cheng, & Mruk, 2013; Xiao, Mruk, Lee, & Cheng, 2011). Thereafter, cells were treated without (control, Ctrl) or with IL-1α (100 pg/ml) for 20 min at 35 °C in a CO$_2$ incubator. Sertoli cells were then harvested by fixing cells in methanol at −20 °C for 5 min and stained for either CAR or JAM-A (red fluorescence) as described in the text. Sertoli cell nuclei were visualized by DAPI (4′,6-diamidino-2-phenylindole) staining. It is noted that treatment of Sertoli cells accelerated the internalization of CAR and JAM-A, with these proteins redistributed from the cell surface into the cell cytosol, confirming data shown in Fig. 10.2. Scale bar, 60 μm, which applies to all other micrographs.

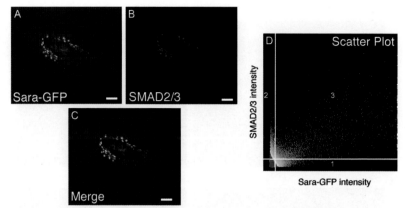

Kristopher D. Marjon and Jennifer M. Gillette, Figure 12.4 Measurement of colocalization within signaling endosomes. Osteoblastic cells (SaOS$_2$) were transfected with Sara-GFP (A) and then fixed and stained with SMAD2/3 (B). The merge image (C) indicates the localization of endogenous SMAD2/3 to the Sara-GFP endosomes. Scale bar—10 μm. The scatter plot (D) indicates the Sara-GFP colocalization with SMAD2/3 (Section 3). Calculations using the Zeiss LSM software indicate an overlap coefficient of 0.9, whereas the correlation $R=0.61$. These data suggest colocalization between the two fluorescent signals.

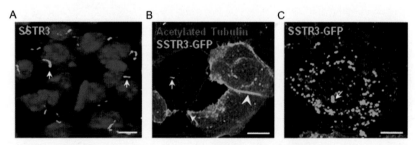

Cristy Tower-Gilchrist et al., Figure 15.1 Localization of SSTR3 in IMCD cells. (A) A frozen section of wild-type adult mouse brain was processed for IF with anti-SSTR3 antibodies (sc-11617, 1:500; Santa Cruz, Dallas, TX). The CA3 region of the hippocampus shows cilia-specific localization of SSTR3. Draq5 nuclear stain is in blue. (B, C) IMCD-3 cells stably expressing SSTR3-GFP were imaged directly (C), or processed for immunofluorescence using an anti-GFP polyclonal antibody (ab290-50, 1:500; Abcam, Cambridge, MA) (to detect SSTR3) and anti-acetylated tubulin monoclonal antibody (T6793, 1:2000; Sigma, Saint Louis, Missouri Sigma) (to detect cilia) (B). SSTR3 localizes to cilia (arrows) and is also detected on the plasma membrane (arrowheads) and in internal endosomes. Scale bars, 10 μm (A, B) and 19 μm (C). *Adapted from Figure 1 of Tower et al. (2011) with permission from Elsevier.*

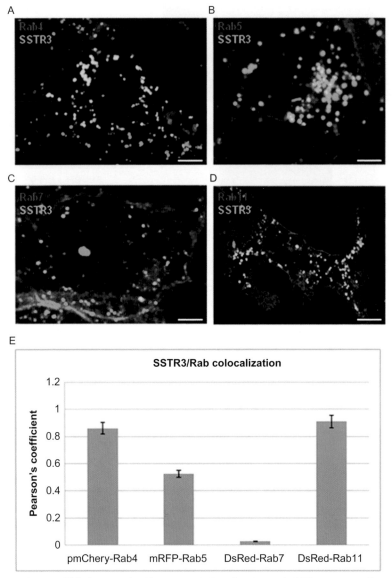

Cristy Tower-Gilchrist et al., Figure 15.2 Localization of SSTR3 to a subset of endosomal compartments. (A–D) IMCD-3 cells stably expressing EGFP-SSTR3 were transfected with mCherry-Rab4 (A), mRFP-Rab5 (B), Ds-Red-Rab7 (C), or Ds-Red Rab11 (D). Cells were imaged at 48 h posttransfection. SSTR3 localizes to Rab4 and Rab5-positive early endosomes, as well as Rab11-positive recycling endosomes, but not to Rab7-positive endosomes. The level of SSTR3 colocalization with each subset of endosomes was measured using Pearson's correlation coefficient (E). Scale bars, 19 μm (A–D). *Adapted from Tower et al. (2011) with permission from Elsevier.*

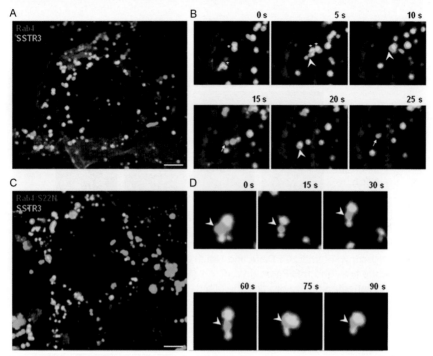

Cristy Tower-Gilchrist *et al.*, **Figure 15.3** Dynamics of SSTR3 movement through endosomal compartments. IMCD-3 cells stably expressing SSTR3 were transfected with mCherry-Rab4 wild-type (A and B) or with mCherry-Rab4/S22N (C and D). Live imaging was carried out, and still images from movies were selected. (B) Arrowheads point to SSTR3 (green) rapidly budding from Rab4-containing (yellow) endosomes. (D) Arrowheads point to SSTR3-only (green) elements that remain in association with Rab4-containing (yellow) elements for an extended time period. Scale bars 19 μm (A and C).

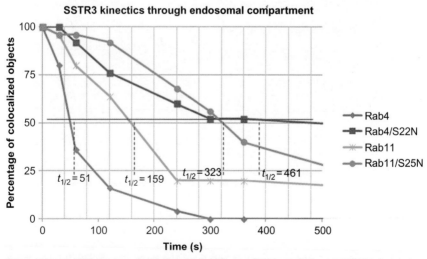

Cristy Tower-Gilchrist *et al.*, **Figure 15.4** Effects of mutant Rabs on SSTR3 dynamics. The Pearson's colocalization coefficients of SSTR3 with wild-type Rab4 and Rab11 or with dominant negative Rab4/S22N and Rab11/S25N were extracted from different times during imaging as in Fig. 15.3 and used to calculate the percentage of colocalized objects at each time point.

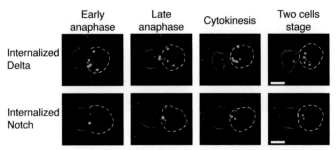

Sylvain Loubéry and Marcos González-Gaitán, Figure 17.1 Anti-Delta and anti-Notch antibody uptake assays. Anti-Delta (top) and anti-Notch (bottom) antibody uptake assays have been performed as described in the text and imaged by spinning-disc confocal microscopy. Internalized Delta and Notch are marked by Alexa488 (green). The dotted white lines and mRFP-Pon (in red), respectively, indicate the parts of the SOP membrane inherited by the pIIa and the pIIb daughter cells. Bars: 5 μm.

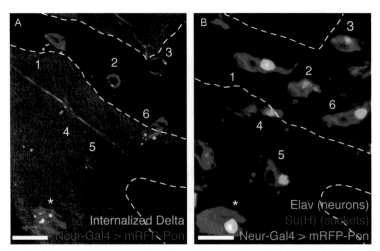

Sylvain Loubéry and Marcos González-Gaitán, Figure 17.3 Correlative antibody uptake assay and immunostaining. The anti-Delta antibody uptake assay (A) and the correlative immunostaining (B) have been performed as described in the text and observed by spinning-disc confocal microscopy. Numbers indicate corresponding SOPs or pIIa/pIIb siblings (in A) and sensory organs (in B); the asterisks indicate a dorso-central macrochaete. Dotted lines indicate folds of the tissue. Bars: 15 μm.

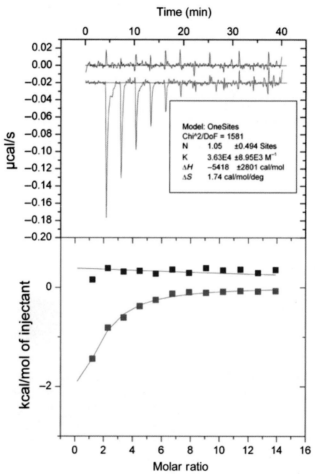

Rajesh Ghai et al., Figure 19.2 Representative ITC experiment showing interaction of the SNX17 PX-FERM protein with a peptide harboring the NPxY/NxxY motif from the APP transmembrane cargo protein (red). The blue curve depicts a blank run (titration of peptide into the buffer).

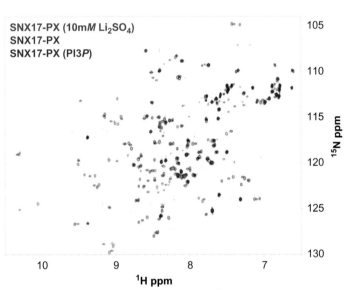

Rajesh Ghai et al., Figure 19.3 Representative ^{1}H–^{15}N two-dimensional HSQC spectra of the SNX17 PX domain under different sample conditions. The spectrum of the unliganded protein (red) is overlaid with spectra of SNX17 in the presence of 10 mM LiSO$_4$ or a molar excess of diC8-PI3P. *This figure is reproduced from Ghai et al. (2011) with permission from the* Proceedings of the National Academy of Sciences USA.